Pediatric /

With an unrestricted educational grant from Pfizer Endocrinecare Pfizer

Workshop, May 16–18, 2010, Turin (Italy)

Pediatric Adrenal Diseases

Volume Editors

Lucia Ghizzoni Turin

Marco Cappa Rome

George Chrousos Athens

Sandro Loche Cagliari

Mohamad Maghnie Genova

55 figures, 14 in color, and 11 tables, 2011

KARGER

Basel · Freiburg · Paris · London · New York · Bangalore · Bangkok · Shanghai · Singapore · Tokyo · Sydney

Endocrine Development

Founded 1999 by Martin O. Savage, London

Lucia Ghizzoni
Division of Endocrinology
Diabetology and Metabolism
Department of Internal Medicine
University of Turin
Turin, Italy

Marco Cappa
Department of Pediatrics
Pediatric Hospital Bambino Gesù
Rome, Italy

George Chrousos
Choremeion Research Laboratory
First Department of Pediatrics
Athens University Medical School
Athens, Greece

Sandro Loche
Regional Hospital for
Microcytaemia
Cagliari, Italy

Mohamad Maghnie
Department of Pediatrics
IRCCS G. Gaslini
University of Genova
Genova, Italy

Library of Congress Cataloging-in-Publication Data

Pediatric adrenal diseases : workshop, May 16-18, 2010, Turin (Italy) / volume editors, Lucia Ghizzoni ... [et al.].
p. ; cm. -- (Endocrine development, ISSN 1421-7082 ; v. 20)
Includes bibliographical references and indexes.
ISBN 978-3-8055-9643-5 (hard cover : alk. paper) -- ISBN 978-3-8055-9644-2 (e-ISBN)
1. Pediatric endocrinology--Congresses. 2.
Children--Diseases--Congresses. I. Ghizzoni, Lucia. II. Series: Endocrine development ; v. 20. 1421-7082
[DNLM: 1. Adrenal Gland Diseases--Congresses. 2. Adrenal
Glands--Congresses. 3. Child. W1 EN3635 v.20 2011 / WK 700]
RJ482.G76P44 2011
618.92'4--dc22

2010044943

Bibliographic Indices. This publication is listed in bibliographic services, including Current Contents® and Index Medicus.

www.karger.com
Printed in Switzerland on acid-free and non-aging paper (ISO 9706) by Reinhardt Druck, Basel
ISSN 1421–7082
ISBN 978–3–8055–9643–5
e-ISBN 978–3–8055–9644–2

Contents

Preface

In the last few years, rapid progress has taken place in our understanding of the adrenal cortex and its pivotal roles in homeostasis. This book emanated from the proceedings of a recent, cutting-edge international meeting in Turin, Italy, and represents the state of the art in the adrenal field. Articles cover the entire area, from the physiology to molecular and cell biology and pathophysiology perspectives, and were written by respected experts and active investigators in each subarea.

The steroidogenesis pathway results in the manufacturing of glucocorticoids, mineralocorticoids and androgens in the layers of the adrenal cortices of humans and/or rodents. The genes, molecules, and cell compartments, such as the endoplasmic reticulum and the mitochondria, that are directly or indirectly involved in this highly complex and vital biochemical path, have been elucidated and presented with clarity in the book. Furthermore, the defects of the genes responsible for relatively common and rare or extremely rare adrenal disorders have been unraveled and presented in individual chapters. Two novel discoveries, one on the P450 oxidoreductase deficiency and the other on P450 side-chain cleavage deficiency, are included in separate chapters.

Our knowledge of the interactions of the adrenal cortices with the adrenal medulla and their importance in the integration of adrenocortical and adrenomedullary function have been further refined. Also, the ability of the adrenal cortex to retain an exquisite balance between adrenocortical stem cell differentiation and growth and steroidogenic cell apoptosis, tightly coupled to the needs of the organism at the resting and stress states, has been better understood. Excellent chapters on these issues have been included in the book. Furthermore, the key adrenal organogenesis transcription factors have been identified and assigned roles in the physiology and pathophysiology of the adrenal cortex, and this area is succinctly and clearly presented in the corresponding chapters.

We have furthered our understanding of the most common disease of the adrenal cortices, congenital adrenal hyperplasia due to 21-hydroxylase deficiency. In this book, an update of the complex molecular pathophysiology of the disorder is presented, while the still controversial antenatal therapy of the condition with dexamethasone is discussed. Furthermore, the long-term effects of the disorder on brain development and function are analyzed.

In the book, the end-hormones of the hypothalamic-pituitary-adrenal (HPA) axis glucocorticoids and androgens and those of the renin-angiotensin system, mineralocorticoids, are discussed from several perspectives. Thus, glucocorticoid therapy refinements in the treatment of adrenal insufficiency and the mechanism of the crucial effects of these hormones on the growth plate are presented. Furthermore, the use of growth hormone to counteract the anti-growth effects of glucocortcoid therapy is discussed. Also, genetic variations in the production of aldosterone of potential clinical importance are described. Finally, novel findings on the mechanisms of adrenarche in the human organism are presented and discussed.

There are three chapters in the book that deal with the actions of glucocorticoids and mineralocorticoids on their target tissues. One of them summarizes the current knowledge on the rare but interesting familial or sporadic glucocorticoid resistance syndrome (Chrousos syndrome). The second provides novel information on the effects of the biological clock on the glucocorticoid receptor throughout the organism, producing a circadian rhythm in the sensitivity of target tissues to glucocorticoids that inversely mirrors the circadian activity of the HPA axis, in effect counteracting its actions. The potential pathologic cardiometabolic effects that uncoupling of this target tissue system with the activity of the HPA axis may have in states such as chronic stress, frequent trans-time zone travel and night-shift work is analyzed in this review article. The third chapter deals with the potential roles of common polymorphisms of the mineralocorticoid receptor gene on the regulation of the human HPA axis and behavior.

Finally, expert review chapters on autoimmune Addison disease and the adrenoleukodystrophy/adrenomyeloneuropathy syndrome are included in the book, as well as a bonus chapter on the ancient history of congenital adrenal hyperplasia by Professor Maria New.

The book will be a useful tool in the hands of anybody that wishes to be up to date with novel research findings in the area of adrenal gland physiology and pathophysiology. Furthermore, the book will be of interest to scientists, physicians and other healthcare professionals who study mitochondrial biology, organogenesis, nuclear receptors, stress biology, glucocorticoid, mineralocorticoid and adrenal androgen actions, pharmacology, and human development.

Lucia Ghizzoni
Marco Cappa
George Chrousos
Sandro Loche
Mohamad Maghnie

Ghizzoni L, Cappa M, Chrousos G, Loche S, Maghnie M (eds): Pediatric Adrenal Diseases.
Endocr Dev. Basel, Karger, 2011, vol 20, pp 1–19

Role of Mitochondria in Steroidogenesis

Walter L. Miller

Department of Pediatrics, University of California, San Francisco, Calif., USA

Abstract

All steroids are made from cholesterol by a complex array of enzymes that includes cytochrome P450 enzymes in both the mitochondria and endoplasmic reticulum. The structures and mechanisms of catalysis by the mitochondrial and endoplasmic reticulum P450s are very similar, but their electron transport mechanisms differ. Nevertheless, mitochondrial P450scc is inactive when targeted to the endoplasmic reticulum with its electron-transfer partners. Adrenals and gonads receive cholesterol from low-density lipoproteins, store it as cholesterol esters, and transport cholesterol to mitochondria by ill-defined mechanisms. The acute quantitative regulation of steroidogenesis is determined by cholesterol import into mitochondria by the steroidogenic acute regulatory protein (StAR). The chronic regulation, determining net steroidogenic capacity, is determined by the amount of P450scc protein present, which is determined by gene transcription. Mutations in StAR cause congenital lipoid adrenal hyperplasia (lipoid CAH), characterized by absent steroidogenesis, potentially lethal salt loss, 46,XY sex reversal and massively enlarged adrenals engorged with cholesterol esters. Non-classic lipoid CAH is a newly described variant caused by StAR mutations that retain 10–25% of function, and characterized by onset of glucocorticoid deficiency after 2 years of age, and mild defects in mineralocorticoid and sex steroid synthesis. Rare mutations in P450scc result in both classic and non-classic hormonal syndromes that are indistinguishable from lipoid CAH, but have small adrenals.

The Cellular Cholesterol Economy

Steroidogenesis occurs in multiple cellular compartments. Low-density lipoproteins (LDL) interact with LDL receptors on the cell surface, resulting in the import of cholesterol. LDL cholesterol esters are taken up by receptor-mediated endocytosis, then stored directly or converted to free cholesterol and used for steroid hormone synthesis. By contrast, rodent adrenals import high-density lipoproteins via a receptor termed SRB1 (scavenger receptor B1); this pathway plays a minor role in human steroidogenesis (fig. 1). Cholesterol is esterified by acyl-CoA:cholesterol transferase (ACAT), stored in lipid droplets, and accessed by activation of hormone-sensitive lipase (HSL) [1]. ACTH stimulates HSL and inhibits ACAT, thus increasing the availability of free

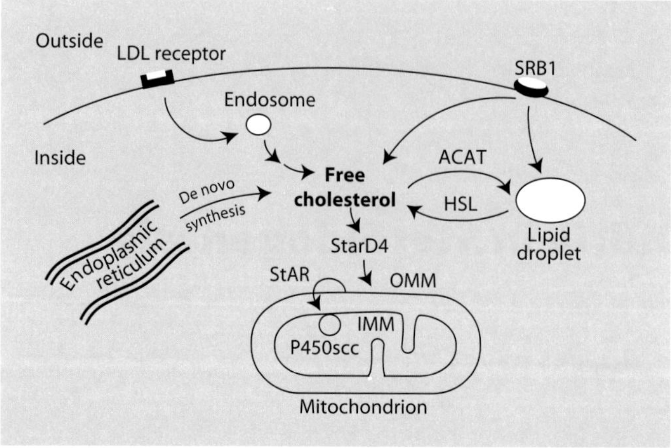

Fig. 1. Import and transfer of cellular cholesterol. Human cells bind circulating low-density lipoproteins (LDL) by receptor-mediated endocytosis, directing the cholesterol to endosomes, whereas rodent cells bind high-density lipoproteins (HDL) via scavenger receptor B1 (SRB1). Cholesterol can also be synthesized de novo in the endoplasmic reticulum. Cholesterol can be esterified by acyl-CoA:cholesterol transferase (ACAT) and stored in lipid droplets as cholesterol esters. Free cholesterol, produced by hormone-sensitive lipase (HSL), may be bound by StarD4 for trans-cytoplasmic transport to membrane destinations, including the outer mitochondrial membrane (OMM). In the adrenals and gonads, the steroidogenic acute regulatory protein, StAR, is responsible for the rapid movement of cholesterol from the OMM to the inner mitochondrial membrane (IMM), where it can be taken up by the cholesterol side-chain cleavage enzyme, P450scc, and converted to pregnenolone. Copyright: W.L. Miller.

cholesterol for steroid hormone synthesis. The first step in steroidogenesis is the cleavage of the cholesterol side chain to yield pregnenolone; this reaction is catalyzed by the cholesterol side-chain cleavage enzyme, P450scc, which resides in the mitochondria. The mechanism(s) by which cholesterol is delivered to the mitochondria remain unclear. Free cholesterol is nearly insoluble (critical micellar concentration ~25–40 nM) [2], thus trans-cytoplasmic cholesterol transport appears to be mediated by cholesterol-binding proteins. Candidate cholesterol-binding proteins include sterol-carrier protein 2 (SCP-2) and its homologue SCP-x, but these appear to be non-specific lipid-binding and transfer proteins that play a minor role in the intracellular cholesterol economy [3]. The steroidogenic acute regulatory protein (StAR) is the first-described member of a family of proteins that contain so-called START (StAR-related lipid transfer) domains, which are found in most metazoan organisms [4]. Fifteen START domain (StarD) proteins appear to serve roles in binding and mediating the intracellular transfer of lipids in mammals [5]. StarD4, 5, and 6 are closely related to StAR [6], and may participate in intracellular cholesterol transport [5].The structure of StarD4 [7] is essentially the same as that of StAR, including a cholesterol-binding pocket, and both StarD4 and D5 can have low levels of StAR-like activity in transfected cells [8]. StarD4, 5 and 6 lack signal sequences that target them to specific subcellular organelles, hence they appear to be confined to the cytoplasm, where they bind insoluble lipids, permitting the lipid to be transported across aqueous cytosol. StarD6, which is primarily expressed in the male germ line, has greater StAR-like activity than StAR itself [9]. Thus these proteins are excellent candidates for the factor(s) that deliver cholesterol to the outer mitochondrial membrane (OMM) from elsewhere in the cell (lipid droplets, endoplasmic reticulum) in most cell types, whereas StAR itself is responsible for delivery from the OMM to the inner mitochondrial membrane (IMM), but only acts in steroidogenic cells. Knockout of the StarD4 gene in mice does not create a disorder of the adrenals or gonads [10], hence the physiologic role of this protein remains uncertain.

Cytochrome P450

Most steroidogenic enzymes are forms of 'cytochrome P450', a term designating a group of oxidative enzymes which have about 500 amino acids and contain a single heme group [11]. P450 means pigment 450, because all of these proteins absorb light at 450 nm in their reduced states. The human genome has genes for 57 cytochrome P450 enzymes. These genes are formally termed *CYP* genes; the encoded proteins may be given the same name without the use of italics (thus the *CYP11A1* gene encodes CYP11A1), but most endocrinologists use the traditional P450 names for the proteins (thus the *CYP11A1* gene encodes P450scc). Seven human cytochrome P450 enzymes are targeted to the mitochondria and are termed 'Type 1'; the other 50 human P450 enzymes are targeted to the endoplasmic reticulum and are termed 'Type 2'. All P450 enzymes activate molecular oxygen using their heme center and electrons from NADPH. The two types of P450 enzymes are distinguished by different mechanisms by which they receive electrons from NADPH, as well as by their intracellular locations. Type 1 enzymes receive electrons from NADPH via a flavoprotein termed ferredoxin reductase (also termed adrenodoxin reductase) [12] and a small iron-sulfur protein termed ferredoxin (also termed adrenodoxin) [13] (fig. 2). Type 2 P450 enzymes receive electrons from NADPH via a single 2-flavin protein termed P450 oxidoreductase (POR) [14]. Each P450 enzyme can catalyze multiple reactions, often with chemically different substrates, hence they are designated by their P450 name (e.g. P450c21) rather than by a reaction-based name (e.g. 21-hydroxylase).

Six P450 enzymes are involved in steroidogenesis. Mitochondrial P450scc is the cholesterol side-chain cleavage enzyme catalyzing the series of reactions formerly termed '20,22-desmolase' in the clinical literature. The two isozymes of mitochondrial P450c11, P450c11β (11β-hydroxylase) and P450c11AS (aldosterone synthase) catalyze 11β-hydroxylase, 18-hydroxylase, and 18-methyloxidase activities. In the endoplasmic reticulum, P450c17 catalyzes both 17α-hydroxylase and 17,20-lyase activities, P450c21 catalyzes the 21-hydroxylation of both glucocorticoids and mineralocorticoids and P450aro catalyzes aromatization of androgens to estrogens.

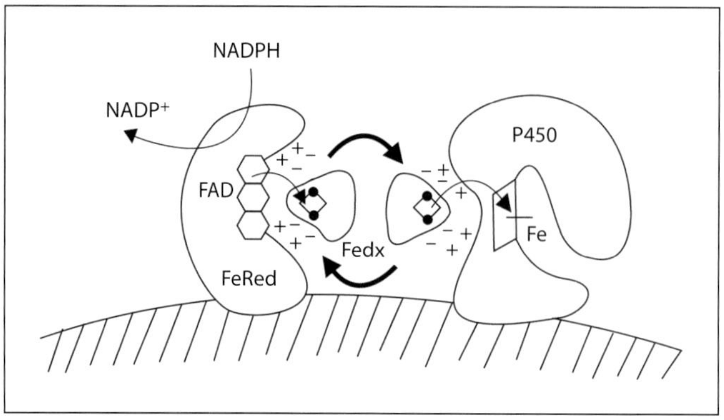

Fig. 2. Organization of Type I P450 enzymes on the inner mitochondrial membrane. NADPH donates electrons to the flavin adenine dinucleotide (FAD) moiety of ferredoxin reductase (FeRed); positively charged residues in ferredoxin reductase interact with negatively charged residues in ferredoxin (Fedx), permitting the electrons to be transferred to the Fe_2S_2 center of ferredoxin. Ferredoxin then dissociates from ferredoxin reductase and diffuses through the mitochrondrial matrix. The same surface of ferredoxin that received the electrons from ferredoxin reductase then interacts with the redox-partner binding-site of a Type I P450, such as P450scc. The electrons from the Fe_2S_2 center of ferredoxin then travel to the heme ring of the P450. The heme iron then mediates catalysis with substrate bound in the P450. Copyright: W.L. Miller.

Mitochondrial Specificity of the Cholesterol Side-Chain Cleavage Reaction

The subcellular localizations on the P450 enzymes are critically linked to their enzymatic activity. For mitochondrial P450scc, which requires electron donation via ferredoxin reductase and ferredoxin, the molar ratios of these two electron-transfer proteins to P450scc are critical determinants of P450scc catalytic activity [15]. Creation of a three-component fusion protein of the general structure H_2N-P450scc-FerredoxinReductase-Ferredoxin-COOH (termed F2) increases the enzyme's maximum velocity (V_{max}) [15]. The requirement for ferredoxin reductase and ferredoxin is not absolute, as fusion of P450scc to POR and targeting of this fusion to the mitochondria yields an active enzyme [16]. However, when P450scc, the F2 fusion, or other P450scc constructs are redirected to the endoplasmic reticulum by genetic engineering, they are wholly inactive [16]. Thus the mitochondrial location is essential for the enzymatic activity of P450scc [17].

Acute versus Chronic Regulation of Steroidogenesis

Steroid secretion requires steroid synthesis. Cells that produce polypeptide hormones, such as pancreatic islet cells or pituitary somatotropes, will first synthesize and then store large amounts of hormone in secretory vesicles ready for rapid release. Thus the secretion of insulin in response to a glucose load entails release of previously synthesized insulin, and is not immediately dependent on insulin biosynthesis. By contrast, steroidogenic cells store very little steroid, so that a rapid steroidogenic response (e.g. adrenal secretion of aldosterone or cortisol in response to stress, or the 'pulsing' of sex steroids in response to an LH surge) requires rapid synthesis of new steroid. ACTH promotes adrenal steroidogenesis at three distinct levels by separate cellular mechanisms. First, over the course of months, long-term exposure to ACTH (as in Cushing's disease or CAH), promotes adrenal growth by promoting cellular hypertrophy and hyperplasia. This growth occurs primarily by ACTH stimulating the production of cAMP, which in turn promotes the synthesis of intra-adrenal insulin-like growth factor 2 (IGF-2) [18, 19], basic fibroblast growth factor [20], and epidermal growth factor [21]. Second, ACTH acting through cAMP in the zona fasciculata and angiotensin II acting through the calcium/calmodulin pathway in the zona reticularis, promote transcription of genes encoding the steroidogenic enzymes, especially the enzymatically rate-limiting enzyme, P450scc [22]. Third, ACTH stimulates phosphorylation of StAR at Ser195 almost immediately [23] and stimulates transcription of the StAR gene within minutes [24], thus increasing the flow of cholesterol from the OMM to IMM where it becomes substrate for the first and rate-limiting enzyme, P450scc.

Steroidogenic Acute Regulatory Protein

Studies in the 1960s showed that the acute steroidogenic response could be blocked by inhibitors of protein synthesis (e.g. puromycin or cycloheximide), indicating that a short-lived protein species mediates this process [25–27]. Studies using two-dimensional gel electrophoresis first showed that this acute steroidogenic response was accompanied by the rapid synthesis of a 37-kDa phosphoprotein [28–30]; Douglas Stocco's laboratory then cloned this factor, which they named the steroidogenic acute regulatory protein – StAR [31]. The history of the discovery of StAR as this long-sought acute trigger of steroidogenesis has been reviewed [32]. StAR is synthesized as a 37-kDa protein that has a typical mitochondrial leader sequence that directs it to the mitochondria; upon mitochondrial entry this leader peptide is cleaved off to yield the 30-kDa intramitochondrial protein. Overexpression of mouse StAR in mouse Leydig MA-10 cells increased their basal rate of steroidogenesis [31], and co-transfection of expression vectors for both StAR and the P450scc system in non-steroidogenic COS-1 cells augmented pregnenolone synthesis above that obtained with the P450scc system alone [33], providing evidence that StAR is the acute regulator of steroidogenesis. The definitive proof of StAR's essential role came when it was shown that StAR mutations cause congenital lipoid adrenal hyperplasia, in which very little steroid is made [33, 34].

Some steroidogenesis is independent of StAR. When non-steroidogenic cells are transfected with the P450scc system, they can convert some cholesterol to pregnenolone in the absence of StAR [15, 16]. Co-transfection with a vector expressing StAR shows that this StAR-independent steroidogenesis occurs at about 14% of the StAR-induced rate [33, 34]. In vivo, the placenta utilizes mitochondrial P450scc to initiate steroidogenesis [35] but does not express StAR [36]. The mechanism of StAR-independent steroidogenesis is unclear; it may occur without a triggering protein, or some other protein may exert StAR-like activity to promote cholesterol flux, but without StAR's rapid kinetics.

StAR Structure and Mechanism of Action

The mechanism of StAR's action has been studied extensively, but remains incompletely understood [37–39]. The short half-life of the 37-kDa cytoplasmic form of StAR suggested it was a 'precursor' and the longer half-life of the 30-kDa intramitochondrial form of StAR initially suggested that this 'mature' 30-kDa form was the biologically active moiety. However, both the 37- and 30-kDa forms of StAR are equally active when expressed in cytoplasm or added to mitochondria in vitro [40]. We built constructs that immobilized StAR on the cytoplasmic surface of the OMM, in the intramembranous space, or on the matrix side of the IMM (fig. 3). StAR was constitutively active on the OMM, but inactive when localized to the intramembranous space or to the IMM [41]. Thus, StAR acts exclusively on the OMM [40, 41]. We also built constructs that would either speed up or slow down the entry of StAR into the mitochondria; faster import decreased activity whereas slower import increased activity [41]. Thus StAR's activity to promote steroidogenesis is proportional to its residency time on the OMM, and it is StAR's cellular localization, and not its cleavage from 37 to 30 kDa, that determines its activity (fig. 3). StAR has a hydrophobic pocket that binds a single molecule of cholesterol [42]. The interaction of StAR with the OMM involves conformational changes [43, 44] that are needed for it to accept and discharge cholesterol molecules. StAR can transfer cholesterol between membranes in vitro [45, 46] suggesting that other protein molecules are not needed for its action; however, the mutant R182L, which is biologically inactive and causes lipoid CAH [47], can also transfer cholesterol to membranes in vitro, and, as described below, other proteins are known to interact with StAR. Thus StAR's action to promote steroidogenesis is distinct from its cholesterol-transfer activity; cholesterol binding is necessary but not sufficient for StAR activity.

Multiple lines of evidence indicate that the action of StAR also requires the peripheral benzodiazepine receptor (PBR, also known as the transporter protein, TSPO) on the OMM [48, 49]. StAR has a functional interaction with PBR [48], and physical interactions with voltage-dependent anion channel 1 (VDAC1) and phosphate carrier protein [50]; these three proteins are found on the OMM. StAR probably interacts with many other proteins as well [49]. StAR is recycled, with each molecule moving hundreds of molecules of cholesterol before the cleavage/inactivation event [51]. Low

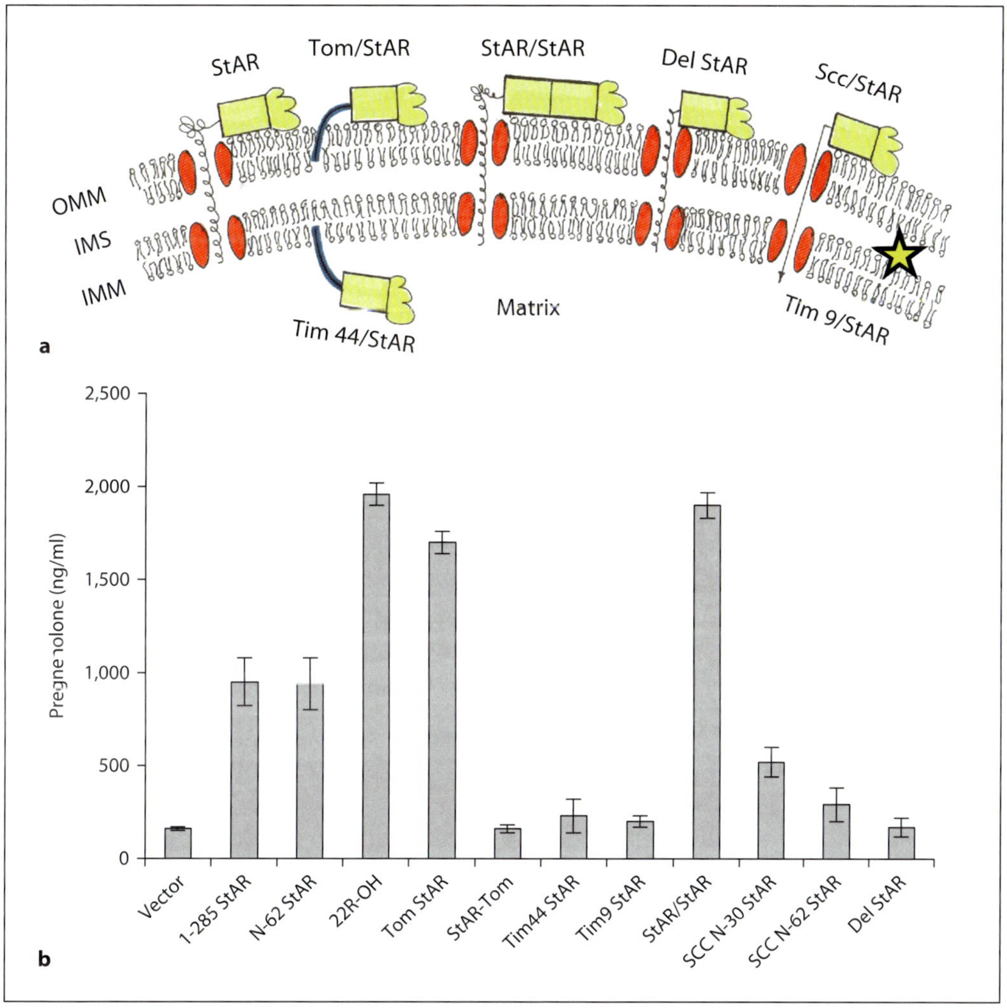

Fig. 3. StAR acts on the outer mitochondrial membrane (OMM). **a** Experimental design, depicting the OMM, intramembranous space (IMS) and inner mitochondrial membrane (IMM). The first 62 residues of StAR are represented as the corkscrew line, the protease-resistant domain (residues 63–188) as the yellow box, and the protease-sensitive carboxyl-terminal domain as the yellow blob. Mitochondrial protein-import proteins are represented by the red ovals. Fusion of N-62 StAR to the C-terminus of Tom20, an OMM protein, generates the Tom/StAR construct, which immobilizes StAR on the OMM; Tim9/StAR similarly immobilizes StAR in the IMS and Tim44/StAR immobilizes StAR on the matrix side of the IMM. StAR/StAR is a construct that fuses 1–188 StAR to N-62 (63–285) StAR, thus dimerizing the protease-resistant domain (residues 63–188); Del StAR retains residues 1–30 but deletes a 'pause' sequence at residues 31–62, and Scc/StAR replaces the StAR leader (either 1–30 or 1–62) with the 39 amino acid leader sequence of human P450scc. **b** Activity of the constructs illustrated in panel **a**, when transfected into COS-1 cells co-transfected with the F2 fusion of the cholesterol side-chain cleavage system. A low level of StAR-independent steroidogenesis is seen with the vector control, but full-length (1–285) or N-62 StAR increase steroidogenesis sixfold. The hydroxysterol 22R-OH-cholesterol bypasses the action of StAR, providing an index of the maximal level of steroidogenesis that can be achieved in this system. Transfection with the Tom/StAR fusion or the StAR/StAR dimer achieves this maximal level of steroidogenesis; by contrast, when StAR is placed at the N-terminus rather than the C-terminus of Tom20 (StAR/Tom), when it is localized to the matrix side of the IMM (Tim44/StAR), or when it is confined to the IMS (Tim9/StAR), it is inactive. Replacing StAR's leader peptide with that of P450scc also reduces activity; cell-free transcription/translation linked to mitochondrial import assays show that the level of activity seen with the leader mutants is inversely proportional to their speed of import [41]. Copyright: W.L. Miller.

levels of steroidogenesis will persist in the absence of StAR at about 14% of the StAR-induced rate [34, 37], accounting for the steroidogenic capacity of tissues that lack StAR (e.g. the placenta). Physical studies and partial proteolysis indicate that residues 63–193 of StAR are protease-resistant and constitute a 'pause-transfer' sequence, which permits the bioactive loosely-folded carboxy-terminal molten globule domain to have increased interaction with the OMM [52].

The three-dimensional structure of StAR has been deduced from the crystal structure of a protein called N-218 MLN64. The term MLN64 refers to metastatic lymph node, clone 64, a 445 amino acid protein cloned from metastatic breast carcinoma [53]. When 218 amino-terminal residues are deleted (N-218), the remaining 227 carboxyl-terminal amino acids are 37% identical and about 50% similar to the sequence of StAR [53]. N-218 MLN64 has about half of the activity of StAR to promote steroidogenesis, both in transfected cells and when purified and added to steroidogenic mitochondria in vitro [54, 55]. MLN64, a related protein called MENTHO [56, 57] and the NPC proteins that are disordered in Niemann-Pick type C disease [58] act in trafficking cholesterol in peroxosomes and lipid droplets. The crystal structure of N-216 MLN64 reveals a globular protein with an α/β helix-grip fold and an elongated hydrophobic pocket measuring about 26 Å deep and 10 Å across at its widest diameter [42]. Modeling indicates that N-216 MLN64 accommodates a single molecule of cholesterol in this pocket, with the 3β-OH group coordinated by the two polar residues at the bottom of the pocket. This structure, the crystal structure of the closely related StarD4 protein [7] and several computational models of StAR [59–61] are all characterized by two long α-helixes at the N- and C-termini, two short α-helixes, and a set of 9 antiparallel β sheets that form a helix-grip fold (fig. 4a). Only the exterior surface of the C-terminal α-helix and small segments of the adjacent Ω-loops interact with the OMM (fig. 4b). The Ω-loops form hydrogen bonds with the C-terminal helix, prohibiting access of cholesterol to StAR's hydrophobic cholesterol-binding pocket. The interaction of StAR with the charged phospholipids head groups on the OMM disrupts these hydrogen bonds, permitting the C-helix to swing open and closed, governing access of cholesterol to the sterol-binding pocket. Immobilizing the C-helix by forming disulfide bonds with the adjacent loops ablates activity while opening these artificial disulfide bonds restores activity, proving that it is the immobilization of the C-helix, and not the amino acid changes, that are responsible for the lost activity [62]. Thus the activity of StAR on the OMM requires an acid-induced disruption of hydrogen bonds and a consequent conformational change in StAR to permit it to bind and release cholesterol.

Disorders of StAR: Classic and Non-Classic Congenital Lipoid Adrenal Hyperplasia

Congenital lipoid adrenal hyperplasia (lipoid CAH) is a very severe disorder of steroidogenesis, characterized by minimal or absent serum concentrations of all steroids,

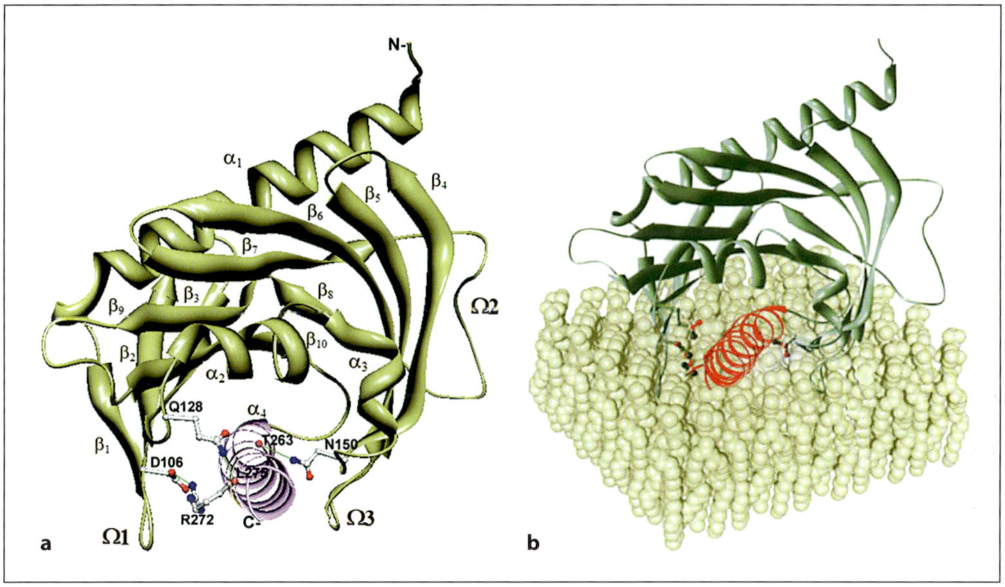

Fig. 4. Model of N-62 StAR. **a** Ribbon diagram showing the N-terminus in the upper right-hand corner and the C-terminal helix in the lower center, extending out of the plane of the diagram. Residues that contribute to the associations between this C-terminal helix and adjacent structures are shown as ball-and-stick representations: carbon atoms are in white, nitrogen are in blue, oxygen in red, and hydrogen bonds in green [reprinted from 60 with permission]. **b** Model showing the orientation of StAR as it interacts with the OMM [cover picture of ref. 60].

high basal ACTH and plasma renin activity, absent steroidal responses to long-term treatment with high doses of ACTH or hCG, and grossly enlarged adrenals filled with cholesterol and cholesterol esters [63–65]. Because these findings indicate a lesion in the conversion of cholesterol to pregnenolone, it was initially thought to result from a disorder in an enzyme involved in this conversion; hence, before the role of P450scc was understood, lipoid CAH was misnamed 20,22-desmolase deficiency [65, 66]. However, the gene for the enzyme that catalyzes this activity, P450scc, is normal in these patients [67, 68], as are the mRNAs for ferredoxin reductase and ferredoxin [68]. Steroidogenesis in the placenta persists in lipoid CAH, permitting the synthesis of progesterone needed for term gestation; this would not be expected to happen if P450scc were involved [69]. The normal P450scc system in conjunction with the accumulation of cholesterol esters in the affected adrenal suggested that the lesion lay in an upstream factor involved in cholesterol transport into mitochondria [68]. The identification of disease-causing StAR mutations in lipoid CAH proved StAR's indispensable role in adrenal and gonadal steroidogenesis [33, 34, 70].

Lipoid CAH is the StAR gene knockout experiment of nature [71]. Studies of lipoid CAH showed that StAR promotes steroidogenesis by increasing the movement of cholesterol into mitochondria [33, 34, 70]. The observation that cells could make some steroids in a StAR-independent fashion led to the two-hit model of lipoid CAH

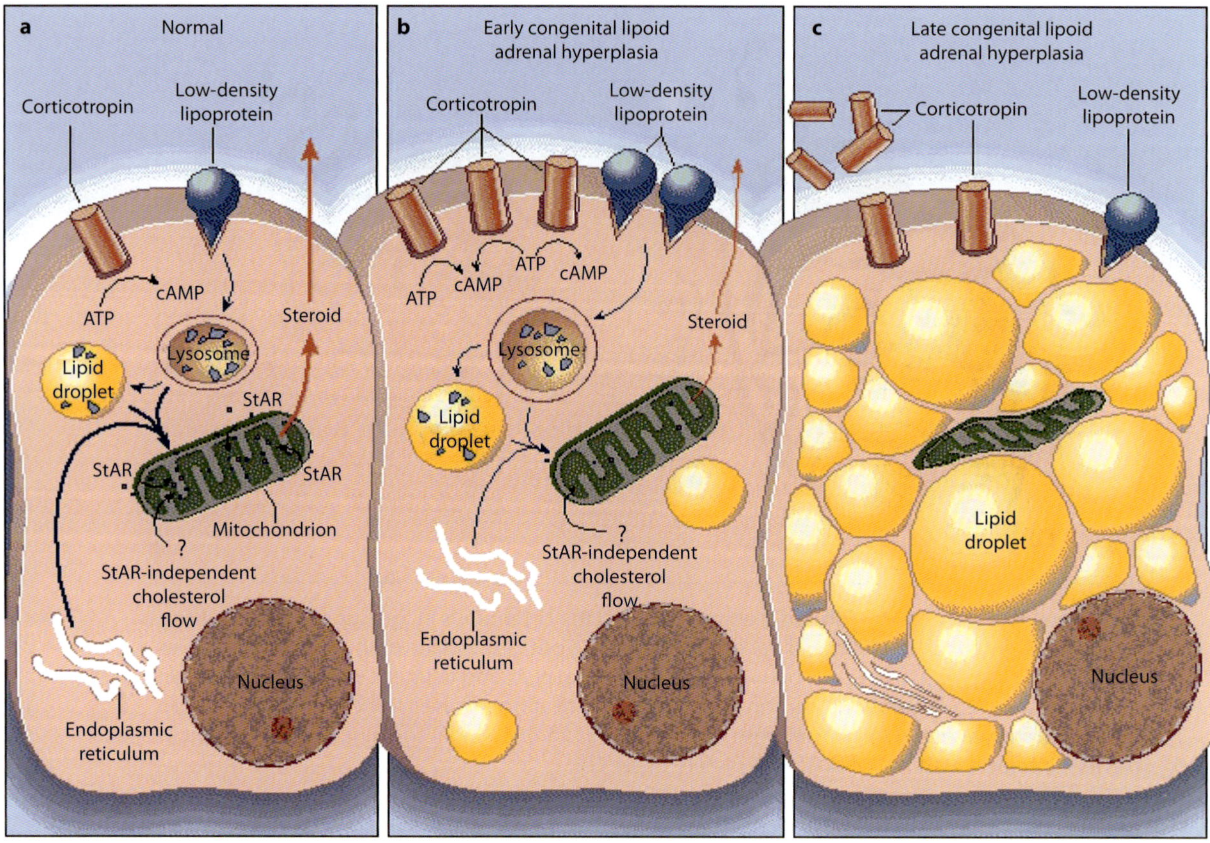

Fig. 5. Two-hit model of lipoid CAH. **a** In a normal adrenal cell, cholesterol is primarily derived from LDL. The rate-limiting step in steroidogenesis is the flow of cholesterol from the outer to the IMM. **b** In early lipoid CAH, StAR-independent mechanisms can still move some cholesterol into the mitochondria, yielding a low level of steroidogenesis; ACTH secretion increases, stimulating further accumulation of cholesterol esters in lipid droplets. **c** As lipid droplets accumulate, they engorge and damage the cell through physical displacement and by the action of cholesterol auto-oxidation products. Steroidogenic capacity is destroyed, and tropic stimulation continues. In the ovary, follicular cells remain unstimulated and undamaged until puberty, when small amounts of estradiol are produced, as in panel **b**, causing partial feminization, with infertility and hypergonadotropic hypogonadism [modified from 34 with permission].

[34] (fig. 5). The first hit is the loss of StAR activity, resulting in a loss of most, but not all, steroidogenesis. Diminished steroidogenesis causes a compensatory rise in ACTH and LH, eliciting an increase intracellular cAMP, which increases biosynthesis of LDL receptors, their consequent uptake of LDL cholesterol, and de novo synthesis of cholesterol. In the absence of StAR, intracellular cholesterol accumulates, causing the second hit, which is the loss of all steroidogenic capacity caused by mitochondrial and cellular damage resulting from the accumulated cholesterol, cholesterol esters, and their auto-oxidation products [34].

The two-hit model explains the clinical findings in lipoid CAH. Because the fetal testis normally makes large amounts of testosterone in fetal life, the Leydig cells are destroyed early in gestation. This eliminates testosterone biosynthesis, so that an affected 46,XY fetus does not undergo virilization and is born with female external genitalia. Nevertheless, Wolffian duct derivatives are developed, indicating the presence of some testosterone synthesis early in fetal life [72], as predicted by the two-hit model. The Sertoli cells are undamaged and produce Müllerian inhibitory hormone normally, so that the phenotypically female 46,XY fetus lacks a cervix, uterus, or fallopian tubes. The steroidogenically active fetal zone of the adrenal is similarly affected early in development, eliminating most DHEA biosynthesis, and hence eliminating the feto-placental production of estriol, so that mid-gestation maternal and fetal estriol levels are very low [69]. By contrast, the definitive zone of the fetal adrenal, which differentiates into the zonae glomerulosa and fasciculata, normally produces very little aldosterone. Furthermore, because fetal salt and water balance are governed by the placenta, stimulation of the glomerulosa by angiotensin II generally does not begin until birth. Consistent with this, many newborns with lipoid CAH do not have a salt-wasting crisis until several months of life, when chronic stimulation then leads to cellular damage [34, 73].

The two-hit model also explains the spontaneous feminization of affected 46,XX females who are treated in infancy and reach adolescence [74, 75]. The fetal ovary makes little or no steroids and lacks detectable mRNAs for the steroidogenic enzymes after the first trimester [76]; consequently the ovary remains largely undamaged until it is stimulated by gonadotropins at the time of puberty, when it then produces small amounts of estrogen by StAR-independent steroidogenesis. However this low level of steroidogenesis cannot generate adequate progesterone in response to the mid-cycle LH surge, and continued stimulation results in cholesterol accumulation and cellular damage, so that biosynthesis of progesterone in the latter part of the cycle is impaired. Gonadotropin stimulation only recruits individual follicles and does not promote steroidogenesis in the whole ovary, hence most follicles remain undamaged and available for future cycles. Monthly cyclicity is determined by the hypothalamic-pituitary axis, and remains normal. A new follicle is recruited with each successive cycle, and more estradiol is produced by StAR-independent steroidogenesis. Although ovarian steroidogenesis is impaired, enough estrogen is produced to induce breast development, especially in the absence of adrenal androgens, hence the patient experiences general feminization, monthly estrogen withdrawal and cyclic vaginal bleeding [34, 74, 75]. Because little progesterone is produced in the latter half of the cycle, the cycles are anovulatory. Measurements of estradiol, progesterone and gonadotropins throughout the cycle in affected adult females with lipoid CAH confirm this model [75]. Similarly, examination of the ovaries of StAR knockout mice confirms the two-hit model [77].

Many StAR mutations have been found in patients with lipoid CAH [34, 78]. About 65–70% of affected Japanese alleles and virtually all affected Korean alleles carry the mutation Q258X [33, 34, 65, 78–80]. The Q258X carrier frequency in these countries is about 1 in 300 [34, 79] so that 1 in every 250,000–300,000 newborns is affected, or about 500 patients in Japan and Korea. Other genetic clusters are found among Palestinian Arabs, carrying the mutation R182L [34], in eastern Saudi Arabia, carrying R182H [73], and in parts of Switzerland, carrying the mutation L260P [81]. Deletion of only 10 carboxy-terminal residues reduces StAR activity by half [40] and deletion of 28 carboxy-terminal residues by the common Q258X mutation eliminates all activity. By contrast, deletion of the first 62 amino-terminal residues has no effect on StAR activity, even though this deletes the entire mitochondrial leader sequence and forces StAR to remain in the cytoplasm [40].

Most patients with lipoid CAH are infants with normal-appearing female genitalia who experience failure to thrive and salt loss in the first weeks of life [34, 65, 78]. Lipoid CAH can present as apparent sudden infant death syndrome [82], and the classic form of the disease my present as late as 1 year of age [73]. By contrast, 'non-classic lipoid CAH' is an attenuated form of the disease caused by mutations that retain about 10–25% of normal StAR activity [83–85]. These patients typically present with mild symptoms of adrenal insufficiency anywhere from 2 years of age to adulthood, and have been mistaken for cases of familial glucocorticoid deficiency [84]. The genetic males have normal external genitalia, indicating normal intrauterine Leydig cell function, but typically have hypergonadotropic hypogonadism. Some patients have very mild disorders of mineralocorticoid secretion, characterized by normal electrolytes and elevated plasma renin activity. Several StAR mutations can cause this phenotype, but the most common is R188C [83–85]. Thus the spectrum of clinical presentation of mutations in the StAR protein is substantially broader than that of classic lipoid CAH, and many affected patients may not be diagnosed correctly [84].

Treatment of lipoid CAH consists of physiologic replacement with glucocorticoids, mineralocorticoids and, in the newborn period, salt [65]. The glucocorticoid requirement is less than in the virilizing congenital adrenal hyperplasias because it is not necessary to suppress excess adrenal androgen production, so that growth in these patients should be normal [65]. Genetic males have female external genitalia and should undergo orchiectomy and be raised as females [34, 65]. Successful pregnancy has been induced in 1 adult female with lipoid CAH by in vitro fertilization using a modified protocol that employed higher doses of progesterone to replace the maternally produced first-trimester progesterone that that the affected mother could not produce [86].

P450scc Deficiency: A Rare New Disease

Conversion of cholesterol to pregnenolone by mitochondrial P450scc is the first, rate-limiting and hormonally regulated step in the synthesis of all steroid hormones [87, 88]. This process involves three distinct chemical reactions, the 22-hydroxylation of cholesterol, 20-hydroxylation of 22(R)-hydroxycholesterol, and oxidative scission of the C20-22 bond of 20(R),22(R)-dihydroxycholesterol (the side-chain cleavage event), yielding pregnenolone and isocaproaldehyde. The reactions catalyzed by P450scc are slow, with a net turnover number of about 6 [89] to 20 [90] molecules of cholesterol per molecule of P450scc per second. Because 20-hydroxycholesterol, 22-hydroxycholesterol, and 20,22-hydroxycholesterol can all be isolated from bovine adrenals in significant quantities, and because 3 mol of NADPH are required per mole of cholesterol converted to pregnenolone, it was initially thought that three separate enzymes were involved. However, protein purification and reconstitution of enzymatic activity in vitro showed that a single protein, P450scc, converted cholesterol to pregnenolone [for reviews, see 91, 92]. These three reactions occur on a single active site that is in contact with the IMM.

The cDNA for human P450scc was cloned in 1986 [35] and the gene was cloned in 1987 [93]. The gene lies on chromosome 15q23-q24, consisting of nine exons spanning about 20 kb. The 2-kb mRNA encodes 521 amino acids, including a 39 amino acid mitochondrial leader peptide that is cleaved off during the entry of P450scc into the mitochondria. A spontaneous deletion of the gene for P450scc in the rabbit [94], knockout of P450scc in the mouse [95], and rare patients with P450scc gene mutations [96, 97] result in the loss of all steroidogenesis, showing that all steroidogenesis is initiated by this one enzyme. Thus it is the presence of P450scc that renders a cell 'steroidogenic' and able to make steroids de novo, as opposed to modifying steroids produced elsewhere, which occurs in many types of cells. Transcription of the CYP11A1 gene encoding P450scc determines how much P450scc enzyme is made, and consequently determines the steroidogenic capacity of a cell. This transcription is regulated by both tissue-specific and hormonally-responsive factors, and can be induced by both the PKA and PKC second messenger systems, which act through different DNA elements in the CYP11A1 promoter [22]. The expression of P450scc and other steroidogenic enzymes in the adrenal and gonad is hormonally inducible [98] and requires the action of the transcription factor, steroidogenic factor 1 (SF1) [99, 100]; by contrast, expression of P450scc in the human placenta is constitutive [101], independent of SF1, and requires LBP proteins [102–105] and TreP-132 [106, 107]. Thus long-term cellular stimulation over the course of days will increase the content of P450scc, and the level of basal steroid produced, as well as the capacity of the cell to mount a steroidogenic response.

At least 6 patients have now been described with severe mutations in P450scc that ablate virtually all activity [102, 103, 108–110]. These patients are clinically and hormonally indistinguishable from those with classical lipoid CAH, except that their StAR genes are normal and they do not have enlarged adrenals. It would seem logical that elimination of P450scc activity would be incompatible with term gestation, as the placenta, a fetal tissue, must produce progesterone in the second half of pregnancy to suppress maternal uterine contractions, thus preventing miscarriage. It is most likely that these few fetuses with P450scc mutations reached term gestation because of unusually protracted maintenance of the maternal corpus luteum of pregnancy, which normally involutes in the second trimester. However, this has not been investigated directly. In addition, just as a non-classical form of lipoid CAH has been described recently [83–85], there is also a non-classical form of P450scc deficiency. These patients have presented at 5–15 years of age, primarily with adrenal insufficiency, and have P450scc mutations that retain 10–20% of normal activity [111, 112]. As with classical P450scc deficiency, these patients do not have the large adrenals typical of lipoid CAH. As these recent discoveries illustrate, more remains to be learned about human steroidogenesis and its disorders.

References

1 Soccio RE, Breslow JL: Intracellular cholesterol transport. Arterioscler Thromb Vasc Biol 2004;24: 1150–1160.
2 Haberland ME, Reynolds JA: Self-association of cholesterol in aqueous solution. Proc Natl Acad Sci USA 1973;70:2313–2316.
3 Gallegos AM, Atshaves BP, Storey SM, Starodub O, Petrescu AD, Huang H, McIntosh AL, Martin GG, Chao H, Kier AB, Schroeder F: Gene structure, intracellular localization, and functional roles of sterol carrier protein-2. Prog Lipid Res 2001;40:498–563.
4 Ponting CP, Aravind L: START: a lipid-binding domain in StAR, HD-ZIP and signalling proteins. Trends Biochem Sci 1999;24:130–132.
5 Soccio RE, Breslow JL: StAR-related lipid transfer (START) proteins: mediators of intracellular lipid metabolism. J Biol Chem 2003;278:22183–22186.
6 Soccio RE, Adams RM, Romanowski MJ, Sehayek E, Burley SK, Breslow JL: The cholesterol-regulated StarD4 gene encodes a StAR-related lipid transfer protein with two closely related homologues, StarD5 and StarD6. Proc Natl Acad Sci USA 2002;99:6943–6948.
7 Romanowski MJ, Soccio RE, Breslow JL, Burley SK: Crystal structure of the *Mus musculus* cholesterol-regulated START protein 4 (StarD4) containing a StAR-related lipid transfer domain. Proc Natl Acad Sci USA 2002;99:6949–6954.
8 Soccio RE, Adams RM, Maxwell KN, Breslow JL: Differential gene regulation of StarD4 and StarD5 cholesterol transfer proteins. J Biol Chem 2005;280: 19410–19418.
9 Bose HS, Whittal RM, Ran Y, Bose M, Baker BY, Miller WL: StAR-like activity and molten globule behavior of StARD6, a male germ-line protein. Biochemistry 2008;47:2277–2288.
10 Riegelhaupt JJ, Waase MP, Garbarino J, Cruz DE, Breslow JL: Targeted disruption of steroidogenic acute regulatory protein D4 leads to modest weight reduction and minor alterations in lipid metabolism. J Lipid Res 2010;51:1134–1143.
11 Gonzalez FJ: The molecular biology of cytochrome P450s. Pharmacol Rev 1998;40:243–288.
12 Solish SB, Picado-Leonard J, Morel Y, Kuhn RW, Mohandas TK, Hanukoglu I, Miller WL: Human adrenodoxin reductase: two mRNAs encoded by a single gene on chromosome 17cena→q25 are expressed in steroidogenic tissues. Proc Natl Acad Sci USA 1988;85:7104–710814.
13 Picado-Leonard J, Voutilainen R, Kao L, Chung B, Strauss JF III, Miller WL: Human adrenodoxin: cloning of three cDNAs and cycloheximide enhancement in JEG-3 cells. J Biol Chem 1988;263:3240–3244.
14 Miller WL: Regulation of steroidogenesis by electron transfer. Endocrinology 2005;146:2544–2550.
15 Harikrishna JA, Black SM, Szklarz GD, Miller WL: Construction and function of fusion enzymes of the human cytochrome P450scc system. DNA Cell Biol 1993;12:371–379.
16 Black SM, Harikrishna JA, Szklarz GD, Miller WL: The mitochondrial environment is required for activity of the cholesterol side-chain cleavage enzyme, cytochrome P450scc. Proc Natl Acad Sci USA 1994;91:7247–7251.
17 Miller WL 1995 Mitochondrial specificity of the early steps in steroidogenesis. J Steroid Biochem Mol Biol 1995;55:607–616.
18 Voutilainen R, Miller WL: Coordinate tropic hormone regulation of mRNAs for insulin-like growth factor II and the cholesterol side-chain cleavage enzyme, P450scc, in human steroidogenic tissues. Proc Natl Acad Sci USA 1987;84:1590–1594.
19 Mesiano S, Mellon SH, Jaffe RB: Mitogenic action, regulation, and localization of insulin-like growth factors in the human fetal adrenal gland. J Clin Endocrinol Metab 1993;76:968–976.
20 Mesiano S, Mellon SH, Gospodarowicz D, Di Blasio AM, Jaffe RB: Basic fibroblast growth factor expression is regulated by corticotropin in the human fetal adrenal: a model for adrenal growth regulation. Proc Natl Acad Sci USA 1991;88:5428–5432.

21 Coulter CL, Read LC, Carr BR; A role for epidermal growth factor in the morphological and functional maturation of the adrenal gland in the fetal rhesus monkey in vivo. J Clin Endocrinol Metab 1996;81: 1254–1260.

22 Moore CCD, Brentano ST, Miller WL: Human P450scc gene transcription is induced through cyclic AMP and repressed by 12-*O*-tetradecanoylph orbol-13-acetate and A23187 by independent *cis*-elements. Mol Cell Biol 1990;10:6013–6023.

23 Arakane F, King SR, Du Y, Kallen CB, Walsh LP, Watari H, Stocco DM, Strauss JF III: Phosphorylation of steroidogenic acute regulatory protein (StAR) modulates its steroidogenic activity. J Biol Chem 1997;272:32656–32662.

24 Stocco DM, Wang XJ, Jo Y, Manna PR: Multiple signaling pathways regulating steroidogenesis and steroidogenic acute regulatory protein expression: more complicated than we thought. Mol Endocrinol 2005;19:2647–2659.

25 Ferguson JJ: Protein synthesis and adrenocorticotropin responsiveness. J Biol Chem 1963;238: 2754–2759.

26 Garren LD, Ney RL, Davis WW: Studies on the role of protein synthesis in the regulation of corticosterone production by ACTH in vivo. Proc Natl Acad Sci USA 1965;53:1443–1450.

27 Davis WW, Garren LD: On the mechanism of action of adrenocorticotropic hormone. The inhibitory site of cycloheximide in the pathway of steroid biosynthesis. J Biol Chem 1968;243:5153–5157.

28 Pon LA, Hartigan JA, Orme-Johnson NR: Acute ACTH regulation of adrenal corticosteroid biosynthesis: rapid accumulation of a phosphoprotein. J Biol Chem 1986;261:13309–13316.

29 Epstein LF, Orme-Johnson NR: Regulation of steroid hormone biosynthesis: identification of precursors of a phosphoprotein targeted to the mitochondrion in stimulated rat adrenal cortex cells. J Biol Chem 1991;266:19739–19745.

30 Stocco DM, Sodeman TC: The 30-kDa mitochondrial proteins induced by hormone stimulation in MA-10 mouse Leydig tumor cells are processed from larger precursors. J Biol Chem 1991;266:19731–19738.

31 Clark BJ, Wells J, King SR, Stocco DM: The purification, cloning and expression of a novel luteinizing hormone-induced mitochondrial protein in MA-10 mouse Leydig tumor cells. Characterization of the steroidogenic acute regulatory protein (StAR). J Biol Chem 1994;269:28314–28322.

32 Stocco DM, Clark BJ: Regulation of the acute production of steroids in steroidogenic cells. Endocr Rev 1996;17:221–244.

33 Lin D, Sugawara T, Strauss JF III, Clark BJ, Stocco DM, Saenger P, Rogol A, Miller WL: Role of steroidogenic acute regulatory protein in adrenal and gonadal steroidogenesis. Science 1995;267:1828–1831.

34 Bose HS, Sugawara T, Strauss JF III, Miller WL: The pathophysiology and genetics of congenital lipoid adrenal hyperplasia. N Engl J Med 1996;335:1870–1878.

35 Chung B, Matteson KJ, Voutilainen R, Mohandas TK, Miller WL: Human cholesterol side-chain cleavage enzyme, P450scc: cDNA cloning, assignment of the gene to chromosome 15, and expression in the placenta. Proc Natl Acad Sci USA 1986;83: 8962–8966.

36 Sugawara T, Holt JA, Driscoll D, Strauss JF III, Lin D, Miller WL, Patterson D, Clancy KP, Hart IM, Clark BJ, Stocco DM: Human steroidogenic acute regulatory protein (StAR): functional activity in COS-1 cells, tissue-specific expression, and mapping of the structural gene to 8p11.2 and an expressed pseudogene to chromosome 13. Proc Natl Acad Sci USA 1995;92:4778–4782.

37 Miller WL, Strauss JF III: Molecular pathology and mechanism of action of the steroidogenic acute regulatory protein, StAR. J Steroid Biochem Mol Biol 1999;69:131–141.

38 Miller WL: StAR search – what we know about how the steroidogenic acute regulatory protein mediates mitochondrial cholesterol import. Mol Endocrinol 2007;21:589–601.

39 Miller WL: Steroidogenic acute regulatory protein (StAR), a novel mitochondrial cholesterol transporter. Biochim Biophys Acta 2007;1771:663–676.

40 Arakane F, Sugawara T, Nishino H, Liu Z, Holt JA, Pain D, Stocco DM, Miller WL, Strauss JF III: Steroidogenic acute regulatory protein (StAR) retains activity in the absence of its mitochondrial targeting sequence: implications for the mechanism of StAR action. Proc Natl Acad Sci USA 1996;93: 13731–13736.

41 Bose HS, Lingappa VR, Miller WL 2002 Rapid regulation of steroidogenesis by mitochondrial protein import. Nature 417:87–91.

42 Tsujishita Y, Hurley JH: Structure and lipid transport mechanism of a StAR-related domain. Nat Struct Biol 2000;7:408–414.

43 Bose HS, Whittal RM, Baldwin MA, Miller WL: The active form of the steroidogenic acute regulatory protein, StAR, appears to be a molten globule. Proc Natl Acad Sci USA 1999;96:7250–7255.

44 Baker BY, Yaworsky DC, Miller WL: A pH-dependent molten globule transition is required for activity of the steroidogenic acute regulatory protein, StAR. J Biol Chem 2005;280:41753–41760.

45 Christinsen K, Bose HS, Harris FM Miller WL, Bell JD: Binding of StAR to synthetic membranes suggests an active molten globule. J Biol Chem 2001; 276:17044–17051.
46 Tuckey RC, Headlam MJ, Bose HS, Miller WL: Transfer of cholesterol between phospholipid vesicles mediated by the steroidogenic acute regulatory protein (StAR). J Biol Chem 2002;277:47123–47128.
47 Baker BY, Epand RF, Epand RM, Miller WL: Cholesterol binding does not predict activity of the steroidogenic acute regulatory protein, StAR. J Biol Chem 2007;282:10223–10232.
48 Hauet T, Yao ZX, Bose HS, Wall CT, Han Z, Li W, Hales DB, Miller WL, Culty M, Papadopoulos V: Peripheral-type benzodiazepine receptor-mediated action of steroidogenic acute regulatory protein on cholesterol entry into Leydig cell mitochondria. Mol Endocrinol 2005;19:540–554.
49 Liu J, Rone MB, Papadopoulos V: Protein-protein interactions mediate mitochondrial cholesterol transport and steroid biosynthesis. J Biol Chem 2006;281:38879–38893.
50 Bose M, Whittal RM, Miller WL, Bose HS: Steroidogenic activity of StAR requires contact with mitochondrial VDAC1 and PCP. J Biol Chem 2008;283:8837–8845.
51 Artemenko IP, Zhao D, Hales DB, Hales KH, Jefcoate CR: Mitochondrial processing of newly synthesized steroidogenic acute regulatory protein (StAR), but not total StAR, mediates cholesterol transfer to cytochrome P450 side-chain cleavage enzyme in adrenal cells. J Biol Chem 2001;276: 46583–46596.
52 Bose HS, Whittal RM, Baldwin MA, Miller WL: The active form of the steroidogenic acute regulatory protein, StAR, appears to be a molten globule. Proc Natl Acad Sci USA 1999;96:7250–7255.
53 Moog-Lutz C, Tomasetto C, Régnier CH, Wendling C, Lutz Y, Muller D, Chenard MP, Basset P, Rio MC: MLN64 exhibits homology with the steroidogenic acute regulatory protein (StAR) and is overexpressed in human breast carcinomas. Int J Cancer 1997;71:183–191.
54 Watari H, Arakane F, Moog-Lutz C, Callen CB, Tomasetto C, Gerton GL, Rio MC, Baker ME, Strauss JF III: MLN64 contains a domain with homology to the steroidogenic acute regulatory protein (StAR) that stimulates steroidogenesis. Proc Natl Acad Sci USA 1997;94:8462–8467.
55 Bose HS, Whittal RM, Huang MC, Baldwin MA, Miller WL: N-218 MLN64, a protein with StAR-like steroidogenic activity, is folded and cleaved similarly to StAR. Biochemistry 2000;39:11722–11731.
56 Alpy F, Wendling C, Rio MC, Tomasetto C: MENTHO, a MLN64 homologue devoid of the START domain. J Biol Chem 2002;277:50780–50787.
57 Alpy F, Latchumanan VK, Kedinger V, Janoshazi A, Thiele C, Wendling C, Rio MC, Tomasetto C: Functional characterization of the MENTAL domain. J Biol Chem 2005;280:17945–17952.
58 Cheruku SR, Xu Z, Dutia R, Lobel P, Storch J: Mechanism of cholesterol transfer from the Niemann-Pick type C2 protein to model membranes supports a role in lysosomal cholesterol transport. J Biol Chem 2006;281:31594–31604.
59 Mathieu AP, Fleury A, Ducharme L, Lavigne P, LeHoux JG: Insights into steroidogenic acute regulatory protein (StAR)-dependent cholesterol transfer in mitochondria: evidence from molecular modeling and structure-based thermodynamics supporting the existence of partially unfolded states of StAR. J Mol Endocrinol 2002;29:327–345.
60 Yaworsky DC, Baker BY, Bose HS, Best KB, Jensen LB, Bell JD, Baldwin MA, Miller WL: pH-dependent interactions of the carboxyl-terminal helix of steroidogenic acute regulatory protein with synthetic membranes. J Biol Chem 2005;280:2045–2054.
61 Murcia M, Faraldo-Gomez JD, Maxfield FR, Roux B: Modeling the structure of the StART domains of MLN64 and StAR proteins in complex with cholesterol. J Lipid Res 2006;47:2614–2630.
62 Baker BY, Yaworsky DC, Miller WL: A pH-dependent molten globule transition is required for activity of the steroidogenic acute regulatory protein, StAR. J Biol Chem 2005;280:41753–41760.
63 Sandison AT: A form of lipoidosis of the adrenal cortex in an infant. Arch Dis Childh 1955;30:538–541.
64 Prader A, Gurtner HP: Das Syndrom des Pseudohermaphroditismus masculinus bei kongenitaler Nebennierenrindenhyperplasie ohne Androgenuberproduktion (adrenaler Pseudohermaphroditismus masculinus). Helv Paediatr Acta 1955;10:397–412.
65 Hauffa BP, Miller WL, Grumbach MM, Conte FA, Kaplan SL: Congenital adrenal hyperplasia due to deficient cholesterol side-chain cleavage activity (20,22-desmolase) in a patient treated for 18 years. Clin Endocrinol 1985;23:481–493.
66 Camacho AM, Kowarski A, Migeon CJ, Brough A: Congenital adrenal hyperplasia due to a deficiency of one of the enzymes involved in the biosynthesis of pregnenolone. J Clin Endocrinol Metab 1968;28: 153–161.

67 Matteson KJ, Chung B, Urdea MS, Miller WL: Study of cholesterol side-chain cleavage (20,22-desmolase) deficiency causing congenital lipoid adrenal hyperplasia using bovine-sequence P450scc oligodeoxyribonucleotide probes. Endocrinology 1986;118: 1296–1305.
68 Lin D, Gitelman SE, Saenger P, Miller WL: Normal genes for the cholesterol side-chain cleavage enzyme, P450scc, in congenital lipoid adrenal hyperplasia. J Clin Invest 1991;88:1955–1962.
69 Saenger P, Klonari Z, Black SM, Compagnone N, Mellon SH, Fleischer A, Abrams CAL, Shackleton CHL, Miller WL: Prenatal diagnosis of congenital lipoid adrenal hyperplasia. J Clin Endocrinol Metab 1995;80:200–205.
70 Tee MK, Lin D, Sugawara T, Holt JA, Guiguen Y, Buckingham B, Strauss JF III, Miller WL: T→A transversion 11 bp from a splice acceptor site in the gene for steroidogenic acute regulatory protein causes congenital lipoid adrenal hyperplasia. Hum Mol Genet 1995;4:2299–2305.
71 Miller WL: Congenital lipoid adrenal hyperplasia: the human gene knockout of the steroidogenic acute regulatory protein. J Mol Endocrinol 1997;19:227–240.
72 Ogata T, Matsuo N, Saito M, Prader A: The testicular lesion and sexual differentiation in congenital lipoid adrenal hyperplasia. Helv Paediatr Acta 1989; 43:531–538.
73 Chen X, Baker BY, Abduljabbar MA, Miller WL: A genetic isolate of congenital lipoid adrenal hyperplasia with atypical clinical findings. J Clin Endocrinol Metab 2005;90:835–840.
74 Bose HS, Pescovitz OH, Miller WL: Spontaneous feminization in a 46,XX female patient with congenital lipoid adrenal hyperplasia caused by a homozygous frame-shift mutation in the steroidogenic acute regulatory protein. J Clin Endocrinol Metab 1997;82:1511–1515.
75 Fujieda K, Tajima T, Nakae J, Sageshima S, Tachibana K, Suwa S, Sugawara T, Strauss JF III: Spontaneous puberty in 46,XX subjects with congenital lipoid adrenal hyperplasia. J Clin Invest 1997;99:1265–1271.
76 Voutilainen R, Miller WL: Developmental expression of genes for the steroidogenic enzymes P450scc (20,22-desmolase), P450c17 (17α-hydroxylase/17,20-lyase) and P450c21 (21-hydroxylase) in the human fetus. J Clin Endocrinol Metab 1986;63:1145–1150.
77 Hasegawa T, Zhao L, Caron KM, Majdic G, Suzuki T, Shizawa S, Sasano H, Parker KL: Developmental roles of the steroidogenic acute regulatory protein (StAR) as revealed by StAR knockout mice. Mol Endocrinol 2000;14:1462–1471.
78 Nakae J, Tajima T, Sugawara T, Arakane F, Hanaki K, Hotsubo T, Igarashi N, Igarashi Y, Ishii T, Koda N, Kondo T, Kohno H, Nakagawa Y, Tachibana K, Takeshima Y, Tsubouchi K, Strauss JF III, Fujieda K: Analysis of the steroidogenic acute regulatory protein (StAR) gene in Japanese patients with congenital lipoid adrenal hyperplasia. Hum Mol Genet 1997;6:571–576.
79 Yoo HW, Kim GH: Molecular and clinical characterization of Korean patients with congenital lipoid adrenal hyperplasia. J Pediatr Endocrinol Metab 1998;11:707–711.
80 Bose HS, Sato S, Aisenberg J, Shalev SA, Matsuo N, Miller WL: Mutations in the steroidogenic acute regulatory protein (StAR) in six patients with congenital lipoid adrenal hyperplasia. J Clin Endocrinol Metab 2000;85:3636–3639.
81 Flück CE, Maret A, Mallet D, Portrat-Doyen S, Achermann JC, Leheup B, Theintz GE, Mullis PE, Morel Y: A novel mutation L260P of the steroidogenic acute regulatory protein gene in three unrelated patients of Swiss ancestry with congenital lipoid adrenal hyperplasia. J Clin Endocrinol Metab 2005;90:5304–5308.
82 Gassner HL, Toppari J, Quinteiro Gonzalez S, Miller WL: Near-miss apparent SIDS from adrenal crisis. J Pediatr 2004;145:178–183.
83 Baker BY, Lin L, Kim CJ, Raza L, Smith CP, Miller WL, Achermann JC: Non-classic congenital lipoid adrenal hyperplasia. A new disorder of the steroidogenic acute regulatory protein with very late presentation and normal male genitalia. J Clin Endocrinol Metab 2006;91:4781–4785.
84 Metherell LA, Naville D, Halaby G, Begeot M, Huebner A, Nurnberg G, Nurnberg P, Green J, Tomlinson JW, Krone NP, Lin L, Racine M, Berney DM, Achermann JC, Arlt W, Clark AJ: Nonclassic lipoid congenital adrenal hyperplasia masquerading as familial glucocorticoid deficiency. J Clin Endocrinol Metab 2009;94:3865–3871.
85 Sahakitrungruang T, Soccio RE, Lang-Muritano M, Walker J, Achermann JC, Miller WL: Clinical, genetic and functional characterization of four patients carrying partial loss-of-function mutations in the steroidogenic acute regulatory protein (StAR). J Clin Endocrinol Metab 2010;95:3352–3359.
86 Khoury K, Barbar E, Ainmelk Y, Ouellet A, LeHoux JG: Gonadal function, first cases of pregnancy, and child delivery in a woman with lipoid congenital adrenal hyperplasia. J Clin Endocrinol Metab 2009; 94:1333–1337.
87 Stone D, Hechter O: Studies on ACTH action in perfused bovine adrenals: aspects of progesterone as an intermediary in cortico-steroidogenesis. Arch Biochem Biophys 1955;54:121–128.

88 Koritz SB, Kumar AM: On the mechanism of action of adrenocorticotropic hormone – the stimulation of the activity of enzymes involved in pregnenolone synthesis. J Biol Chem 1970;245: 152–157.
89 Kuwada M, Kitajima R, Suzuki H, Horie S: Purification and properties of cytochrome P-450 (SCC) from pig testis mitochondria. Biochem Biophy Res Com 1991;176:1501–1508.
90 Tuckey RC, Cameron KJ: Catalytic properties of cytochrome P-450scc. Biochim Biophys Acta 1993; 1163:185–194.
91 Simpson ER: Cholesterol side-chain cleavage, cytochrome P450, and the control of steroidogenesis. Mol Cell Endocrinol 1979;13:213–227.
92 Miller WL: Molecular biology of steroid hormone synthesis. Endocr Rev 1988;9:295–318.
93 Morohashi K, Sogawa K, Omura T, Fujii-Kuriyama Y: Gene structure of human cytochrome P-450(scc), cholesterol desmolase. J Biochem 1987;101:879–887.
94 Yang X, Iwamoto K, Wang M, Artwohl J, Mason JI, Pang S: Inherited congenital adrenal hyperplasia in the rabbit is caused by a deletion in the gene encoding cytochrome P450 cholesterol side-chain cleavage enzyme. Endocrinology 1993;132:1977–1982.
95 Hu MC, Hsu NC, El Hadj NB, Pai CI, Chu HP, Wang CKL, Chung BC: Steroid deficiency syndromes in mice with targeted disruption of Cyp11a1. Mol Endocrinol 2002;16:1943–1950.
96 Tajima T, Fujieda K, Kouda N, Nakae J, Miller WL: Heterozygous mutation in the cholesterol side-chain cleavage enzyme (P450scc) gene in a patient with 46,XY sex reversal and adrenal insufficiency. J Clin Endocrinol Metab 2001;86:3820–3825.
97 Kim CJ, Lin L, Huang N, Quigley CA, AvRuskin TW, Achermann JC, Miller WL: Severe combined adrenal and gonadal deficiency caused by novel mutations in the cholesterol side-chain cleavage enzyme, P450scc. J Clin Endocrinol Metab 2008;93: 696–702.
98 Hum DW, Miller WL: Transcriptional regulation of human genes for steroidogenic enzymes. Clin Chem 1993;39:333–340.
99 Parker KL, Schimmer BP: Steroidogenic factor 1: a key determinant of endocrine development and function. Endocr Rev 1997;18:361–377.
100 Morohashi K: Gonadal and extragonadal functions of Ad4BP/SF-1: developmental aspects. Trends Endocrinol Metab 1999;10:169–173.
101 Moore CCD, Hum DW, Miller WL: Identification of positive and negative placental-specific basal elements, a transcriptional repressor, and a cAMP response element in the human gene for P450scc. Mol Endocrinol 1992;6:2045–2058.
102 Huang N, Miller WL: Cloning of factors related to HIV-inducible LBP proteins that regulate steroidogenic factor-1-independent human placental transcription of the cholesterol side-chain cleavage enzyme, P450scc. J Biol Chem 2000;275:2852–2858.
103 Huang N, Miller WL: LBP proteins mediate SF1-independent expression of P450scc in human placental JEG-3 cells. Mol Endocrinol 2005;19: 409–420.
104 Henderson YC, Frederick MJ, Jayakumar Choi AY, Wang MT, Kang Y, Evans R, Spring PM, Uesugi M, Clayman GL: Human LBP-32/MGR is a repressor of the P450scc in human choriocarcinoma cell line JEG-3. Placenta 2007;28:152–160.
105 Henderson YC, Frederick MJ, Wang MT, Hollier LM, Clayman GL: LBP-1b, LBP-9, and LBP-32/MGR detected in syncytiotrophoblasts from first-trimester human placental tissue and their transcriptional regulation. DNA Cell Biol 2008;27: 71–79.
106 Gizard F, Lavallee B, DeWitte F, Hum DW: A novel zinc finger protein TReP-132 interacts with CPB/p300 to regulate human CYP11A1 gene expression. J Biol Chem 2001;276:33881–33892.
107 Gizard F, Lavallee B, DeWitte F, Teissier E, Staels B, Hum DW: The transcriptional regulating protein of 132 kDa (TReP-132) enhances P450scc gene transcription through interaction with steroidogenic factor-1 in human adrenal cells. J Biol Chem 2002; 277:39144–39155.
108 Katsumata N, Ohtake M, Hojo T, Ogawa E, Hara T, Sato N, Tanaka T: Compound heterozygous mutations in the cholesterol side-chain cleavage enzyme gene (*CYP11A*) cause congenital adrenal insufficiency in humans. J Clin Endocrinol Metab 2002;87: 3808–3813.
109 Hiort O, Holterhus P-M, Werner R, Marschke C, Hoppe U, Partsch J, Riepe FG, Achermann JC, Struve D: Homozygous disruption of P450 side-chain cleavage (*CYP11A1*) is associated with prematurity, complete 46,XY sex reversal, and severe adrenal failure. J Clin Endocrinol Metab 2005;90: 538–541.
110 Al Kandari H, Katsumata N, Alexander S, Rasoul MA: Homozygous mutation of P450 side-chain cleavage enzyme gene (CYP11A1) in 46,XY patient with adrenal insufficiency, complete sex reversal and agenesis of corpus callosum. J Clin Endocrinol Metab 2006;91:2821–2826.

111 Rubtsov P, Karmanov M, Sverdlova P, Spirin P, Tiulpakov A: A novel homozygous mutation in CYP11A1 gene is associated with late-onset adrenal insufficiency and hypospadias in a 46,XY patient. J Clin Endocrinol Metab 2009;94:936–939.

112 Sahakitrungruang T, Blackett PR, Miller WL: Partial defect in the cholesterol side-chain cleavage enzyme, P450scc (CYP11A1) presenting as non-classic congenital lipoid adrenal hyperplasia (submitted).

Prof. Walter L. Miller
Department of Pediatrics, Room 672-S, University of California
San Francisco, CA 94143 0978 (USA)
Tel. +1 415 476 2598, E-Mail wlmlab@ucsf.edu

Ghizzoni L, Cappa M, Chrousos G, Loche S, Maghnie M (eds): Pediatric Adrenal Diseases.
Endocr Dev. Basel, Karger, 2011, vol 20, pp 20–27

The Physiology and Biochemistry of Adrenarche

Richard J. Auchus

Division of Endocrinology and Metabolism, Department of Internal Medicine, University of Texas Southwestern Medical Center, Dallas, Tex., USA

Abstract

Adrenarche refers to the rise in adrenal 19-carbon steroid production, primarily dehydroepiandrosterone (DHEA) and DHEA sulfate (DHEAS). Adrenarche, which occurs only in humans and a few higher primates, is independent of puberty and manifests clinically as the appearance of axillary and pubic hair, usually about age 8. The physiologic correlate of adrenarche is the development of the adrenal zona reticularis, which has many of the biochemical properties of the fetal adrenal. Recent studies have shown that adrenarche is not an abrupt and 'signaled' process occurring in mid-childhood but rather a continuous process since birth. In addition, some differences in gene expression between the zona reticularis and the fetal adrenal gland have begun to emerge. The purpose of adrenarche in human physiology remains enigmatic, although disease states associated with abnormalities in DHEA(S) production provide some clues.

What Is Adrenarche?

We are captivated by those aspects of our humanity regarded has unique to the human species. Cognition, religion, morality, and sport are some of these items, and all have become major topics of research, philosophy, and popular attention. Perhaps this attraction is why adrenarche has held the interest of endocrinologists and developmental biologists for decades. Adrenarche refers to the rise in 19-carbon (C_{19}) steroid production by the adrenal gland, primarily dehydroepiandrosterone (DHEA) and its sulfate (DHEAS) [1]. This increase in adrenal steroid output was traditionally recognized at about 6–8 years of age [2], and the clinical correlate is the development of axillary and pubic hair (pubarche). Adrenarche occurs only in human beings and a few other old world monkeys, such as chimpanzees and gorillas [3, 4], hence my digression above. In contrast, rodents and small mammals make little or no DHEAS, and rhesus macaques make high amounts of DHEAS at birth with little change throughout life [4]. Adrenarche occurs independent of gonads and puberty [5], a uniquely adrenal and (almost) uniquely human event.

For years, a 'trigger' for adrenarche was sought, similar to the rise in gonadotropins as the initiator of puberty. Although some fragments of pro-opiomelanocortin were proposed [6], none have been conclusively implicated [7, 8], although several probably participate in the process. Irrespective of its mechanism, the histologic correlate of adrenarche is the expansion of the zona reticularis (ZR), the innermost zone of the adrenal cortex. The ZR continues to grow throughout young adulthood, reaches its maximum development in the third decade of life, and then gradually involutes with aging [9]. This involution, which parallels an age-related decline in circulating DHEAS concentrations, has implicated the DHEAS decline as component of the aging process. Clinical trials have not consistently found benefit for DHEAS replacement in the elderly, as discussed later.

Adrenocorticotropin (ACTH) is certainly required for adrenarche to occur, as demonstrated by patients with hypopituitarism, who do not experience adrenarche. Conversely, women with Cushing's disease often suffer from high DHEAS production and hyperandrogenism. In fact, patients with Cushing's disease who are cured and experience a phase of central adrenal insufficiency for many months have low DHEAS, which can remain suppressed for years [10, 11], as if they must experience adrenarche again.

Perhaps the assumption that adrenarche is an event heralded in mid-childhood has clouded our thinking about adrenarche. This limitation derives from the prodigious synthesis of DHEAS in early adulthood, with typical values of 400 μg/dl (10 μmol/l). Immunoassays for DHEAS, as a result, generally cut off at a lower limit of 10–35 μg/dl, as if there is no point in calibrating the assay lower. Integrated measures of urinary C_{19} steroid, which primarily derive from the adrenal in prepubertal children, provide a different view of adrenarche. Urinary C_{19} steroid excretion rises continuously and exponentially from birth [12], suggesting that adrenarche is a gradual process that evolves throughout childhood rather than punctuating a specific phase. It is best to think of adrenarche as starting at birth but not having clinical manifestations until mid-childhood, which, coincidentally, is about the time commercial assays measure DHEAS above the limit of detection for the assays.

We have only a rudimentary understanding of the molecular mechanisms driving the zonation of the adrenal cortex. Although all cells of the adrenal cortex express the cholesterol side chain cleavage enzyme (CYP11A1, P450scc) and thus convert cholesterol to pregnenolone, the regulation of this process and the repertoire of downstream enzymes varies. The zona glomerulosa (ZG) produces aldosterone under the control of the renin-angiotensin system, while the zona fasciculata (ZF) and ZR respond to ACTH by producing cortisol and DHEAS, respectively (fig. 1). Curiously, the ZR resembles the fetal adrenal gland (FA) both histologically and biochemically, yet the FA involutes after birth and probably does not give rise to the ZR. The C_{19} steroids of the FA provide substrate for the massive production of estriol [13], primarily in the

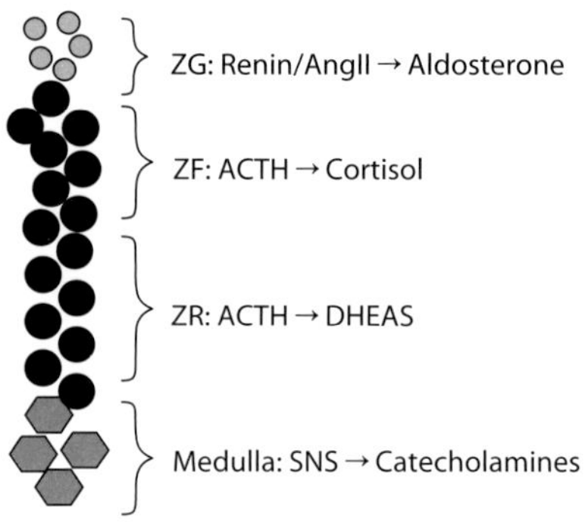

Fig. 1. Schematic of adrenal zonation. The zonae glomerulosa (ZG), fasciculata (ZF), and reticularis (ZR) form the adrenal cortex, and the medulla produces catecholamines under the stimulation of the sympathetic nervous system (SNS). The renin-angiotensin II (AngII) system regulates aldosterone production in the ZG, while ACTH stimulates cortisol and DHEAS production by the ZF and ZR, respectively.

third trimester of pregnancy, attesting to the biosynthetic capacity of the ZR. Some markers such as the enzyme AKR1C3 (17β-hydroxysteroid dehydrogenase type 5, 17β-HSD5) are expressed more highly in the ZR than in the FA [14], suggesting that the two tissues differ in other respects as well. Regardless, the genesis of the ZR and the mechanisms responsible for its growth remain unknown. Besides ACTH, insulin-like growth factors 1 and 2, fibroblast growth factor, and other growth factors stimulate growth of adrenal cells [15], but patients with growth hormone deficiency still experience adrenarche, so these other factors appear not essential, unlike ACTH.

Biochemistry and Enzymology of Adrenarche

The conversion of DHEAS from pregnenolone ostensibly requires only two enzymes, cytochrome P450c17 (CYP17A1, steroid 17-hydroxylase/-lyase) and DHEA sulfotransferase (SULT2A1), but the molecular machinery extends beyond these two proteins (fig. 2). The complexities of adrenarche and conditions that perturb adrenarche derive in part from the biochemical characteristics of human P450c17 and the participation of additional proteins in these reactions. First of all, human P450c17 is about equally active in 17α-hydroxylating pregnenolone and progesterone, although about 25% of progesterone is 16α-hydroxylated [16]. The 17-hydroxylase reaction requires the flavoprotein P450 oxidoreductase (POR) to transfer electrons from reduced nicotinamide adenine dinucleotide phosphate (NADPH) to the P450 heme moiety, and POR performs this function for all microsomal cytochromes P450 in the adrenal, testis, liver, and other tissues. The 17,20-lyase reaction, in contrast, is much more efficient in the pregnenolone pathway than the progesterone pathway [17]. The 17,20-lyase reaction also requires POR, but in addition, the presence of roughly one molar equivalent of cytochrome b_5 (b5) using in vitro systems with P450c17 and POR increases the rate of this reaction tenfold [18, 19]. The mechanism of b5 action

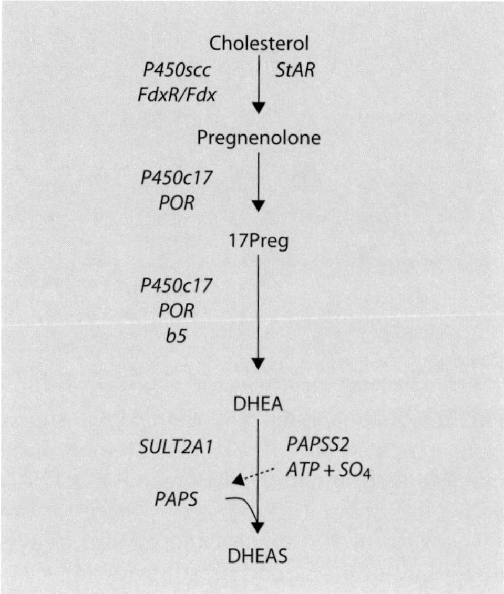

Fig. 2. Pathway and enzymes involved in adrenarche. Cholesterol is converted to pregnenolone by P450scc using ferridoxin and ferridoxin reductase (Fdx, FdxR) as electron transfer proteins. The steroidogenic acute regulatory protein (StAR) mobilizes cholesterol into steroidogenesis. P450c17 requires POR for both the 17-hydroxylase and 17,20-lyase reactions, and P450c17 also requires b5 for optimal 17,20-lyase activity. SULT2A1 converts DHEA to DHEAS using PAPS, which is synthesized from ATP and sulfate by PAPSS2.

remains unclear, although electron transfers to and from b5 are probably not involved [17], and b5 residues E48 and E49 are critical for this action [20]. Genetic evidence supports the need for all three proteins for normal 17,20-lyase activity, as isolated 17,20-lyase deficiency can be caused by mutations in the genes encoding P450c17 [21, 22], POR [23], or b5 [24]. P450c17 phosphorylation on serine and threonine is also involved in maximizing activity [25].

The consequences derived from these properties of human P450c17 are illustrated in figure 3. Human adrenal microsomes contain P450c17, P450c21, POR, b5, and 3β-hydroxysteroid dehydrogenase/isomerase type 2 (3β-HSD2) but not the mitochondrial CYP11 enzymes. Incubation with steroid and NADPH enables catalysis by P450c17 and P450c21 but not 3β-HSD2, which uses NAD^+ as cofactor. Pregnenolone is metabolized to 17α-hydroxypregnenolone, and 17α-hydroxypregnenolone is readily converted to DHEA. Progesterone, in contrast, is 17α- and 16α-hydroxylated via P450c17, but little androstenedione, the 17,20-lyase product, is formed from 17α-hydroxyprogesterone. Instead, 17α-hydroxyprogesterone is 21-hydroxylated by P450c21, yielding 11-deoxycortisol. In the ZG, this pathway generates cortisol, whereas absence of P450c21 in the ZR limits flux to DHEA.

Conversion of pregnenolone to DHEA would also be thwarted if 3β-HSD2 was abundant in the ZR. Human 3β-HSD2 catalyzes the conversion of many 3β-hydroxy-5-ene (Δ^5) steroids (pregnenolone, 17α-hydroxypregnenolone, DHEA, and androst-5-ene-3β,17β-diol) to their 3-keto-4-ene (Δ^4) congeners (progesterone, 17α-hydroxyprogesterone, androstenedione, and testosterone, respectively) [26, 27]. This conversion is essentially irreversible, unlike other HSD reactions [28], since the

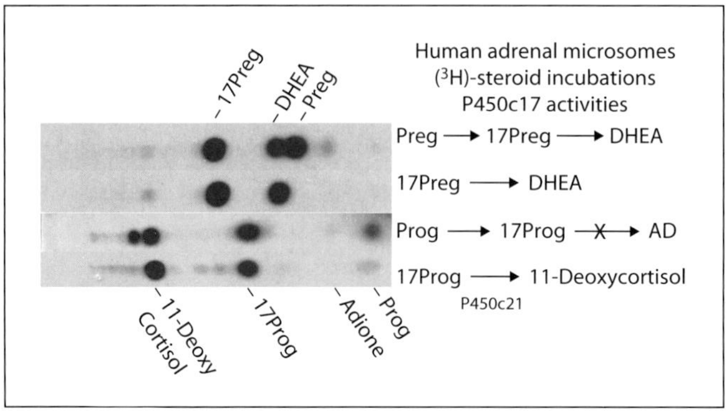

Fig. 3. Human adrenal microsomal P450 steroid metabolic pathways. Human adrenal microsomes were incubated with [^{3}H]-labeled steroids and NADPH, and the metabolites were extracted, chromatographed, and imaged. Human P450c17 17α-hydroxylates both pregnenolone (Preg to 17Preg) and progesterone (Prog to 17Prog, as well as 16α-hydroxyprogesterone, spot beneath 11-deoxycortisol). Human P450c17 also cleaves 17Preg to DHEA, but 17Prog is a poor substrate; instead, 17Prog is a good substrate for P450c21, yielding 11-deoxycortisol.

isomerase activity creates a very stable α,β-unsaturated ketone group. Adrenarche is associated with a decline in 3β-HSD2 expression specifically in the ZR [29, 30], which resembles the enzymology of the FA.

Immunohistochemical experiments have elegantly demonstrated the optimal pattern of gene expression in the ZR for DHEAS production. POR is expressed in all zones, P450c17 in the ZF and ZR, and 3β-HSD2 in the ZG and ZF but poorly in the ZR [31]. In addition, the ZR alone shows high expression of SULT2A1 and b5 [31], which augments 17,20-lyase activity and enables sulfation of DHEA. Curiously, b5 is widely expressed in many tissues, so the mechanism of relatively suppressed b5 expression in the ZR is unknown.

SULT2A1 catalyzes the sulfonation of many Δ^5 steroids, but given the enzymology of the ZR, DHEA is the major substrate. SULT2A1 requires phosphoadenosine phosphosulfate (PAPS) as the sulfate donor, and PAPS is synthesized in the adrenal primarily by PAPS synthase type 2 (PAPSS2) [32] using ATP and sulfate. Consequently, metabolism of pregnenolone to DHEAS requires the presence of at least three additional proteins beyond P450c17 and SULT2A1: POR, b5, and PAPSS2. In addition, low 3β-HSD2 expression is essential to keeping steroidogenesis along the Δ^5 pathway to DHEAS.

What Is the Purpose of Adrenarche?

While this is the most important question regarding adrenarche and its study, this is an issue about which, alas, we have perhaps the least insight. Patients with disorders preventing adrenarche, such as 17-hydroxylase deficiency and adrenal hypoplasia, do not appear to have any consistent major health deficits after glucocorticoid and mineralocorticoid replacement, so the production of DHEAS per se does not appear to be essential for life. Disorders in which adrenarche occurs early and/or with higher than normal DHEAS production, in contrast, are associated with certain health problems, particularly in women. Over half of testosterone in women normally derives from adrenal DHEAS as precursor, so androgen excess is an expected consequence of conditions such as premature adrenarche [33], classical and nonclassical 21-hydroxylase deficiency (21-OHD) [34], and tumors that produce DHEAS in girls and women. The sulfonation of DHEA itself is another protective mechanism of mitigating androgen excess in women despite high C_{19} steroid synthesis, because DHEAS is not a substrate for 3β-HSD enzymes, unlike DHEA. A girl with PAPSS2 deficiency, for example, had high circulating androstenedione and testosterone with premature pubarche despite low DHEAS [32]. In contrast, excess DHEAS production rarely harms men unless extreme, as exemplified by men with nonclassical 21-OHD being rarely ascertained [35], whereas men with classical 21-OHD can have infertility and hypogonadism [36] but few other problems related to adrenal C_{19} steroid production [37]. Could adrenarche be simply a necessary evil and consequence of human adrenal biochemistry, development, and zonation? This pessimistic view implies that adrenarche is best when least prominent, and only ill can derive from augmentation of DHEAS production.

A causative role of the age-related decline in DHEAS in the aging process has been suggested, but the larger and well-designed trials have not supported this notion [38]. In contrast, women with adrenal insufficiency have been shown to benefit from DHEAS replacement in some studies, mainly in their sexuality and sense of well-being [39]. Similar results have been obtain in hypopituitary women treated with low doses of testosterone [40], so it is not clear if the benefits of DHEA treatment derive from its conversion to testosterone or from direct effects of DHEA on the brain.

Conclusions and Future Directions

Adrenarche is not the renaissance of the fetal adrenal in mid-childhood as previously thought. Subtle differences between the fetal adrenal and the ZR are now recognized, and the genesis of the ZR is a gradual, exponentially progressive process that begins in infancy. The molecular mechanisms governing adrenal zonation and ZR development remain obscure, and the purposes of DHEAS in human physiology are controversial. Until we solve these last problems, we will not understand the meaning of adrenarche.

References

1 Miller WL: Androgen synthesis in adrenarche. Rev Endocr Metab Disord 2009;10:3–17.

2 Lashansky G, Saenger P, Fishman K, Gautier T, Mayes D, Berg G, DiMartino-Nardi J, Reiter E: Normative data for adrenal steroidogenesis in a healthy pediatric population: age- and sex-related changes after ACTH stimulation. J Clin Endocrinol Metab 1991;73:674–686.

3 Smail PJ, Faiman C, Hobson WC, Fuller GB, Winter JSD: Further studies on adrenarche in nonhuman primates. Endocrinology 1982;111:844–848.

4 Cutler GB, Glenn M, Bush M, Hodgen GD, Graham CE, Loriaux DL: Adrenarche: a survey of rodents, domestic animals and primates. Endocrinology 1978;103:2112–2118.

5 Sklar CA, Kaplan SL, Grumbach MM: Evidence for dissociation between adrenarche and gonadarche: studies in patients with idiopathic precocious puberty, gonadal dysgenesis, isolated gonadotropin deficiency, and constitutionally delayed growth and adolescence. J Clin Endocrinol Metab 1980;51:548–556.

6 Parker LN, Odell WD: Evidence for the existence of cortical androgen stimulating hormone. Am J Physiol 1979;236:616.

7 Mellon SH, Shively JE, Miller WL: Human proopiomelanocortin (79-96), a proposed androgen stimulatory hormone, does not affect steroidogenesis in cultured human fetal adrenal cells. J Clin Endocrinol Metab 1991;72:19–22.

8 Penhoat A, Sanchez P, Jaillard C, Langlois D, Begeot M, Saez JM: Human proopiomelanocortin (79–96), a proposed cortical androgen stimulating hormone, does not affect steroidogenesis in cultured human adult adrenal cells. J Clin Endocrinol Metab 1991;72:23–26.

9 Orentreich N, Brind JL, Rizer RL, Vogelman JH: Age changes and sex differences in serum dehydroepiandrosterone sulfate concentrations throughout adulthood. J Clin Endocrinol Metab 1984;59:551–555.

10 Klose M, Jorgensen K, Kristensen LO: Characteristics of recovery of adrenocortical function after treatment for Cushing's syndrome due to pituitary or adrenal adenomas. Clin Endocrinol (Oxf) 2004;61:394–399.

11 Kleiber H, Rey F, Temler E, Gomez F: Dissociated recovery of cortisol and dehydroepiandrosterone sulphate after treatment for Cushing's syndrome. J Endocrinol Invest 1991;14:489–492.

12 Remer T, Boye KR, Hartmann MF, Wudy SA: Urinary markers of adrenarche: reference values in healthy subjects, aged 3–18 years. J Clin Endocrinol Metab 2005;90:2015–2021.

13 Miller WL: Steroid hormone biosynthesis and actions in the materno-feto-placental unit. Clin Perinatol 1998;25:799–817.

14 Nakamura Y, Hornsby PJ, Casson P, Morimoto R, Satoh F, Xing Y, Kennedy MR, Sasano H, Rainey WE: Type 5 17β-hydroxysteroid dehydrogenase (AKR1C3) contributes to testosterone production in the adrenal reticularis. J Clin Endocrinol Metab 2009;94:2192–2198.

15 Townsend SF, Dallman MF, Miller WL: Insulin-like growth factor-I and -II mRNAs after two stimuli of adrenal growth; in Spencer E (ed): Modern Concepts of Insulin-Like Growth Factors. New York, Elsevier, 1991; pp 241–251.

16 Swart P, Swart AC, Waterman MR, Estabrook RW, Mason JI: Progesterone 16α-hydroxylase activity is catalyzed by human cytochrome P450 17α- hydroxylase. J Clin Endocrinol Metab 1993;77: 98–102.

17 Auchus RJ, Lee TC, Miller WL: Cytochrome b_5 augments the 17,20-lyase activity of human P450c17 without direct electron transfer. J Biol Chem 1998; 273:3158–3165.

18 Katagiri M, Kagawa N, Waterman MR: The role of cytochrome b_5 in the biosynthesis of androgens by human P450c17. Arch Biochem Biophys 1995;317: 343–347.

19 Yanagibashi K, Hall PF: Role of electron transport in the regulation of the lyase activity of C-21 side-chain cleavage P450 from porcine adrenal and testicular microsomes. J Biol Chem 1986;261: 8429–8433.

20 Naffin-Olivos JL, Auchus RJ: Human cytochrome b_5 requires residues E48 and E49 to stimulate the 17,20-lyase activity of cytochrome P450c17. Biochemistry 2006;46:100–108.

21 Geller DH, Auchus RJ, Mendonça BB, Miller WL: The genetic and functional basis of isolated 17,20-lyase deficiency. Nat Genet 1997;17:201–205.

22 Sherbet DP, Tiosano D, Kwist KM, Hochberg Z, Auchus RJ: CYP17 mutation E305G causes isolated 17,20-lyase deficiency by selectively altering substrate binding. J Biol Chem 2003;278:48563–48569.

23 Hershkovitz E, Parvari R, Wudy SA, Hartmann MF, Gomes LG, Loewental N, Miller WL: Homozygous mutation G539R in the gene for P450 oxidoreductase in a family previously diagnosed as having 17,20-lyase deficiency. J Clin Endocrinol Metab 2008;93:3584–3588.

24 Kok RC, Timmerman MA, Wolffenbuttel KP, Drop SL, de Jong FH: Isolated 17,20-lyase deficiency due to the cytochrome b_5 mutation W27X. J Clin Endocrinol Metab 2010;95:994–999.

25 Zhang L, Rodriguez H, Ohno S, Miller WL: Serine phosphorylation of human P450c17 increases 17,20-lyase activity: implications for adrenarche and for the polycystic ovary syndrome. Proc Natl Acad Sci USA 1995;92:10619–10623.
26 Thomas JL, Berko EA, Faustino A, Myers RP, Strickler RC: Human placental 3β-hydroxy-5-ene-steroid dehydrogenase and steroid 5/4-ene-isomerase: purification from microsomes, substrate kinetics, and inhibition by product steroids. J Steroid Biochem 1988;31:785–793.
27 Lee TC, Miller WL, Auchus RJ: Medroxyprogesterone acetate and dexamethasone are competitive inhibitors of different human steroidogenic enzymes. J Clin Endocrinol Metab 1999;84:2104–2110.
28 Mizrachi D, Auchus RJ: Androgens, estrogens, and hydroxysteroid dehydrogenases. Mol Cell Endocrinol 2009;301:37–42.
29 Dardis A, Saraco N, Rivarola MA, Belgorosky A: Decrease in the expression of the 3β-hydroxysteroid dehydrogenase gene in human adrenal tissue during prepuberty and early puberty: implications for the mechanism of adrenarche. Pediatr Res 1999;45:384–388.
30 Gell JS, Carr BR, Sasano H, Atkins B, Margraf L, Mason JI, Rainey WE: Adrenarche results from development of a 3β-hydroxysteroid dehydrogenase-deficient adrenal reticularis. J Clin Endocrinol Metab 1998;83:3695–3701.
31 Suzuki T, Sasano H, Takeyama J, Kaneko C, Freije WA, Carr BR, Rainey WE: Developmental changes in steroidogenic enzymes in human postnatal adrenal cortex: immunohistochemical studies. Clin Endocrinol (Oxf) 2000;53:739–747.
32 Noordam C, Dhir V, McNelis JC, Schlereth F, Hanley NA, Krone N, Smeitink JA, Smeets R, Sweep FC, Claahsen-van der Grinten HL, Arlt W: Inactivating PAPSS2 mutations in a patient with premature pubarche. N Engl J Med 2009;360:2310–2318.
33 Ibañez L, Virdis R, Potau N, Zampolli M, Ghizzoni L, Albisu MA, Carrascosa A, Bernasconi S, Vicens-Calvet E: Natural history of premature pubarche: an auxological study. J Clin Endocrinol Metab 1992;74:254–257.
34 White PC, Speiser PW: Congenital adrenal hyperplasia due to 21-hydroxylase deficiency. Endocr Rev 2000;21:245–291.
35 Speiser PW, New MI, White P: Molecular genetic analysis of nonclassical steroid 21-hydroxylase deficiency associated with HLA-B14DR1. N Engl J Med 1988;319:19–23.
36 Claahsen-van der Grinten HL, Otten BJ, Sweep FC, Hermus AR: Repeated successful induction of fertility after replacing hydrocortisone with dexamethasone in a patient with congenital adrenal hyperplasia and testicular adrenal rest tumors. Fertil Steril 2007;88:705 e5–e8.
37 Auchus RJ: Congenital adrenal hyperplasia in adults. Curr Opin Endocrinol Diabet Obes 2010;17:210–216.
38 Nair KS, Rizza RA, O'Brien P, Dhatariya K, Short KR, Nehra A, Vittone JL, Klee GG, Basu A, Basu R, Cobelli C, Toffolo G, Dalla Man C, Tindall DJ, Melton LJ 3rd, Smith GE, Khosla S, Jensen MD: DHEA in elderly women and DHEA or testosterone in elderly men. N Engl J Med 2006;355:1647–1659.
39 Arlt W, Callies F, van Vlijmen JC, Koehler I, Reincke M, Bidlingmaier M, Huebler D, Oettel M, Ernst M, Schulte HM, Allolio B: Dehydroepiandrosterone replacement in women with adrenal insufficiency. N Engl J Med 1999;341:1013–1020.
40 Miller KK, Biller BM, Beauregard C, Lipman JG, J. J, Schoenfeld D, Sherman JC, Swearingen B, Loeffler J, Klibanski A: Effects of testosterone replacement in androgen-deficient women with hypopituitarism: a randomized, double-blind, placebo-controlled study. J Clin Endocrinol Metab 2006;91:1683–1690.

Dr. Richard J. Auchus, MD, PhD
Division of Endocrinology and Metabolism, Department of Internal Medicine
University of Texas Southwestern Medical Center
5323 Harry Hines Blvd, Dallas, TX 75390-8857 (USA)
Tel. +1 214 648 6751, Fax +1 214 648 8917, E-Mail richard.auchus@UTSouthwestern.edu

Ghizzoni L, Cappa M, Chrousos G, Loche S, Maghnie M (eds): Pediatric Adrenal Diseases.
Endocr Dev. Basel, Karger, 2011, vol 20, pp 28–37

Update on the Corticomedullary Interaction in the Adrenal Gland

Matthias Haase[a] · Holger S. Willenberg[a] · Stefan R. Bornstein[b]

[a]Department of Endocrinology, Diabetes and Rheumatology, University Hospital Duesseldorf, Duesseldorf, and [b]Department of Medicine III, Carl Gustav Carus Medical School, University of Technology Dresden, Dresden, Germany

Abstract

The adrenal gland is formed by the adrenal medulla and the adrenal cortex. Both tissues descend from different origins during embryonal development. While the chromaffin cells are derived from the neural crest, the adrenocortical cells stem from a cell condensation in the celomic epithelium. Already during adrenal organogenesis, close interactions between the two tissue types are necessary for the differentiation, morphogenesis and survival of the adrenal gland. Moreover, the communication between the chromaffin and adrenocortical cells ensures a regular function of the adult adrenal gland including the regulation of hormone synthesis and responses to stress. This is even more important since the cortical-chromaffin crosstalk is also relevant for the pathogenesis of different diseases. In the past decade, significant progress in the understanding of the cortical-chromaffin communication has been made. Here, we summarize the insights gained from in vitro studies, from animal models and from clinical observations.

The adult human adrenal gland is composed of two different tissue types, namely the adrenal medulla and the adrenal cortex. Both tissues have different origins during embryonal development and produce different types of hormones. At an early stage of development, cells within the celomic epithelium condense and form the adrenogonadal primordium. A short time later during organogenesis, the adrenogonadal primordium divides and neural crest-derived chromaffin precursor cells migrate towards the adrenal primordium where they start forming the adrenal medulla. This represents the first contact between steroidogenic cells of the adrenal cortex and the chromaffin cell type and marks the beginning of an interaction for lifetime. However, the morphology that underlies the adrenocortical-medullary interaction changes during development. First, chromaffin cells form dispersed islets within the developing adrenal gland. Not until birth, the adrenal medullary cells start to coalesce and constitute the adult adrenal medulla. At this time, the fetal adrenocortical zones start to

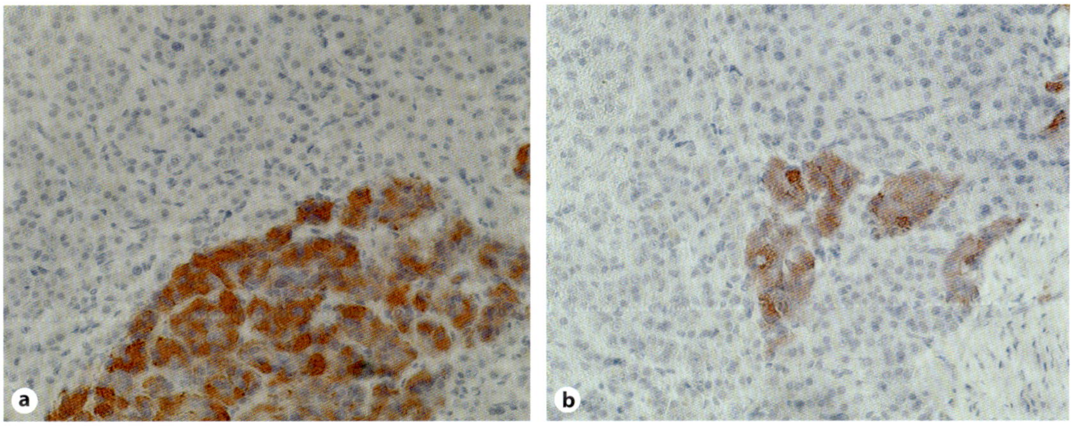

Fig. 1. **a** Chromaffin cells are stained red with an antibody to chromogranin A using immunohistochemistry technique. The adjacent adrenocortical cells belong to the zona reticularis. **b** Chromaffin cells are stained red with an antibody to chromogranin A using immunohistochemistry technique. The chromaffin cells are completely surrounded by adrenocortical cells.

differentiate into adult zones, including *zona fasciculata* and *zona glomerulosa*. Later on, the *zona reticularis* develops during a process called adrenarche and gives rise to an increase in adrenal androgen secretion [1]. Three main structural characteristics facilitate the interaction between the adrenal cortex and medulla: (1) The blood flow is directed centripedally from an outer arterial plexus to the central vein. The blood in the cortical capillaries collects adrenal steroid hormones from each zone for delivery to the chromaffin tissue. (2) Although the structure of the adrenal gland is in change during life, a close contact between chromaffin cells and adrenocortical cells is continuously maintained. The cortical-medullary junction is mainly composed of medullary cells and adrenocortical cells of the *zona reticularis* (fig. 1a). (3) Also, in the adult adrenal gland, islets of cortical cells are regularly found between adrenal medullary cells and vice versa chromaffin cells can be found in all zones of the adrenal cortex (fig. 1b) [2].

During the last years, significant progresses in the understanding of the adrenocortical-medullary interactions have been made.

Evidence from in vitro Studies

How the Adrenal Cortex Actuates the Function of Chromaffin Cells

Glucocorticoids can induce enzymes necessary for catecholamine biosynthesis, including the phenylethanolamine N-methyltransferase (PNMT) that converts norepinephrine to epinephrine, the tyrosine hydroxylase (TH) that metabolizes tyrosine to dihydroxy-phenylalanine (L-DOPA) and the dopamine-β-hydroxylase that converts dopamine to norepinephrine. Evidence for this important action of glucocorticoids comes primarily from cell culture models that employ chromaffin cells.

In 1965, Wurtman and Axelrod [3] demonstrated impressively that the expression of PNMT as the key enzyme of epinephrine synthesis and consequently the synthesis of epinephrine in rats depends on the presence of glucocorticoids. Therefore, the adrenal medulla can be seen as a 'target organ' of the adrenal cortex, controlled by the pituitary-adrenocortical axis [3]. Anderson and Axel [4] demonstrated in 1986 that the fate of bipotential neuroendocrine adrenomedullary precursors of the rat is determined by glucocorticoids. In these studies, the nerve growth factor (NGF) induces a neuronal phenotype in these cells, while the coincubation of the precursors with dexamethasone directs cell differentiation towards an endocrine phenotype. In the following, cell culture experiments confirmed the important role of glucocorticoids for the differentiation of sympathoadrenal precursor cells by demonstrating that glucocorticoids are essential and may be sufficient to induce an endocrine phenotype of sympathoadrenal precursors that express a functional glucocorticoid receptor system [5]. However, the work of other groups demonstrated that glucocorticoids alone may not be sufficient for the determination of a chromaffin cell type. PC12 cells for example may require a combination of glucocorticoids, neuropeptides, growth factors, and extracellular matrix for a functional chromaffin phenotype [6].

Interestingly, more and more data suggest that adrenal androgens such as dehydroepiandrosterone (DHEA) and its sulfate ester (DHEA-S) also interfere with the physiology of chromaffin cells. Adrenal androgens are synthesized within the human fetal adrenal gland as well as in the zona reticularis of the adult adrenal gland in close proximity to the adrenal medulla and show highest serum levels at the age around 25 years [1, 7]. DHEA induces a neuroendocrine phenotype in NGF-stimulated PC12 cells and the expression of neuroendocrine markers such as chromogranin A is stimulated by the coincubation with DHEA-S [8, 9]. Furthermore, DHEA and DHEA-S stimulate catecholamine release and synthesis by chromaffin cells and neuronal differentiation is inhibited by DHEA-S as shown by a decrease in neurite outgrowth and a decrease in the expression of neuronal markers [10].

Proliferation and survival of chromaffin cells are also affected by adrenal androgens. In PC12 cells, DHEA and DHEA-S reduce NGF-promoted cell survival and lead to an inhibition of the mitogen-activated protein kinases ERK1/2 [9, 11]. DHEA also inhibits insulin-like growth factor II (IGF-II)-induced proliferation in young bovine chromaffin cells, while basic fibroblast growth factor-induced proliferation was not affected [12]. Other studies confirmed the antiproliferative effects of DHEA after stimulation with leukemia inhibitory factor and epidermal growth factor in young and adult bovine chromaffin cells [13]. In contrast, DHEA-S increased the proliferative effects of growth factors such as epidermal growth factor in adult bovine cells but not in young bovine cells, indicating that not only the serum level of DHEA-S but also the activity of DHEA-S is age-dependent [13].

Recently, it has been shown that the mineralocorticoid receptor is expressed in rat PC12 cells and that adrenal mineralocorticoids enhance dopamine synthesis and TH expression by the activation of the mineralocorticoid receptor and in part also by non-genomic effects [14].

How Chromaffin Cells Actuate the Function of the Adrenal Cortex
Several studies have examined the effects of chromaffin cell products on adrenocortical cells, including catecholamines and neuropeptides [15, 16]. The catecholamines epinephrine and norepinephrine are one of the main products of human and animal chromaffin cells. Coculture experiments with chromaffin and adrenocortical cells demonstrate that chromaffin cell products stimulate the steroidogenic activity of adrenocortical cells and that catecholamines contribute in part to this observation [17]. Other studies show that catecholamines stimulate the cyclic adenosine monophosphate synthesis, expression of steroidogenic enzymes and steroid release, including aldosterone, cortisol and androstenedione in different adrenocortical cell types [15, 18–20]. Moreover, a vast number of neuropeptides that originate from chromaffin cells or sympathoadrenal nerve endings have been shown to stimulate or to inhibit adrenocortical function including enkephalins, dynorphin, neuropeptide Y, substance P and related peptides, neurokinin A, vasoactive intestinal peptide (VIP), galanin, atrial natriuretic peptide, somatostatin, β-endorphin, adrenomedullin, corticotropin-releasing hormone, bombesin, vasopressin and others [15]. For example, VIP is expressed within the adrenal medulla and has been identified as a potent stimulator of DHEA, testosterone, androstendione, cortisol, and aldosterone secretion in rat, pig and human adrenocortical cells, while the effects on adrenocortical cells seem to be mediated in part indirectly by catecholamines [21–24]. The regulation of VIP secretion in the adrenal gland is in part controlled by splanchnic nerves but also by proinflammatory cytokines such as interleukin-1α [22, 25]. The neuropeptide galanin is produced in the adrenal medulla and increases glucocorticoid secretion by rat adrenocortical cells through the direct activation of galanin receptors on adrenocortical cells and also indirectly through the stimulation of catecholamine release from medullary cells [26]. Recent studies confirm the stimulatory effects of proinflammatory cytokines including tumor necrosis factor-α on the VIP and galanin expression in chromaffin cells, suggesting also a role of the cortical-medullary interaction for inflammatory responses [27]. However, the fine tuning of the cortical-medullary interaction and especially the signaling network of neuropeptides is complex and poorly understood.

Evidence from Animal Models

Animal models have significantly contributed to the exploration of the cortical-medullary interactions in the adrenal gland. Initially, Conn and Fajans [28] assumed

Table 1. Animal models that provide relevant data on the cortical-medullary interaction

Model	Primarily affected tissue	Consequence for adjacent tissue	Related literature
Steroidogenic factor-1 (SF-1) k.o.	no adrenal cortex	reduced number of chromaffin cells, delayed expression of PNMT in SF-1 heterozygous mice, impaired chromaffin cell migration	– Luo X et al: Cell 1994;77:481–490 – Gut P et al: Development 2005;132:4611–4619 – Lohr J et al: Cell Tissue Res 2006;325:437–444
Glucocorticoid receptor k.o.	chromaffin cells affected, steroid effects impaired	lack of the PNMT enzyme and secretogranin II in chromaffin cells	– Finotto S et al: Development 1999;126:2935–2944
21-Hydroxylase k.o.	no zonation of the adrenal cortex, low corticosterone	incomplete formation of adrenal medulla, low PNMT expression, low dopamine, epinephrine and norepinephrine concentrations	– Bornstein SR et al: FASEB J 1999;13:1185–1194 – Tajima T et al: Gene Ther 1999;6:1898–1903
Corticotropin-releasing hormone type 1 receptor (CRH1R) k.o.	pituitary and adrenal affected, low corticosterone	low expression of PNMT, low adrenal epinephrine levels, depletion of epinephrine-containing secretory granules	– Yoshida-Hiroi M et al: Mol Psychiatry 2002;7:967–974
β-Catenin – conditional inactivation in adrenocortical cells	adrenal cortex smaller, decrease in SF-1-positive cells	adrenocortical and chromaffin precursor cells disappear at embryonic day E18.5	– Kim AC et al: Development 2008;135:2593–6202
Sonic hedgehog (SHH) – conditional inactivation in adrenocortical cells	hypoplastic adrenal cortex	failure of encapsulation of the adrenal medulla by the adrenal cortex, impaired morphogenesis	– Ching S et al: Genesis 2009;47:628–637
CITED2 k.o.	adrenal cortex	complete adrenal agenesis	– Bamforth SD et al: Nat Genet 2001;29:469–474 – Haase M et al: J Endocrinol 2007;192:459–465
Interleukin-6 overexpression	adrenocortical enlargement	hyperplasia of adrenal medulla	– Raber J et al: J Neurosci 1997;17: 9473–9480

Table 1. Continued

Model	Primarily affected tissue	Consequence for adjacent tissue	Related literature
Erbb3 receptor k.o.	lack of chromaffin cells	disorganization of the adrenal cortex	– Britsch S et al: Genes Dev 1998;12:1825–1836
TH k.o.	adrenal medulla	tubular mitochondria in steroidogenic cells with a reduction in internal membranes, low corticosterone, increase in liposomes	– Bornstein SR et al: Proc Natl Acad Sci USA 2000;97:14742–14747
Ascorbic acid transporter (SVCT2) k.o.	dopamine-β-hydroxylase deficiency, low tissue levels of norepinephrine, apoptotic chromaffin cells	low corticosterone (following low ACTH)	– Bornstein SR et al: FASEB J 2003;17:1928–1930 – Patak P et al: Endocr Res 2004;30:871–875

from the literature in the early 1950s that adrenalin acts on the pituitary to lead to adrenocortical hormone secretion. This was later confirmed to some extent in patients with Shy-Drager syndrome [29]. However, Conn also began to discuss that adrenaline may act directly on the cortex in a model of insulin-induced hypoglycemia. Much later, glucocorticoids were shown to increase the catecholamine release from perfused canine adrenal glands [30]. However, the development of genetic knockout models allowed for studying the contribution of specific enzymes and factors to the integrity of adrenal function. Many of the studies support the data obtained from in vitro studies, whereas some studies change existing concepts and give direction for new research. We have summarized the most important studies in table 1. Of note, the systemic application of glucocorticoids for restauration of eucortisolemia in glucocorticoid deficiency does not prevent the changes in the medulla, most likely because high local concentrations are not achieved this way.

Evidence from Clinical Observations

A number of studies confirmed the clinical supposition that patients with adrenal insufficiency have defective catecholamine secretion. Lower plasma levels of catecholamines are observed in patients with secondary and primary adrenal insufficiency despite regular regiments for substitution of glucocorticoids and mineralocorticoids [31, 32]. Glucocorticoid deficiency due to a loss-of function mutation in the ACTH receptor or hypoadrenalism following prolonged glucocorticoid administration for control of inflammatory or autoimmune diseases also result in decreased adrenomedullary function [33, 34]. It has been shown that the decrease in plasma and urinary epinephrine concentrations represents a dose-dependent function of adrenocortical secretory capacity [35]. In addition to the biochemical data, a look into histology reveals that the adrenal chromaffin cells are depleted of secretory granules and that the medulla is not intact in patients with 21-hydroxylase deficiency [35, 36]. This leads to a lack of adequate catecholamine supply to the body which is of functional relevance since glucose homeostasis is not intact in patients with 21-hydroxylase deficiency as shown in high-intensity exercise and the development of hypoglycemia can not be overcome by increasing hydrocortisone doses [37]. On the hyperfunctional side of the spectrum of hormone secretion, it was found that chromaffin tissue adjacent to hypersecretory adrenocortical tumor cells may become hyperplastic and rarely appears as an ipsilateral pheochromocytoma [38, 39]. Although the small number of such cases rather suggests mere coincidence of ipsilateral adrenocortical and adrenomedullary lesions, the occurrence of pheochromocytoma crisis following glucocorticoid therapy has also been reported [40].

On the other hand, the adrenal steroidogenesis may come under direct control of chromaffin tissue when these neuroendocrine cells aberrantly secrete ACTH or when adrenal cortical cells ectopically express β-adrenoceptors [41–43]. Interestingly, adrenocortical tumors seem to preferentially develop from cortical islets within the medulla [44]. In patients with adrenomedullary hypofunction due to the Shy-Drager syndrome, cortisol responses to insulin-induced hypoglycemia are abolished and are diminished after metopirone administration although the ACTH rise is augmented [29].

Conclusion

There is vast evidence from in vitro and in vivo studies that the adrenal medullary and the adrenal cortical tissues interact with each other in multiple ways. There is an interexchange of classical hormones, bioactive peptides and there are cell-cell communications via direct contacts. The knowledge from classic cell culture and organ-perifusion models as well as from classic surgical animal studies was recently complemented by investigations in genetic knockout animals. Impressively, they support old data of the interaction between the chromaffin and adrenocortical tissues in large part but also give rise to new concepts. Since clinical studies in affected patients also confirm the hypothesis of such an interaction, this has consequences for the design of new in vitro and in vivo experiments but also for the diagnosis and treatment of diseases that affect the function of the adrenal cortex or/and the adrenal medulla.

Acknowledgement

This work was supported by a grant from the Dr. Robert Pfleger-Stiftung, Bamberg, Germany, to H.S.W.

References

1 Belgorosky A, Baquedano MS, Guercio G, Rivarola MA. Adrenarche: postnatal adrenal zonation and hormonal and metabolic regulation. Horm Res 2008;70:257–267.

2 Bornstein SR, Gonzalez-Hernandez JA, Ehrhart-Bornstein M, Adler G, Scherbaum WA: Intimate contact of chromaffin and cortical cells within the human adrenal gland forms the cellular basis for important intraadrenal interactions. J Clin Endocrinol Metab 1994;78:225–232.

3 Wurtman RJ, Axelrod J. Adrenaline synthesis: control by the pituitary gland and adrenal glucocorticoids. Science. 1965;150:1464–1465.

4 Anderson DJ, Axel R. A bipotential neuroendocrine precursor whose choice of cell fate is determined by NGF and glucocorticoids: Cell 1986;47:1079–1090.

5 Seidl K, Unsicker K. The determination of the adrenal medullary cell fate during embryogenesis: Dev Biol 1989;136:481–490.

6 Beaujean D, Rosenbaum C, Müller HW, Willemsen JJ, Lenders J, Bornstein SR: Combinatorial code of growth factors and neuropeptides define neuroendocrine differentiation in PC12 cells. Exp Neurol 2003;184:348–358.

7 Mesiano S, Jaffe RB: Developmental and functional biology of the primate fetal adrenal cortex. Endocr Rev 1997;18:378–403.

8 Ziegler CG, Sicard F, Lattke P, Bornstein SR, Ehrhart-Bornstein M, Krug AW: Dehydroepiandrosterone induces a neuroendocrine phenotype in nerve growth factor-stimulated chromaffin pheochromocytoma PC12 cells. Endocrinology 2008;149:320–328.

9 Krug AW, Langbein H, Ziegler CG, Bornstein SR, Eisenhofer G, Ehrhart-Bornstein M: Dehydroepiandrosterone-sulphate (DHEA-S) promotes neuroendocrine differentiation of chromaffin pheochromocytoma PC12 cells. Mol Cell Endocrinol 2009;300:126–131.

10 Charalampopoulos I, Dermitzaki E, Vardouli L, Tsatsanis C, Stournaras C, Margioris AN, Gravanis A: Dehydroepiandrosterone sulfate and allopregnanolone directly stimulate catecholamine production via induction of tyrosine hydroxylase and secretion by affecting actin polymerization. Endocrinology 2005;146:3309–3318.

11 Ziegler CG, Sicard F, Sperber S, Ehrhart-Bornstein M, Bornstein SR, Krug AW: DHEA reduces NGF-mediated cell survival in serum-deprived PC12 cells. Ann N Y Acad Sci 2006;1073:306–311.

12 Sicard F, Krug AW, Ziegler CG, Sperber S, Ehrhart-Bornstein M, Bornstein SR: Role of DHEA and growth factors in chromaffin cell proliferation. Ann N Y Acad Sci 2006;1073:312–316.

13 Sicard F, Ehrhart-Bornstein M, Corbeil D, Sperber S, Krug AW, Ziegler CG, Rettori V, McCann SM, Bornstein SR: Age-dependent regulation of chromaffin cell proliferation by growth factors, dehydroepiandrosterone (DHEA), and DHEA sulfate. Proc Natl Acad Sci USA 2007;104:2007–2012.

14 Goto J, Otsuka F, Yamashita M, Suzuki J, Otani H, Takahashi H, Miyoshi T, Mimura Y, Ogura T, Makino H: Enhancement of aldosterone-induced catecholamine production by bone morphogenetic protein-4 through activating Rho and SAPK/JNK pathway in adrenomedullar cells. Am J Physiol Endocrinol Metab 2009;296:E904–E916.

15 Ehrhart-Bornstein M, Hinson JP, Bornstein SR, Scherbaum WA, Vinson GP: Intraadrenal interactions in the regulation of adrenocortical steroidogenesis. Endocr Rev 1998;19:101–143.

16 Willenberg HS, Schinner S, Ansurudeen I: New mechanisms to control aldosterone synthesis. Horm Metab Res 2008;40:435–441.

17 Haidan A, Bornstein SR, Glasow A, Uhlmann K, Lübke C, Ehrhart-Bornstein M: Basal steroidogenic activity of adrenocortical cells is increased 10-fold by coculture with chromaffin cells. Endocrinology 1998;139:772–780.

18 Ehrhart-Bornstein M, Bornstein SR, Trzeclak WH, Usadel H, Güse-Behling H, Waterman MR, Scherbaum WA: Adrenaline stimulates cholesterol side-chain cleavage cytochrome P450 mRNA accumulation in bovine adrenocortical cells. J Endocrinol 1991;131:R5–R8.

19 Güse-Behling H, Ehrhart-Bornstein M, Bornstein SR, Waterman MR, Scherbaum WA, Adler G. Regulation of adrenal steroidogenesis by adrenaline: expression of cytochrome P450 genes: J Endocrinol 1992;135:229–237.

20 Haidan A, Bornstein SR, Liu Z, Walsh LP, Stocco DM, Ehrhart-Bornstein M: Expression of adrenocortical steroidogenic acute regulatory (StAR) protein is influenced by chromaffin cells. Mol Cell Endocrinol 2000;165:25–32.

21 Nussdorfer GG, Mazzocchi G: Vasoactive intestinal peptide (VIP) stimulates aldosterone secretion by rat adrenal glands in vivo. J Steroid Biochem 1987; 26:203–205.

22 Ehrhart-Bornstein M, Bornstein SR, Scherbaum WA, Pfeiffer EF, Holst JJ: Role of the vasoactive intestinal peptide in a neuroendocrine regulation of the adrenal cortex. Neuroendocrinology 1991;54: 623–628.

23 Bornstein SR, Haidan A, Ehrhart-Bornstein M: Cellular communication in the neuro-adrenocortical axis: role of vasoactive intestinal polypeptide (VIP). Endocr Res 1996;22:819–829.

24 Cobb VJ, Williams BC, Mason JI, Walker SW: Direct stimulation of cortisol secretion from the human NCI H295 adrenocortical cell line by vasoactive intestinal polypeptide. J Hypertens 1997;15:1735–1738.

25 Eskay RL, Eiden LE: Interleukin-1α and tumor necrosis factor-α differentially regulate enkephalin, vasoactive intestinal polypeptide, neurotensin, and substance P biosynthesis in chromaffin cells. Endocrinology 1992;130:2252–2258.

26 Andreis PG, Tortorella C, Ziolkowska A, Spinazzi R, Malendowicz LK, Neri G, Nussdorfer GG: Evidence for a paracrine role of endogenous adrenomedullary galanin in the regulation of glucocorticoid secretion in the rat adrenal gland. Int J Mol Med 2007;19:511–515.

27 Ait-Ali D, Stroth N, Sen JM, Eiden LE: PACAP-cytokine interactions govern adrenal neuropeptide biosynthesis after systemic administration of LPS. Neuropharmacology 2010;58:208–214.

28 Conn JW, Fajans SS: The pituitary-adrenal system. Annu Rev Physiol 1952;14:453–480.

29 Sasaki K, Aoyagi K, Murabayashi S, Baba T, Kitaoka M, Mastunaga M, Takebe K: Pituitary-adrenocortical system in patients with Shy-Drager syndrome. Horm Metab Res 1983;15:143–146.

30 Critchley JA, Henderrson CG,Moffat LE, Ungar A,Waite J, West CP: Proceedings: the release of catecholamines from perfused canine adrenal glands by corticoglucocorticoids. Journal of Physiology 1976;254:30P–31P.

31 Rudman D, Moffitt SD, Fernhoff PM, Blackston RD, Faraj BA: Epinephrine deficiency in hypocorticotropic hypopituitary children. J Clin Endocrinol Metab 1981;53:722–729.

32 Bornstein SR, Breidert M, Ehrhart-Bornstein M, Kloos B, Scherbaum WA: Plasma catecholamines in patients with Addison's disease. Clin Endocrinol (Oxf) 1995;42:215–218.

33 Zuckerman-Levin N, Tiosano D, Eisenhofer G, Bornstein S, Hochberg Z: The importance of adrenocortical glucocorticoids for adrenomedullary and physiological response to stress: a study in isolated glucocorticoid deficiency. J Clin Endocrinol Metab 2001;86:5920–5924.

34 Bornstein SR, Briegel J: A new role for glucocorticoids in septic shock: balancing the immune response. Am J Respir Crit Care Med 2003;167:485–486.

35 Merke DP, Chrousos GP, Eisenhofer G, Weise M, Keil MF, Rogol AD, Van Wyk JJ, Bornstein SR: Adrenomedullary dysplasia and hypofunction in patients with classic 21-hydroxylase deficiency. N Engl J Med 2000;343:1362–1368.

36 Charmandari E, Eisenhofer G, Mehlinger SL, Carlson A, Wesley R, Keil MF, Chrousos GP, New MI, Merke DP: Adrenomedullary function may predict phenotype and genotype in classic 21-hydroxylase deficiency. J Clin Endocrinol Metab 2002;87:3031–3037.

37 Weise M, Mehlinger SL, Drinkard B, Rawson E, Charmandari E, Hiroi M, Eisenhofer G, Yanovski JA, Chrousos GP, Merke DP: Patients with classic congenital adrenal hyperplasia have decreased epinephrine reserve and defective glucose elevation in response to high-intensity exercise. J Clin Endocrinol Metab 2004;89:591–597.

38 Wolf A, Willenberg HS, Cupisti K, Schott M, Geddert H, Raffel A, Bornstein SR, Scherbaum WA, Knoefel WT: Adrenal pheochromocytoma with contralateral cortisol-producing adrenal adenoma: diagnosic and therapeutic management. Horm Metab Res 2005;37:391–395.

39 Sakamoto N, Tojo K, Saito T, Fujimoto K, Isaka T, Tajima N, Ikeda K, Yamada H, Furuta N, Sasano H: Coexistence of aldosterone-producing adrenocortical adenoma and pheochromocytoma in an ipsilateral adrenal gland. Endocr J 2009;56:213–219.

40 Rosas AL, Kasperlik-Zaluska AA, Papierska L, Bass BL, Pacak K, Eisenhofer G: Pheochromocytoma crisis induced by glucocorticoids: a report of four cases and review of the literature. Eur J Endocrinol 2008; 158:423–429.

41 Vrezas I, Willenberg HS, Mansmann G, Hiroi N, Fritzen R, Bornstein SR: Ectopic adrenocorticotropin (ACTH) and corticotropin-releasing hormone (CRH) production in the adrenal gland: Basic and clinical aspects. Microsc Res Techniq 2003;61:308–314.

42 Willenberg HS, Feldkamp J, Lehmann R, Schott M, Goretzki PE, Scherbaum WA: A case of catecholamine-excess and Cushing's syndrome due to a corticotropin-secreting paraganglioma. Ann NY Acad Sci 2006;1073:52–58.

43 Lacroix A, Tremblay J, Rousseau G, Bouvier M, Hamet P: Propranolol therapy for ectopic beta-adrenergic receptors in adrenal Cushing's syndrome. N Engl J Med 1997;337:1429–1434.

44 Bornstein SR, Stratakis CA, Chrousos GP: Adrenocortical tumors: recent advances in basic concepts and clinical management. Ann Intern Med 1999;130:757–765.

Holger S. Willenberg, MD
Department of Endocrinology, Diabetes and Rheumatology
University Hospital Duesseldorf, Moorenstrasse 5
DE–40225 Duesseldorf (Germany)
Tel. +49 211 811 7810, Fax +49 211 811 7860, E-Mail Holger.Willenberg@uni-duesseldorf.de

Ghizzoni L, Cappa M, Chrousos G, Loche S, Maghnie M (eds): Pediatric Adrenal Diseases.
Endocr Dev. Basel, Karger, 2011, vol 20, pp 38–46

Role of DAX-1 (*NR0B1*) and Steroidogenic Factor-1 (*NR5A1*) in Human Adrenal Function

Ranna El-Khairi · Alejandro Martinez-Aguayo · Bruno Ferraz-de-Souza · Lin Lin · John C. Achermann

Developmental Endocrinology Research Group, Clinical and Molecular Genetics Unit, UCL Institute of Child Health, University College London, London, UK

Abstract

The nuclear receptor transcription factors DAX-1 *(NR0B1)* and SF-1 *(NR5A1)* regulate many aspects of adrenal and reproductive development and function. Disruption of the genes encoding these factors can be associated with pediatric adrenal disease. DAX-1 mutations are classically associated with X-linked adrenal hypoplasia congenita, hypogonadotropic hypogonadism and impaired spermatogenesis. However, other phenotypes are also being reported, such as isolated mineralocorticoid insufficiency, premature sexual development, primary adrenal insufficiency in a 46,XX patient and late-onset X-linked adrenal hypoplasia congenita and/or hypogonadotropic hypogonadism. SF-1 mutations have also been associated with primary adrenal insufficiency, together with 46,XY disorders of sex development. However it is emerging that SF-1 changes are a relatively rare cause of primary adrenal failure in humans, and most individuals with SF-1 mutations have a spectrum of 46,XY disorders of sex development phenotypes. These conditions range from 46,XY females with streak gonads and müllerian structures, through children with ambiguous genitalia and inguinal testes, to severe penoscrotal hypospadias with undescended testes. Therefore, the human gonad appears to be more sensitive than the adrenal gland to loss of SF-1 function. This review will focus on the expanding range of phenotypes associated with DAX-1 and SF-1 mutations.

The human adrenal gland first develops from an area of intermediate mesoderm at around 4 weeks postconception and undergoes rapid growth and differentiation during fetal life into a fully functional organ capable of releasing cortical and medullary hormones. Defects in several different transcriptional, signaling and mitogenic pathways can result in underdevelopment of the adrenal gland, leading to the clinical condition congenital adrenal hypoplasia (or adrenal hypoplasia congenita (AHC)) [1]. Many of these processes also regulate adrenal function postnatally and throughout the lifespan, potentially resulting in primary adrenal insufficiency when disruption occurs.

In this chapter we will focus on the role of the nuclear receptor transcription factors DAX-1 *(NR0B1)* and steroidogenic factor-1 (SF-1, *NR5A1*, Ad4BP) in human adrenal development and function. The importance of these factors in human adrenal disease is now well established, but the spectra of phenotypes associated with these conditions continue to expand.

DAX-1 (*NR0B1*)

DAX-1 (dosage-sensitive sex reversal – AHC critical region on the X chromosome 1) is an atypical nuclear receptor first reported in 1994 to be associated with X-linked AHC (OMIM 300200) [2]. The genetic locus for this transcription factor was mapped to the short arm of the X chromosome (Xp21) as X-linked AHC can occur as part of a contiguous gene deletion syndrome with glycerol kinase deficiency and Duchenne muscular dystrophy. *DAX1* (officially called *NR0B1*) encodes an atypical nuclear receptor with a ligand-like binding domain at the carboxyl-terminus and an unusual amino-terminal region containing 3.5 repeats of an LXXLL motif-containing sequence (fig. 1). DAX-1/*NR0B1* is expressed in key areas of the adrenal gland, gonad, and central reproductive axis during development and postnatal life, consistent with its role in the development and function of these endocrine systems.

The 'classical' clinical features of X-linked AHC are: (1) primary adrenal failure presenting in early infancy or childhood; (2) hypogonadotropic hypogonadism presenting as absent or arrested puberty, and (3) impaired spermatogenesis [2, 3]. Boys with this condition usually need glucocorticoid and mineralocorticoid replacement throughout life and induction of secondary sex characteristics with testosterone. In general, GnRH pumps or recombinant gonadotropins are not effective at reversing azoospermia [4, 5], although TESE/ICSI following prolonged gonadotropin treatment has been successful [Christin-Maitre, unpubl. data].

Approximately one-third of boys with X-linked AHC have a deletion of *DAX1 (NR0B1)* [6]. A contiguous gene deletion syndrome affecting another Xp locus gene occurs in about half of these cases. Approximately two-thirds of individuals have point mutations in the *DAX1 (NR0B1)* gene. These changes can be insertions, deletions or nonsense changes scattered throughout the two exons. The most common point mutations are frameshift changes (approx. 50%) and nonsense mutations (approx. 30%) (fig. 1, upper panel). Missense mutations have been reported in 20% of patients with point mutations and are mostly located within the putative ligand-binding region of DAX-1 (fig. 1, lower panel) [6]. These changes can cause several different defects such as altered protein structure, reduced protein-cofactor interactions, or disrupted nuclear localization [7, 8]. Missense changes in the amino-terminus of DAX-1 are very rare.

In addition to the classic phenotype of X-linked AHC described above, several variant or partial phenotypes have now been reported (table 1). These include predominant mineralocorticoid insufficiency [9], hyperandrogenism in early childhood

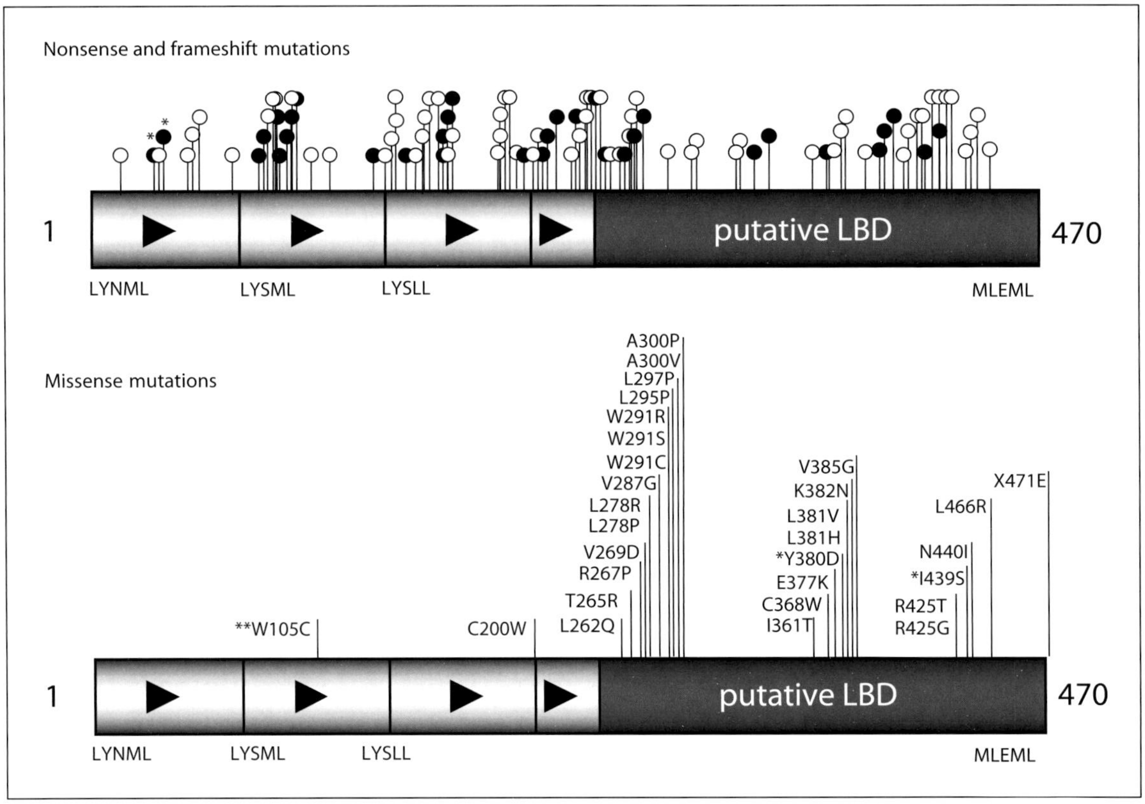

Fig. 1. DAX-1 structure and a selection of the nonsense (●), frameshift (○) and missense changes reported. Those changes associated with a milder phenotype (*) or isolated mineralcorticoid deficiency (**) are indicated. The potential LXXLL domains are shown below the protein structure. LBD = Ligand-binding domain. Modified with permission from Lin et al. [6]. Copyright: The Endocrine Society, 2006.

[10], early puberty with later pubertal arrest [11], and adrenal insufficiency in a girl with an Xp deletion and extreme skewed X-inactivation [12]. Furthermore, late-onset X-linked AHC has been described in several men who presented with primary adrenal failure or hypogonadotropic hypogonadism in late adolescence or early adulthood [13, 14]. In some cases, these milder conditions may be due to missense mutations with partial loss-of-function, or to premature stop codons at the amino-terminus of the protein with subsequent translation of an alternative isoform of the DAX-1 protein that retains partial function due to preserved LXXLL domains [13, 14].

The precise functional role of DAX-1 in adrenal development and function remains unclear. Many studies have shown that DAX-1 can act as a repressor of gene transcription through its effect on the related nuclear receptor steroidogenic factor-1 (SF-1, *NR5A1*). Indeed, crystallography of DAX-1 bound to the related receptor LRH-1 (NR5A2) predicts that two DAX-1 proteins bind to the ligand-like binding domain

Table 1. Phenotypes associated with changes in DAX-1 *(NR0B1)*

Clinical phenotype	Example ref.
Primary adrenal insufficiency, HH, infertility	2, 3
Isolated mineralocorticoid insufficiency	9
Premature sexual development	10, 11
Adrenal insufficiency or delayed puberty in girls with skewed X inactivation or gene conversion	12
Late-onset X-linked AHC or HH in early adulthood	13, 14, 30
Presymptomatic diagnosis	31

AHC = Adrenal hypoplasia congenita, HH = hypogonadotropic hypogonadism.

of the partner, thereby potentially altering its activation capacity [15]. However, such a repressor role for DAX-1 is paradoxical given the fact that loss-of-function changes in DAX-1 result in adrenal hypoplasia. Two alternative hypotheses have been suggested. Firstly, DAX-1 may be needed to prevent adrenal stem cell differentiation, so that expansion of a pool of progenitor stem cells can occur before these cells mature into a mature steroidogenic lineage [7]. Loss of DAX-1 repression would result in premature differentiation without prior expansion, so that the total number of cells is reduced and adrenal hypoplasia results. Alternatively, DAX-1 may actually function as an activator of gene transcription in certain cases, such as with the steroid receptor RNA activator or on specific promoters (e.g. pre-B-cell leukemia transcription factor 1). These activator effects may be cell- or time-specific [16, 17].

Although the exact mechanisms underlying X-linked AHC and its associated reproductive consequences remain to be fully elucidated, understanding the underlying pathogenic mechanism is important to gain new insight into this condition and if new treatment approaches are to be developed.

Steroidogenic Factor-1 (*NR5A1*)

SF-1 is a related nuclear receptor that is expressed in the adrenal gland, gonads, ventromedial hypothalamus, gonadotropes and spleen during development and into postnatal life. Like DAX-1, SF-1 has a carboxyl-terminal ligand-like binding domain, but its amino-terminal DNA-binding domain is more typical of the nuclear receptor superfamily (fig. 2). This region contains a two zinc finger DNA-binding domain (with a 'P-box' motif), an accessory region ('A-box') that supports monomeric binding, and a hinge region that can undergo posttranslational modification (e.g. phosphorylation, SUMOylation). SF-1 was originally thought to be an 'orphan' nuclear receptor as no naturally-occurring ligand was known. Recent data following crystallization of the ligand-binding domain suggest that this region may contain a small phospholipid

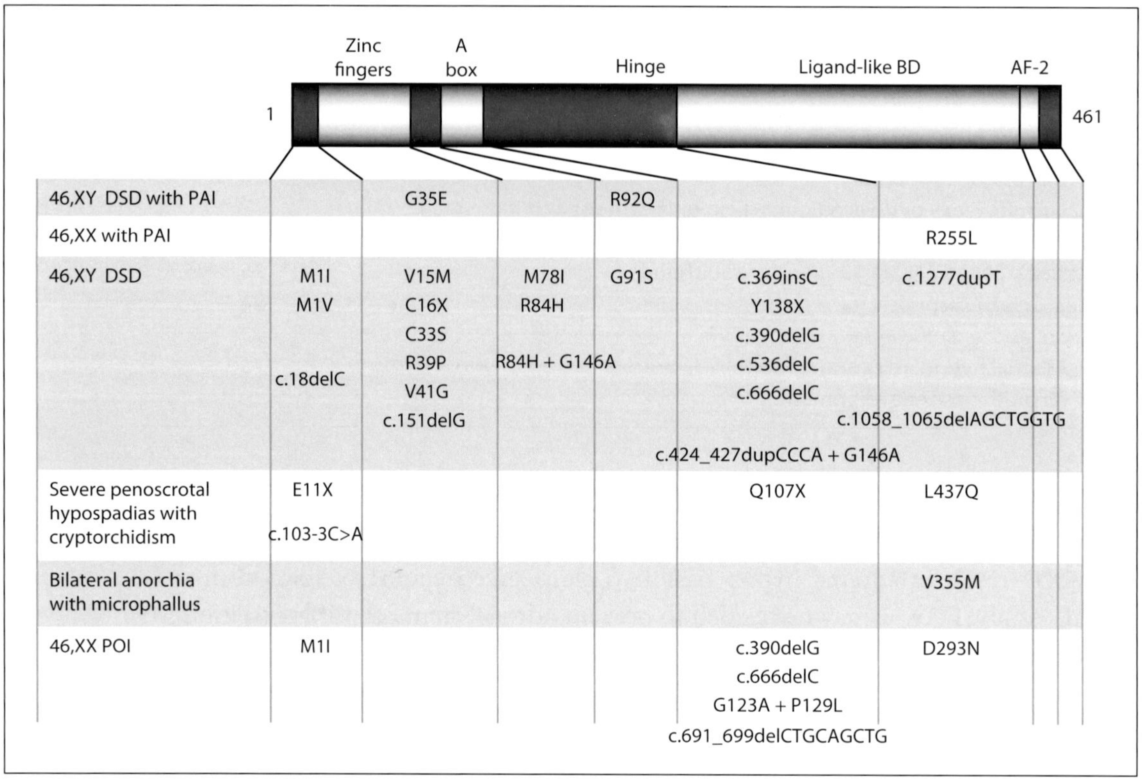

Fig. 2. SF-1 structure and a selection of the changes reported in patients with adrenal insufficiency and/or reproductive phenotypes. Nonsense, frameshift and missense changes are shown. BD = Binding domain, PAI = primary adrenal insufficiency, DSD = disorder of sex development, POI = primary ovarian insufficiency. Modified with permission from Lin et al. [6]. Copyright: The Endocrine Society, 2006.

molecule [18]. Although it is still not know if a high-specificity natural ligand exists for SF-1, this finding raises the possibility that SF-1 function could be modulated by signaling pathways, environmental chemicals or pharmacological agents (table 2).

SF-1 regulates an array of target genes involved in adrenal and reproductive development and in steroidogenic and metabolic function. Complete deletion of the gene encoding SF-1 in the mouse causes adrenal and gonadal agenesis, a female phenotype and müllerian structures in XY mice, partial gonadotropin deficiency, and late-onset obesity in animals rescued by adrenal transplantation [19].

Initial attempts to identify SF-1 changes in humans focused on individuals with a similar collection of features (i.e., primary salt-losing adrenal insufficiency, 46,XY disorders of sex development (DSD) and müllerian structures). This is a very rare phenotype in humans and in more than a decade only 2 such patients have been reported who harbor SF-1 changes. The first individual described has a de novo heterozygous p.G35E change in the 'P-box' of SF-1 [20]. This motif forms the principle DNA-

Table 2. Phenotypes associated with changes in SF-1 *(NR5A1)*

Clinical phenotype	Example ref.
Primary adrenal insufficiency, 46,XY female, streak gonads, uterus	20, 21
Primary adrenal insufficiency in 46,XX girl	22
46,XY DSD, streak gonads, uterus	23
46,XY DSD, clitoromegaly/ambiguous, inguinal testes, no uterus	23
Severe penoscrotal hypospadias, small penis, undescended testes	24
Bilateral anorchia, small penis	25
Male factor infertility	26
Primary ovarian insufficiency	27

DSD = Disorder of sex development.

binding structure of SF-1 by interfacing with the major groove of DNA. The p.G35E may have a mild competitive or dominant negative effect in certain assay systems but the effect on different promoters can be variable and the phenotype may be the result of reduced SF-1 transactivation on multiple targets throughout the genome and at different stages of development; therefore, a potentially complex and non-linear system. The importance of SF-1 in regulating human adrenal and gonad development and function was confirmed following the report of an infant with a similar phenotype (primary salt-losing adrenal failure, 46,XY DSD, müllerian structures) who had inherited a homozygous p.R92Q alteration in SF-1 in a recessive fashion [21]. This change lies within the 'A-box' of SF-1 and interferes with monomeric DNA-binding stability. Again, the effects of this homozygous change are complex and variable, but in a range of promoter assays mean functional activity was in the order of 30–40% of wild-type. Although a 46,XX girl with primary adrenal failure was reported to have a heterozygous point mutation in SF-1 (p.R255L) [22], a systematic study of children and adults with a range of potential adrenal hypoplasia phenotypes did not reveal any additional mutations in SF-1 in a cohort of more than 70 DAX-1-negative individuals tested [6]. Therefore, SF-1 mutations are a well-established but relatively rare cause of primary adrenal failure in humans. No complete loss of function changes or deletions of SF-1 have been described.

More recently, an association between SF-1 changes and 46,XY DSD has emerged [e.g. see 23]. These children typically harbor heterozygous nonsense or frameshift mutations in SF-1 that are predicted to cause haploinsufficiency of the protein, or heterozygous missense changes with severe loss of function but no clear dominant negative activity. Affected individuals have a spectrum of 46,XY DSD phenotypes ranging from 46,XY females with streak gonads and müllerian structures, through children with ambiguous genitalia and inguinal testes, to severe penoscrotal hypospadias with undescended testes [24]. In these cases there seems to be a predominant androgen biosynthesis defect with variable defects in testis development and/or integrity.

SF-1 changes often occur de novo, but can also be transmitted in a recessive pattern or from the mother in a sex-limited dominant manner. Other missense mutations in SF-1 have be found in association with bilateral anorchia (vanishing testis syndrome) with micropenis [25], male factor infertility [26] and in women with primary ovarian insufficiency or premature menopause [27]. Of note, where data are available, adrenal function appears to be normal in all these individuals. Therefore, the human gonad appears to be more sensitive than the adrenal gland to loss of SF-1 function, although this group of patients will need long-term follow-up to establish whether there is a risk of developing adrenal failure with time. The potential metabolic effects of SF-1 disruption also need addressing in the long term, as ventromedial hypothalamic defects and late-onset obesity have been reported in mice [28], and obesity and hypertension have been noted in some patients. A polymorphic variability in SF-1 may influence some of these processes in the population.

Finally, overexpression of SF-1 may act as a drive to pediatric adrenal tumorigenesis, as somatic duplication or multiple copy numbers of the locus containing SF-1 have been found in some children with adrenal tumors on a background of loss of heterozygosity for *TP53* [29]. In such cases, SF-1 could be a future therapeutic target for treatment of this rare but important group of conditions.

Acknowledgements

J.C.A. is a Wellcome Trust Senior Fellow in Clinical Science (079666). A.M.A. holds a scholarship from Becas-Chile (www.becaschile.cl). B.FdS. holds a scholarship from Capes/Brazil (4798066). We thank the physicians and families who have been involved in our studies over the years and recognize the invaluable contribution Keith L. Parker has made in this field.

References

1 Else T, Hammer GD: Genetic analysis of adrenal absence: agenesis and aplasia. Trends Endocrinol Metab 2005;16:458–468.

2 Muscatelli F, Strom TM, Walker AP, Zanaria E, Recan D, Meindl A, Bardoni B, Guioli S, Zehetner G, Rabl W, Schwarz HP, Kaplan J, Camerino G, Meitinger T, Monaco AP: Mutations in the DAX-1 gene give rise to both X-linked adrenal hypoplasia congenita and hypogonadotropic hypogonadism. Nature 1994;372:672–676.

3 Reutens AT, Achermann JC, Ito M, Ito M, Gu WX, Habiby RL, Donohoue PA, Pang S, Hindmarsh PC, Jameson JL: Clinical and functional effects of mutations in the DAX-1 gene in patients with adrenal hypoplasia congenita. J Clin Endocrinol Metab 1999;84:504–511.

4 Mantovani G, Mancini M, Gazzano G, Spada A, Colpi GM, Beck-Peccoz P, Persani L: Somatic mutational analysis of DAX1 in testes from men with idiopathic azoospermia. Fertil Steril 2005;84:1542–1544.

5 Seminara SB, Achermann JC, Genel M, Jameson JL, Crowley WF Jr: X-linked adrenal hypoplasia congenita: a mutation in *DAX1* expands the phenotypic spectrum in males and females. J Clin Endocrinol Metab 1999;84:4501–4509.

6 Lin L, Gu WX, Ozisik G, To WS, Owen CJ, Jameson JL, Achermann JC: Analysis of DAX1 (NR0B1) and steroidogenic factor-1 (NR5A1) in children and adults with primary adrenal failure: ten years' experience. J Clin Endocrinol Metab 2006;91:3048–3054.

7 Lalli E, Sassone-Corsi P: DAX-1, an unusual orphan receptor at the crossroads of steroidogenic function and sexual differentiation. Mol Endocrinol 2003;17: 1445–1453.

8 Lehmann SG, Wurtz JM, Renaud JP, Sassone-Corsi P, Lalli E: Structure-function analysis reveals the molecular determinants of the impaired biological function of DAX-1 mutants in AHC patients. Hum Mol Genet 2003;12:1063–1072.

9 Wiltshire E, Couper J, Rodda C, Jameson JL, Achermann JC: Variable presentation of X linked adrenal hypoplasia congenita. J Pediatr Endocrinol Metab 2001;14:1093–1096.

10 Landau Z, Hanukoglu A, Sack J, Goldstein N, Weintrob N, Eliakim A, Gillis D, Sagi M, Shomrat R, Kosinovsky EB, Anikster Y: Clinical and genetic heterogeneity of congenital adrenal hypoplasia due to *NR0B1* gene mutations. Clin Endocrinol (Oxf) 2010;72:448–454.

11 Domenice S, Latronico AC, Brito VN, Arnhold IJ, Kok F, Mendonca BB: Adrenocorticotropin-dependent precocious puberty of testicular origin in a boy with X-linked adrenal hypoplasia congenita due to a novel mutation in the *DAX1* gene. J Clin Endocrinol Metab 2001;86:4068–4071.

12 Shaikh MG, Boyes L, Kingston H, Collins R, Besley GT, Padmakumar B, Ismayl O, Hughes I, Hall CM, Hellerud C, Achermann JC, Clayton PE: Skewed X inactivation is associated with phenotype in a female with adrenal hypoplasia congenita. J Med Genet 2008;45:e1.

13 Tabarin A, Achermann JC, Recan D, Bex V, Bertagna X, Christin-Maitre S, Ito M, Jameson JL, Bouchard P: A novel mutation in DAX1 causes delayed-onset adrenal insufficiency and incomplete hypogonadotropic hypogonadism. J Clin Invest 2000;105:321–328.

14 Ozisik G, Mantovani G, Achermann JC, Persani L, Spada A, Weiss J, Beck-Peccoz P, Jameson JL: An alternate translation initiation site circumvents an amino-terminal *DAX1* nonsense mutation leading to a mild form of X-linked adrenal hypoplasia congenita. J Clin Endocrinol Metab 2003;88:417–423.

15 Sablin EP, Woods A, Krylova IN, Hwang P, Ingraham HA, Fletterick RJ: The structure of corepressor Dax-1 bound to its target nuclear receptor LRH-1. Proc Natl Acad Sci USA 2008;105:18390–18395.

16 Ferraz-de-Souza B, Martin F, Mallet D, Hudson-Davies RE, Cogram P, Lin L, Gerrelli D, Beuschlein F, Morel Y, Huebner A, Achermann JC: CBP/p300-interacting transactivator, with Glu/Asp-rich C-terminal domain, 2, and pre-B-cell leukemia transcription factor 1 in human adrenal development and disease. J Clin Endocrinol Metab 2009;94: 678–683.

17 Xu B, Yang WH, Gerin I, Hu CD, Hammer GD, Koenig RJ: Dax-1 and steroid receptor RNA activator (SRA) function as transcriptional coactivators for steroidogenic factor 1 in steroidogenesis. Mol Cell Biol 2009;29:1719–1734.

18 Krylova IN, Sablin EP, Moore J, Xu RX, Waitt GM, MacKay JA, Juzumiene D, Bynum JM, Madauss K, Montana V, Lebedeva L, Suzawa M, Williams JD, Williams SP, Guy RK, Thornton JW, Fletterick RJ, Willson TM, Ingraham HA: Structural analyses reveal phosphatidylinositols as ligands for the NR5 orphan receptors SF-1 and LRH-1. Cell 2005;120: 343–355.

19 Schimmer BP, White PC: Steroidogenic factor 1: its roles in differentiation, development, and disease. Mol Endocrinol 2010;24:1322–1337.

20 Achermann JC, Ito M, Ito M, Hindmarsh PC, Jameson JL: A mutation in the gene encoding steroidogenic factor-1 causes XY sex reversal and adrenal failure in humans. Nat Genet 1999;22:125–126.

21 Achermann JC, Ozisik G, Ito M, Orun UA, Harmanci K, Gurakan B, Jameson JL: Gonadal determination and adrenal development are regulated by the orphan nuclear receptor steroidogenic factor-1, in a dose-dependent manner. J Clin Endocrinol Metab 2002;87:1829–1833.

22 Biason-Lauber A, Schoenle EJ: Apparently normal ovarian differentiation in a prepubertal girl with transcriptionally inactive steroidogenic factor 1 (NR5A1/SF-1) and adrenocortical insufficiency. Am J Hum Genet 2000;67:1563–1568.

23 Lin L, Philibert P, Ferraz-de-Souza B, Kelberman D, Homfray T, Albanese A, Molini V, Sebire NJ, Einaudi S, Conway GS, Hughes IA, Jameson JL, Sultan C, Dattani MT, Achermann JC: Heterozygous missense mutations in steroidogenic factor 1 (SF1/Ad4BP, NR5A1) are associated with 46,XY disorders of sex development with normal adrenal function. J Clin Endocrinol Metab 2007;92:991–999.

24 Köhler B, Lin L, Mazen I, Cetindag C, Biebermann H, Akkurt I, Rossi R, Hiort O, Gruters A, Achermann JC: The spectrum of phenotypes associated with mutations in steroidogenic factor 1 (SF-1, NR5A1, Ad4BP) includes severe penoscrotal hypospadias in 46,XY males without adrenal insufficiency. Eur J Endocrinol 2009;161:237–242.

25 Philibert P, Zenaty D, Lin L, Soskin S, Audran F, Leger J, Achermann JC, Sultan C: Mutational analysis of steroidogenic factor 1 (NR5a1) in 24 boys with bilateral anorchia: a French collaborative study. Hum Reprod 2007;22:3255–3261.

26 Bashamboo A, Ferraz-de-Souza B, Lourenco D, Lin L, Sebire NJ, Montjean D, Bignon-Topalovic J, Mandelbaum J, Siffroi JP, Christin-Maitre S, Radha krishna U, Rouba H, Ravel C, Seeler J, Achermann JC, McElreavey K: Human male infertility associated with mutations in NR5A1 encoding steroidogenic factor 1. Am J Hum Genet 2010;87:505–512.
27 Lourenco D, Brauner R, Lin L, De PA, Weryha G, Muresan M, Boudjenah R, Guerra-Junior G, Maciel-Guerra AT, Achermann JC, McElreavey K, Bashamboo A: Mutations in NR5A1 associated with ovarian insufficiency. N Engl J Med 2009;360:1200–1210.
28 Majdic G, Young M, Gomez-Sanchez E, Anderson P, Szczepaniak LS, Dobbins RL, McGarry JD, Parker KL: Knockout mice lacking steroidogenic factor 1 are a novel genetic model of hypothalamic obesity. Endocrinology 2002;143:607–614.
29 Almeida MQ, Soares IC, Ribeiro TC, Fragoso MC, Marins LV, Wakamatsu A, Ressio RA, Nishi MY, Jorge AAL, Lerario AM, Alves VAF, Mendonca BB, Latronico AC: Steroidogenic factor 1 overexpression and gene amplification are more frequent in adrenocortical tumors from children than from adults. J Clin Endocrinol Metab 2010;95:1458–1462.
30 Guclu M, Lin L, Erturk E, Achermann JC, Cangul H: Puberty, stress, and sudden death. Lancet 2010; 376:1512.
31 Achermann JC, Silverman BL, Habiby RL, Jameson JL: Presymptomatic diagnosis of X-linked adrenal hypoplasia congenita by analysis of DAX1. J Pediatr 2000;137:878–881.

Dr. John C. Achermann
Developmental Endocrinology Research Group, Clinical and Molecular Genetics Unit
UCL Institute of Child Health, University College London
30 Guilford Street, London WC1N 1EH (UK)
Tel. +44 207 905 2887, Fax +44 207 404 6191, E-Mail j.achermann@ich.ucl.ac.uk

Ghizzoni L, Cappa M, Chrousos G, Loche S, Maghnie M (eds): Pediatric Adrenal Diseases.
Endocr Dev. Basel, Karger, 2011, vol 20, pp 47–53

Functional and Physiological Consequences of StAR Deficiency: Role in Lipoid Congenital Adrenal Hyperplasia

Steven R. King[a] · Amrit Bhangoo[b] · Douglas M. Stocco[a]

[a]Department of Cell Biology and Biochemistry, Texas Tech University Health Sciences Center, Lubbock, Tex., and [b]Pediatric Endocrinology Division of Infants and Children's Hospital of Brooklyn at Maimonides, Brooklyn, N.Y., USA

Abstract

The steroidogenic acute regulatory (StAR) protein is essential for all hormone-stimulated steroid biosynthesis. Accordingly, its absence gives rise to the most severe form of congenital adrenal hyperplasia (CAH), lipoid CAH. This life-threatening condition typically manifests itself in the perinatal period. Partial loss-of-function StAR mutations incompletely manifest the condition later in life and are a cause of familial glucocorticoid deficiency type 3. Here, we discuss StAR, its expression pattern and the clinical consequences of the loss of its activity.

StAR and Congenital Lipoid Adrenal Hyperplasia

All steroids derive from pregnenolone, a product of the cholesterol side-chain cleavage reaction catalyzed by cytochrome P450scc (CYP11A1) [reviewed in 1]. The supply of substrate cholesterol from the outer mitochondrial membrane to this inner mitochondrial membrane enzyme is regulated by the steroidogenic acute regulatory (StAR) protein. The transfer of cholesterol to P450scc constitutes the rate-limiting step in steroidogenesis. Proof of StAR's function comes from in vitro cell culture and biochemical studies. The most compelling evidence was the discovery that mutations in the StAR gene cause congenital lipoid adrenal hyperplasia (lipoid CAH).

This potentially fatal inherited autosomal recessive disorder is the rarest and most severe form of CAH [reviewed in 2]. It results from a general loss of all steroid production and presents as primary adrenal insufficiency or Addison's disease. Patient symptoms include low body weight, severe dehydration, skin hyperpigmentation, respiratory distress and vomiting. The underlying cause is salt-wasting, hyperkalemia, hypotension and hypoglycemia as a result of low levels of steroids as can be observed in serum and urinary samples. Due to the loss of gonadal steroid production, 46,XY patients also present with a spectrum of disorders of sexual differentiation.

The genetic basis for the disease involves compromised P450scc activity. In a landmark 1995 study, the reason for the drop in activity was traced to inactivating mutations in the StAR gene [3]. While lipoid CAH is now recognized to be usually caused by StAR mutations, subsequent research has determined that rare cases arise from changes in P450scc [2].

How StAR Stimulates Steroidogenesis

Levels of StAR respond to trophic hormone stimulation chiefly through cAMP signaling [reviewed in 4, 5]. StAR is produced as a 285-amino-acid protein with an N-terminal signal sequence that targets it to the interior of the mitochondria. During import into the mitochondria, the signal sequence is clipped, giving rise to mature 30-kDa forms of the protein that differ by posttranslational modifications [6, 7]. One modification, the phosphorylation of Ser195 by protein kinase A, upregulates StAR activity [8, 9]. Mutation of the serine to alanine contributes to lipoid CAH [10].

Once inside the mitochondria, StAR no longer influences steroid synthesis. Therefore, StAR catalyzes cholesterol transfer to P450scc in the short time it is outside the mitochondria. Continual synthesis of steroids then requires continual new synthesis of StAR [11, 12]. Removal of the signal sequence blocks import of StAR into the mitochondria without impairing StAR function in vitro [13]. However, only 40% of StAR knockout mice rescued with a StAR mutant lacking the signal sequence survive past birth and the survivors exhibit some symptoms of lipoid CAH [14]. Thus, specific targeting of StAR to the mitochondria is essential in vivo.

All lipoid CAH mutations affect the C-terminal StAR-related lipid transfer (START) domain of StAR (fig. 1). This domain is essential for function and forms a pocket that may bind cholesterol [15].

Not all steroid production is lost in lipoid CAH. Minor alternative pathways of cholesterol transfer to P450scc exist and mitigate the severity of the condition. Candidate factors responsible for this include the StAR homolog MLN64 [16] and hydroxylated cholesterol molecules (oxysterols).

Pathogenesis of Lipoid CAH due to StAR Deficiency

StAR is expressed in nearly all steroidogenic tissues except the human placenta (for a more exhaustive review, see Bhangoo et al. [2]). A description of the role of StAR in other tissues and the effects of its loss in lipoid CAH are summarized below.

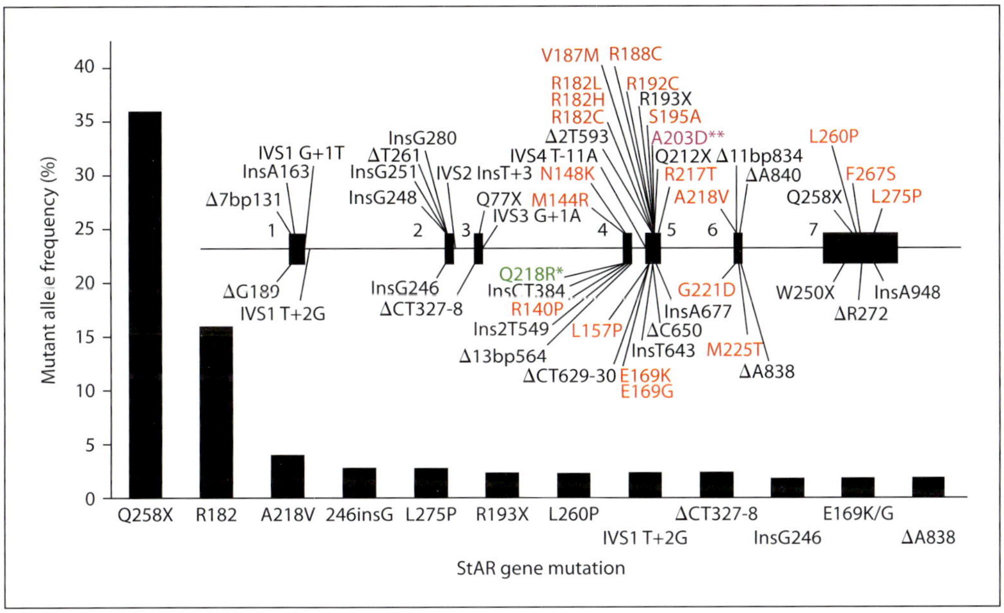

Fig. 1. Frequency of the common StAR gene mutations in reported lipoid CAH patients up to 2010. *Inset:* Approximate locations of reported StAR gene mutations in lipoid CAH patients to date. Specific amino acid changes are highlighted in red. Shown are exons 1–7 (boxes) and the intervening introns. *Gain of function. **Shown in vitro to have lowered activity; patient evidence indicates it does not cause lipoid CAH by itself.

Adrenals

The synthesis of adrenal steroids is controlled by StAR [2]. In turn, StAR production is controlled by ACTH and angiotensin II [4]. The fetal adrenal cortex also generates StAR and steroids by 14 weeks of gestation. StAR appears in the outer definitive zone after 23 weeks. The consequences of the loss of StAR activity in utero seem limited to effects on lung maturation. After birth, adrenal insufficiency occurs within the first 2 months of life in the majority of patients.

With compromised adrenal steroidogenesis, feedback regulation of pituitary ACTH release is lost. The resulting persistently high ACTH levels drive cholesterol uptake, leading to adrenal hyperplasia. Steroid replacement is a highly successful treatment strategy and normalizes ACTH levels, but unlike other forms of CAH, adrenal size responds more slowly. Due to the loss of negative feedback regulation, StAR–/– mice have elevated ACTH and corticotropin-releasing hormone levels with extensive lipid deposits in the adrenal cortex [17]. The age of onset of adrenal crisis varies from postnatal day 1 to 16. Mice that die within the first 2 days have difficulty breathing likely due to the loss of glucocorticoids in utero and less exposure to maternal steroids than longer-lived siblings. Differences in the level of maternal steroids persisting in the circulation postnatally or genetic background may further determine the time of death.

The events leading to adrenal crisis can be described by the 'two-hit' hypothesis [18]. The first 'hit' is the immediate loss of adrenal steroids due to mutations in StAR. Some steroids are still produced due to the low level of activity mutant StAR may retain and StAR-independent cholesterol transfer to P450scc. The second 'hit' occurs when the buildup of cholesteryl esters physically and oxidatively damages the cell, completely eliminating steroidogenesis and triggering the crisis. Empirically, this model is not completely satisfactory since few patients actually exhibit adrenal hyperplasia reflective of pathologic lipid accumulation at the time of adrenal crisis, irrespective of age of onset.

Testis

Testicular fetal and adult Leydig cells at all stages of differentiation express StAR primarily under the control of LH postnatally [2]. The loss of testosterone production in utero in lipoid CAH causes undervirilization or complete feminization, with the majority of 46,XY patients exhibiting female external genitalia. In the classic form of lipoid CAH, testes are located either intra-abdominally or inguinally with defects ranging from spermatogenic arrest to Sertoli cell only, but are nondysgenetic. The removal of the testes is recommended due to the risk of development of carcinoma in situ [19, 20].

With the loss of StAR, lipids accumulate more slowly in Leydig cells than in the adrenal, reflecting a lower demand for steroid synthesis. In fact, abnormal lipid deposits are not detectable in the perinatal testes of lipoid CAH patients, but are usually observed by 14 months. In StAR–/– mice, postnatal testes are inguinal and seminal vesicles are smaller [17]. In pubertal mice sustained by steroid supplementation, the prostate and seminal vesicles are small and hypoplastic [21]. Spermatogenesis occurs with increased germ cell death and sperm maturation delay. As in the human, Leydig cells accumulate lipids at a slower rate than adrenal tissue.

Ovary

The prepubertal ovary is nonsteroidogenic, possessing a very low level of P450scc mRNA and no StAR [2]. With puberty, FSH stimulates StAR, P450scc and androgen production in theca and interstitial cells associated with developing follicles. Granulosa cells acquire the ability to produce steroids with the onset of P450scc expression following the peak in estrogen, but do not synthesize StAR until after the LH surge. Expression of StAR remains high in luteal cells, but is lost in corpora albicantia.

Since steroidogenesis is delayed until the time of puberty, the ovary is initially spared damage in lipoid CAH [reviewed in 22]. Lipids accumulate in theca cells associated with maturing follicles. Although this pathology may eventually halt steroid production in these cells, it does not directly affect cells associated with new follicles recruited in subsequent cycles. While progesterone levels are typically undetectable, the rise of estrogens at the commencement of puberty is typically sufficient to initiate the development of secondary sexual characteristics and regular menses for years. Still, there is a wide variability in the pubertal development of lipoid CAH patients.

Surprisingly, estrogen levels are largely unaffected. The source of estradiol (E_2) is StAR-independent androgen production in growing follicles. Still, patients ultimately present a high LH/FSH ratio and infertility. Patients may cycle for several years before suffering primary ovarian failure. Current evidence indicates that these cycles are anovulatory [22].

The high levels of LH instigate the formation of large, possibly life-threatening ovarian cysts similar to polycystic ovarian syndrome without an increase in androgen levels. The majority of patients who spontaneously enter puberty eventually develop irregular periods and/or these ovarian cysts. Given the possible complications of this condition, such as infertility, early menopause and acute emergencies like cyst torsion, hormonal interventions should be considered to mitigate these problems.

Similar to the human case, the ovary of the StAR knockout mouse is morphologically normal until puberty [21]. At that time, lipid deposits appear in concert with stromal cell luteinization and incomplete follicular maturation. Ovaries enlarge with age, displaying few follicles and vacuolated stromal cells, indicative of premature ovarian failure. Still, estrogen levels are maintained at wild-type levels.

While 46,XX females are inherently infertile, successful pregnancies were achieved in 2 patients through active clinical management of folliculogenesis through to early pregnancy [23, 24]. A key part of the strategy is the supplementation of progesterone to make up for the lost luteal production of the hormone during pregnancy.

Variabilities in Lipoid CAH Presentation

Less severe forms of the disorder, termed nonclassic, present later in life without many features of the disease and correspond to at least one mutant StAR allele maintaining a significant level of activity. Partial loss-of-function StAR mutations can present as familial glucocorticoid deficiency (FGD) type 3 [25]. While mutations in StAR that cause an incomplete loss of function generally predict a later presentation of the disorder, the converse is not always true. Thus, other uncharacterized factors further shape the age of adrenal crisis aside from the severity of the StAR mutation.

Conclusions

Increased awareness of lipoid CAH and FGD3 due to StAR deficiency and the ease of genetic screening have increased accurate diagnosis, allowing for timely and life-saving treatment of patients. The two-hit model explains the findings in many patients. Moreover, the observation that FGD and some lipoid CAH patients exhibit normal or hypoplastic adrenal glands and the differences in age of adrenal crisis compel us to ask what variables determine the clinical presentation. The type of mutation, the level of StAR-independent steroidogenesis and exposure to maternal steroids in utero all may contribute.

References

1 Stocco DM, Clark BJ: Regulation of the acute production of steroids in steroidogenic cells. Endocr Rev 1996;17:221–244.

2 Bhangoo A, Anhalt H, Ten S, King SR: Phenotypic variations in lipoid congenital adrenal hyperplasia. Pediatr Endocrinol Rev 2006;3:258–271.

3 Lin D, Sugawara T, Strauss JF III, Clark BJ, Stocco DM, Saenger P, Rogol A, Miller WL: Role of steroidogenic acute regulatory protein in adrenal and gonadal steroidogenesis. Science 1995;267:1828–1831.

4 Lavoie HA, King SR: Transcriptional regulation of steroidogenic genes: STARD1, CYP11A1 and HSD3B. Exp Biol Med (Maywood) 2009;234:880–907.

5 Stocco DM, Wang X, Jo Y, Manna PR: Multiple signaling pathways regulating steroidogenesis and steroidogenic acute regulatory protein expression: more complicated than we thought. Mol Endocrinol 2005;19:2647–2659.

6 Clark BJ, Wells J, King SR, Stocco DM: The purification, cloning, and expression of a novel luteinizing hormone-induced mitochondrial protein in MA-10 mouse Leydig tumor cells. Characterization of the steroidogenic acute regulatory protein (StAR). J Biol Chem 1994;269:28314–28322.

7 King SR, Ronen-Fuhrmann T, Timberg R, Clark BJ, Orly J, Stocco DM: Steroid production after in vitro transcription, translation, and mitochondrial processing of protein products of complementary deoxyribonucleic acid for steroidogenic acute regulatory protein. Endocrinology 1995;136:5165–5176.

8 Arakane F, King SR, Du Y, Kallen CB, Walsh LP, Watari H, Stocco DM, Strauss JF III: Phosphorylation of steroidogenic acute regulatory protein (StAR) modulates its steroidogenic activity. J Biol Chem 1997;272:32656–32662.

9 Krueger RJ, Orme-Johnson NR: Acute adrenocorticotropic hormone stimulation of adrenal corticosteroidogenesis. Discovery of a rapidly induced protein. J Biol Chem 1983;258:10159–10167.

10 Noriyuki K, Ayako T, Naoko S, Reiko H, Toshiaki T: Adrenal gland. Significance of new StAR gene mutation of S195A-StAR phosphorylation found in congenital adrenal lipoid hyperplasia. Clin Endocrinol 2000;48:141–143.

11 Epstein LF, Orme-Johnson NR: Regulation of steroid hormone biosynthesis. Identification of precursors of a phosphoprotein targeted to the mitochondrion in stimulated rat adrenal cortex cells. J Biol Chem 1991;266:19739–19745.

12 Stocco DM, Sodeman TC: The 30-kDa mitochondrial proteins induced by hormone stimulation in MA-10 mouse Leydig tumor cells are processed from larger precursors. J Biol Chem 1991;266:19731–19738.

13 Arakane F, Sugawara T, Nishino H, Liu Z, Holt JA, Pain D, Stocco DM, Miller WL, Strauss JF III: Steroidogenic acute regulatory protein (StAR) retains activity in the absence of its mitochondrial import sequence: implications for the mechanism of StAR action. Proc Natl Acad Sci USA 1996;93:13731–13736.

14 Sasaki G, Ishii T, Jeyasuria P, Jo Y, Bahat A, Orly J, Hasegawa T, Parker KL: Complex role of the mitochondrial targeting signal in the function of steroidogenic acute regulatory protein revealed by bacterial artificial chromosome transgenesis in vivo. Mol Endocrinol 2008;22:951–964.

15 Tsujishita Y, Hurley JH: Structure and lipid transport mechanism of a StAR-related domain. Nat Struct Biol 2000;7:408–414.

16 Tomasetto C, Regnier C, Moog-Lutz C, Mattei MG, Chenard MP, Lidereau R, Basset P, Rio MC: Identification of four novel human genes amplified and overexpressed in breast carcinoma and localized to the q11-q21.3 region of chromosome 17. Genomics 1995;28:367–376.

17 Caron KM, Soo SC, Wetsel WC, Stocco DM, Clark BJ, Parker KL: Targeted disruption of the mouse gene encoding steroidogenic acute regulatory protein provides insights into congenital lipoid adrenal hyperplasia. Proc Natl Acad Sci USA 1997;94:11540–11545.

18 Bose HS, Sugawara T, Strauss JF III, Miller WL: The pathophysiology and genetics of congenital lipoid adrenal hyperplasia. N Engl J Med 1996;335:1870–1878.

19 Korsch E, Peter M, Hiort O, Sippell WG, Ure BM, Hauffa BP, Bergmann M: Gonadal histology with testicular carcinoma in situ in a 15-year-old 46,XY female patient with a premature termination in the steroidogenic acute regulatory protein causing congenital lipoid adrenal hyperplasia. J Clin Endocrinol Metab 1999;84:1628–1632.
20 Abdulhadi-Atwan M, Jean A, Chung WK, Meir K, Ben NZ, Stratigopoulos G, Oberfield SE, Fennoy I, Hirsch HJ, Bhangoo A, Ten S, Lerer I, Zangen DH: Role of a founder c.201_202delCT mutation and new phenotypic features of congenital lipoid adrenal hyperplasia in Palestinians. J Clin Endocrinol Metab 2007;92:4000–4008.
21 Hasegawa T, Zhao L, Caron KM, Majdic G, Suzuki T, Shizawa S, Sasano H, Parker KL: Developmental roles of the steroidogenic acute regulatory protein (StAR) as revealed by StAR knockout mice. Mol Endocrinol 2000;14:1462–1471.
22 Bhangoo A, Buyuk E, Oktay K, Ten S: Phenotypic features of 46,XX females with StAR protein mutations. Pediatr Endocrinol Rev 2007;5:633–641.
23 Khoury K, Barbar E, Ainmelk Y, Ouellet A, LeHoux JG: Gonadal function, first cases of pregnancy, and child delivery in a woman with lipoid congenital adrenal hyperplasia. J Clin Endocrinol Metab 2009;94:1333–1337.
24 Sertedaki A, Pantos K, Vrettou C, Kokkali G, Christofidou C, Kanavakis E, Dacou-Voutetakis C: Conception and pregnancy outcome in a patient with 11-bp deletion of the steroidogenic acute regulatory protein gene. Fertil Steril 2009;91:934–938.
25 Metherell LA, Naville D, Halaby G, Begeot M, Huebner A, Nurnberg G, Nurnberg P, Green J, Tomlinson JW, Krone NP, Lin L, Racine M, Berney DM, Achermann JC, Arlt W, Clark AJ: Nonclassic lipoid congenital adrenal hyperplasia masquerading as familial glucocorticoid deficiency. J Clin Endocrinol Metab 2009;94:3865–3871.

Douglas M. Stocco, PhD
Department of Cell Biology and Biochemistry
Texas Tech University Health Sciences Center
3601 4th Street, Lubbock, TX 79430 (USA)
Tel. +1 806 743 2505, Fax +1 806 743 2990, E-Mail doug.stocco@ttuhsc.edu

Ghizzoni L, Cappa M, Chrousos G, Loche S, Maghnie M (eds): Pediatric Adrenal Diseases.
Endocr Dev. Basel, Karger, 2011, vol 20, pp 54–62

P450 Side-Chain Cleavage Deficiency – A Rare Cause of Congenital Adrenal Hyperplasia

Berthold Hauffa[a] · Olaf Hiort[b]

[a]Division of Pediatric Endocrinology and Diabetes, Department of Pediatrics, University of Lübeck, Lübeck, and [b]Division of Pediatric Endocrinology and Diabetes, Department of Pediatrics, University of Duisburg-Essen, Duisburg-Essen, Germany

Abstract

Mitochondrial cytochrome P450 side-chain cleavage enzyme (P450scc) converts cholesterol to pregnenolone and is the initiating enzyme for steroidogenesis. It is encoded by a single-copy gene on chromosome 15. For a long time it was thought that deficiency of P450scc was not compatible with life due to lack of progesterone for maintenance of pregnancy. However, recently a total of 8 patients with missense or even nonsense mutations of CYP11A1 have been described. Depending on the severity of the enzyme dysfunction, patients present with mild to severe early-onset adrenal failure. In 46,XY patients also a disorder of sex development is prevalent with hypospadias to complete female phenotype. P450scc deficiency is the differential diagnosis of congenital lipoid adrenal hyperplasia caused by mutations in the steroidogenic acute regulatory protein, however, in contrast to the latter, these patients do not have adrenal hyperplasia but small adrenals and gonads.

In 1951, Stone and Hechter [1] proposed and later provided evidence that conversion of cholesterol to pregnenolone is the first, enzymatically rate-limiting step in the synthesis of all adrenal and gonadal steroids. The conversion requires the presence of a single cholesterol side-chain cleavage enzyme called cytochrome P450scc [2].

Cytochrome P450 proteins constitute a large group of oxidative enzymes containing a single heme group. P450 stands for 'pigment 450', indicating that these proteins absorb light at 450 nm in their reduced states. P450scc, sometimes termed CYP11A1 according to an updated nomenclature [3], is the cholesterol-specific terminal oxidase in a mitochondrial electron-transfer system also containing a flavoprotein, ferredoxin reductase, and an iron-sulfur protein, ferredoxin [4].

Defective P450scc activity interferes with normal steroid production in the adrenal glands and the gonads. It predicts a biochemical and clinical phenotype with severe glucocorticoid and mineralocorticoid deficiency associated with excess cholesterol storage in steroid-producing cells (lipoid adrenal hyperplasia), massively elevated ACTH and renin plasma concentrations and a female appearance of 46,XY individuals, with absent puberty and hypergonadotropic hypogonadism in both male and female karyotypes. Indeed, already in 1955, Prader [5, 6] had described children with lipoid congenital adrenal hyperplasia. Their clinical course suggested a P450scc defect following an autosomal-recessive pattern of inheritance [7]. Up to 1986, a total of 32 individuals believed to have P450scc deficiency were identified [8], indicating that P450scc deficiency would be a rare disease. In 1 patient in whom a postmortem examination had been carried out, the P450scc enzymatic activity of adrenal mitochondrial homogenates was significantly reduced to 10% of normal [9].

In 1986, human P450scc cDNA was cloned, sequenced full-length, and assigned to one single gene on chromosome 15 [10]. The human P450scc gene was later mapped to 15q23-q24 [11]. When this information was used for the molecular genetic analysis of patients with clinical signs suggestive of P450scc deficiency, it came as a surprise that in almost all of these individuals lesions of the P450scc gene could be excluded [12]. It was argued that homozygous P450scc deficiency in humans is incompatible with life since fetuses would spontaneously abort at 6–7 weeks of gestation, at a time when the defective fetal placental tissue P450scc is unable to take over the progesterone production from the maternal corpus luteum needed for pregnancy maintenance [13].

The search continued for a gene whose malfunction would only compromise adrenal and gonadal steroid production but leave placental steroidogenesis intact. Success came with the identification of the steroidogenic acute regulatory (StAR) protein needed for translocation of cholesterol to the inner mitochondrial membrane where the P450scc enzyme resides [14]. In the human, the StAR gene was mapped to chromosome 8p11.2 [15]. Most patients with lipoid congenital adrenal hyperplasia turned out to have mutations in StAR indicating that this protein is indispensable for normal adrenal and gonadal steroidogenesis [16, 17]. The StAR gene is not expressed in the fetal placenta tissue, where progesterone biosynthesis occurs independently, explaining why fetuses with StAR defects may survive [15, 18].

Biochemistry

A cell is called 'steroidogenic', meaning capable of de novo steroid synthesis, if it expresses P450scc. Steroidogenic cells do not store significant quantities of steroid hormones. Therefore, steroid secretion from these cells directly depends on steroid synthesis [19]. The side-chain cleavage reaction as the enzymally rate-limiting step of steroid synthesis is hormonally regulated by ACTH and angiotensin II in the adrenal gland, and by LH in the gonads, and is mediated in these organs by the steroid acute

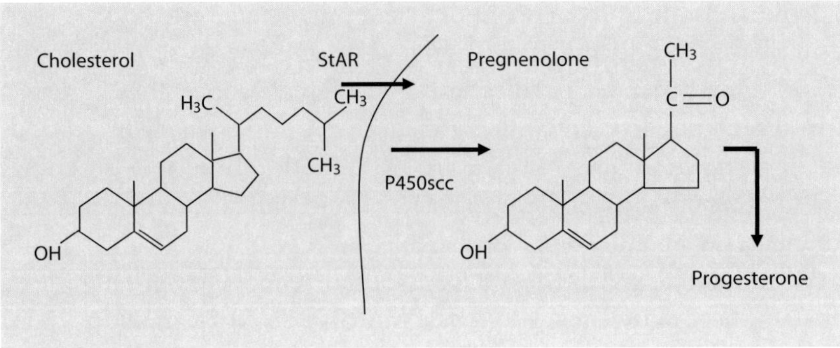

Fig. 1. Synthesis of pregnenolone from cholesterol.

regulatory protein (StAR) [20]. In the placenta, side-chain cleavage occurs independently from StAR.

Cholesterol side-chain cleavage is conceived as a sequential process of three distinct chemical reactions: 20α-hydroxylation, 22-hydroxylation and interruption of the C20-C22 carbon-carbon bond, thereby shortening the cholesterol side-chain by six carbon atoms (fig. 1). This occurs without significant accumulation of the intermediate products, yielding pregnenolone and isocaproic acid. The exact mechanism of these reactions is unknown. The resulting pregnenolone is considered an obligate precursor of all other steroid biosynthetic pathways [21].

The amount of steroid a cell will produce is determined by acute and chronic regulatory mechanisms. P450scc is a slow enzyme, converting only about six molecules of cholesterol to pregnenolone per molecule per minute [22]. Thus, the chronic steroidogenic capacity of a cell is determined by the transcription of the P450scc gene providing the intracellular amount of the enzyme. In nonplacental tissues, acute regulation in response to stimuli occurs via the StAR protein that accelerates the movement of cholesterol from the outer to the inner mitochondrial membrane [23].

In the Leydig cells of the testis, biosynthesis of testosterone is stimulated by LH binding to its receptor. This results in increased intracellular concentrations of cAMP and activation of cAMP-dependent pathways, which increases the conversion of cholesterol into pregnenolone. Expression of P450scc in the human fetal testis occurs early and is most abundant between 14 and 16 weeks of gestation, diminishing by the 20th to 26th gestational week [24]. Intact testosterone biosynthesis in the male fetus is necessary for male sexual differentiation, and later in life for maintaining reproductive capacity.

In the ovary, LH stimulates the theca cells to produce aromatizable androgens [25]. In the corpus luteum, the P450scc gene is expressed relatively constantly throughout the luteal cycle, not explaining the cyclical changes in luteal steroidogenesis. Luteal steroidogenesis is rather associated with changes in StAR gene expression [25]. In the placenta, steroid biosynthesis occurs without StAR. The mechanism of the StAR-independent mechanism of progesterone synthesis in the placenta is unclear. The presence of some other protein with StAR-like activity has been proposed [26].

Molecular Genetics and Biology

The human full-length cDNA of CYP11A1 was cloned in 1986 and assigned to chromosome 15 [10]. The gene is at least 20 kb long and is encodes 9 exons divided by 8 introns. The sequence analysis revealed that the nine separated exons code for a primary structure consisting of 521 amino acids which shows 72% homology with that of bovine P-450(SCC) [27]. To exert its action of monooxygenation, the enzyme requires electrons to activate molecular oxygen. These are provided by NADPH via a mitochondrial electron transfer system, consisting of adrenodoxin reductase, a flavin-adenine-dinucleotide (FAD)-containing reductase, and adrenodoxin, a soluble iron sulfur ferrodoxin type electron transfer protein. Adrenodoxin acts as an electron shuttle between adrenodoxin reductase and P450scc as well as other mitochondrial cytochrome P450s.

The crystal structure of CYP11A1 has not been determined, but homology modeling studies demonstrated the major functional regions. With site-directed mutagenesis, selected residues of the enzyme were evaluated for their specific properties. The residues R425 and R426 are close to the heme binding moiety and K403 and K405 are associated with the adrenodoxin interaction [28]. P450scc is very closely associated with the inner mitochondrial membrane, where it is actively translocated in a pre-protein state. There, a 39-amino-acid sequence is cleaved to yield the mature enzyme. Thus, the enzyme is only functional in the mitochondria, because of its processing but also because of the requirement of mitochondrion-specific electron transfer [29].

Naturally occurring mutations of CYP11A1 have been described and investigated in vitro. The first naturally occurring mutation was described by Tajima et al. [30] in a heterozygous state in a patient with clinical symptoms of delayed-onset congenital adrenal hyperplasia. Sequencing had revealed a six-base pair insertion in exon 4, which leads to an insertion of an additional glycine and asparagine between codons 271 and 272. When this mutation was recreated in vitro and expressed as a single catalytically active fusion protein, the mutant was not able to convert 20α-hydroxycholesterol to pregnenolone. However, cotransfection of the wild-type and the mutant containing expression vectors did not reveal a significant effect on the wild-type's enzymatic activity. This would be a proof for the autosomal-recessive mode of inheritance and the need for two affected alleles to induce P450scc deficiency. The authors argued, however, that a dominant negative effect might play a role.

A total of 8 families with P450scc deficiency have been described to this date. Accept for the initial case description by Tajima et al. [30], they all bear homozygous or compound heterozygous mutations in CYP11A1A, mostly inherited from their parents. These parents are described as healthy and fertile, despite the heterozygous

Table 1. Naturally occurring mutations in CYP11A1 and their retained activity in vitro

Mutation	Retained activity
p.L141W	38.5%
p.A189V	Unclear, some aberrant splicing
p.L222P	7%
p.271_272+GR	0
c.835delA = p.L288X	0
p.R353W	8.1%
p.A359V	11.7%
p.V415E	0
IVS3+(2-3)insT	0

carrier status. This clinically supports the fact that P450scc deficiency is an autosomal-recessive disorder.

Three mutations are associated with a complete failure of enzymatic activity in vitro. This includes the premature termination at codon position 288 due to missense mutation c.835delA = p.L288X [31, 32]. But also the splice site mutation IVS3+(2-3) insT as well as the amino acid substitution mutation V415E did not suggest any remaining enzyme function in in vitro studies [32]. However, the amino acid substitutions R353W [34], A359V [35], L141W [33], and L222P [35] retained some residual activity, therefore allowing pregnenolone formation to some extent. Moreover, the A189V substitution does not affect P450scc activity, but might affect a cryptic splice site [33]. The mutations and their enzymatic activities are summarized in table 1.

Clinical Findings in P450scc Deficiency

Initially it was thought that congenital lipoid adrenal hyperplasia as originally described by Prader [5, 6] would correspond to an enzyme deficiency affecting the first step of pregenolone synthesis. However, these patients were mostly found to bear defects in the StAR protein, and due to its estimated profound effects on placental progesterone synthesis, P450scc deficiency was thought not to occur in humans. With the first case descriptions of relevant CYP11A1 mutations, it became clear that fetuses may survive pregnancy. Apparently, however, some families report several abortions, suggestive of premature termination of pregnancies in some cases due to diminished progesterone formation [31, 32]. It has been reported that the mothers of affected children may have very low estrone levels during pregnancy [31]. This may be the cause also for the early onset of labor and the prematurity of this patient with a deleterious mutation.

Table 2. Mutations and clinical findings in patients with CYP11A1 mutations reported so far

No.	Mutation	Genital aspect	Adrenal failure	Gestation	Ref.
1	p.L288X homozygous	female	9 days	31 weeks	32
2	p.L288X IVS3+(2-3)insT	female	8 days	term	33
3	p.271 272+GR none	female virilized	2–4 years	term	31
4	p.L141W p.V415E	female	1–9 days	term	33
5	p.R353W p.A189V	female	7–9 months	term	34
6	p.A359V homozygous	female	21 months	term	35
7	p.A359V homozygous	female virilized	3 years	term	37
8	p.L222P homozygous	male hypospadias	9 years	term	36

Depending on the mutations in CYP11A1, manifestation of adrenal failure may be at different ages. In the case of deleterious mutations, onset of adrenal failure is with birth or shortly thereafter (table 2), in the case of remaining enzymatic activity the disorder may not come to attention as late as 8 years [35]. All patients had a highly elevated ACTH and low or even immeasurable adrenal and gonadal steroids. The degree of adrenal failure seems to correlate also with the amount of embryonal testicular androgen synthesis in affected 46,XY children. In patients with the deleterious mutations, a completely female phenotype has been described, while in patients with remaining enzymatic activity, the phenotype may vary from female with slight virilization to male with hypospadias. The testes in these patients are small in size. Histology reveals immature testicular tissue without germ cells (own observation of the patient reported in Hiort et al. [31]). In contrast to congenital lipoid adrenal hyperplasia, the adrenals are also small in size and do not show enlargement due to lipoid degeneration (fig. 2). In one patient, additional agenesis of the corpus callosum has been described. As the family was highly consanguineous, a direct association of CYP11A1 deficiency to other malformations is not proven.

Very little information is available about 46,XX patients with P450scc deficiency. Grabensee et al. [36] reported a consanguineous family with several affected members. One is a 46,XX female who presented with adrenal insufficiency. After adequate

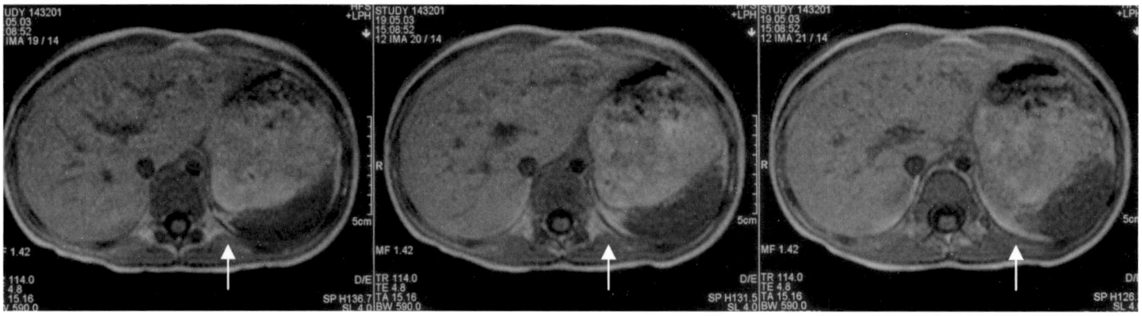

Fig. 2. Magnetic resonance imaging of the renal-adrenal region of a patient with CYP11A1A mutation, demonstrating small or even absent adrenals (arrows).

glucocorticoid replacement therapy, she apparently entered puberty spontaneously, but did not develop stable menstrual cycles. Four in vitro fertilization attempts ended with an early loss of pregnancy. This patient bore the mutation A359V, which is described with residual enzyme activity. This might therefore be the reason for some ovarian function.

Practical Summary

P450scc deficiency is a very rare severe defect in steroid biosynthesis. It is associated with early to intermediate onset of adrenal insufficiency and, in the case of 46,XY karyotype, a variable degree of undervirilization. All steroid measurements will be low to undetectable. P450scc deficiency should be thought of in the differential diagnosis to StAR deficiency when the adrenals are small and not suggestive of congenital lipoid adrenal hyperplasia.

References

1 Stone D, Hechter O: Studies on ACTH action in perfused bovine adrenals: aspects of progesterone as an intermediary in corticosteroidogenesis. Arch Biochem 1955;54:121–128.

2 Shikita M, Hall PF: Cytochrome P-450 from bovine adrenocortical mitochondria: an enzyme for the side chain cleavage of cholesterol. I. Purification and properties. J Biol Chem 1973;248:5598–5604.

3 Nelson DR, Kamataki T, Waxman DJ, Guengerich FP, Estabrook RW, Feyereisen R, Gonzalez FJ, Coon MJ, Gunsalus IC, Gotoh O, Okuda K, Nebert DW: The P450 superfamily: update on new sequences, gene mapping, accession numbers, early trivial names of enzymes, and nomenclature. DNA Cell Biol 1993;12:1–51.

4 Omura T, Sato R, Cooper DY, Rosenthal O, Estabrook RW: Function of cytochrome P-450 of microsomes. Fed Proc 1965;24:1181–1189.

5 Prader A, Siebenmann RE: Nebenniereninsuffizienz bei kongenitaler Lipoidhyperplasie der Nebennieren. Helv Paediatr Acta 1957;12:569–595.

6 Prader A, Gurtner HP: Das Syndrom des Pseudohermaphroditismus masculinus bei kongenitaler Nebennierenhyperplasie ohne Androgenüberproduktion (adrenaler Pseudohermaphroditismus masculinus). Helv Paediatr Acta 1955;10:397–412.
7 Prader A, Anders GJ: Zur Genetik der kongenitalen Lipoidhyperplasie der Nebennieren. Helv Paediatr Acta 1962;17:285–289.
8 Hauffa BP, Miller WL, Grumbach MM, Conte FA, Kaplan SL: Congenital adrenal hyperplasia due to deficient cholesterol side-chain cleavage activity (20,22-desmolase) in a patient treated for 18 years. Clin Endocrinol (Oxf) 1985;23:481–493.
9 Koizumi S, Kyoya S, Miyawaki TM, Kidani H, Funabashi T: Cholesterol side-chain cleavage enzyme activity and cytochrome P-450 content in adrenal mitochondria of a patient with congenital lipoid adrenal hyperplasia (Prader disease). Clin Chim Acta 1977;77:301–306.
10 Chung BC, Matteson KJ, Voutilainen R, Mohandas TK, Miller WL: Human cholesterol side-chain cleavage enzyme, P450scc: cDNA cloning, assignment of the gene to chromosome 15, and expression in the placenta. Proc Natl Acad Sci USA 1986;83: 8962–8966.
11 Sparkes RS, Klisak I, Miller WL: Regional mapping of genes encoding human steroidogenic enzymes: P450scc to 15q23-q24, adrenodoxin to 11q22; adrenodoxin reductase to 17q24-q25; and P450c17 to 10q24-q25. DNA Cell Biol 1991;10:359–365.
12 Lin D, Gitelman SE, Saenger P, Miller WL: Normal genes for the cholesterol side chain cleavage enzyme, P450scc, in congenital lipoid adrenal hyperplasia. J Clin Invest 1991;88:1955–1962.
13 Miller WL: Why nobody has P450scc (20,22 desmolase) deficiency. J Clin Endocrinol Metab 1998; 83:1399–1400.
14 Clark BJ, Wells J, King SR, Stocco DM: The purification, cloning, and expression of a novel luteinizing hormone-induced mitochondrial protein in MA-10 mouse Leydig tumor cells. Characterization of the steroidogenic acute regulatory protein (StAR). J Biol Chem 1994;269:28314–28322.
15 Sugawara T, Holt JA, Driscoll D, Strauss JF III, Lin D, Miller WL, Patterson D, Clancy KP, Hart IM, Clark BJ, Stocco DM: Human steroidogenic acute regulatory protein: functional activity in COS-1 cells, tissue-specific expression, and mapping of the structural gene to 8p11.2 and a pseudogene to chromosome 13. Proc Natl Acad Sci USA 1995;92:4778–4782.
16 Lin D, Sugawara T, Strauss JF III, Clark BJ, Stocco DM, Saenger P, Rogol A, Miller WL: Role of steroidogenic acute regulatory protein in adrenal and gonadal steroidogenesis. Science 1995;267:1828–1831.
17 Bose HS, Sugawara T, Strauss JF III, Miller WL: The pathophysiology and genetics of congenital lipoid adrenal hyperplasia. N Engl J Med 1996;335:1870–1878.
18 Saenger P, Klonari Z, Black SM, Compagnone N, Mellon SH, Fleischer A, Abrams CA, Shackelton CH, Miller WL: Prenatal diagnosis of congenital lipoid adrenal hyperplasia. J Clin Endocrinol Metab 1995;80:200–205.
19 Miller WL: Disorders of androgen synthesis – from cholesterol to dehydroepiandrosterone. Med Princ Pract 2005;14(suppl 1):58–68.
20 Miller WL: Androgen synthesis in adrenarche. Rev Endocr Metab Disord 2009;10:3–17.
21 Lieberman S, Lin YY: Reflections on sterol side-chain cleavage process catalyzed by cytochrome P450(scc). J Steroid Biochem Mol Biol 2001;78:1–14.
22 Kuwada M, Kitajima R, Suzuki H, Horie S: Purification and properties of cytochrome P-450 (SCC) from pig testis mitochondria. Biochem Biophys Res Commun 1991;176:1501–1508.
23 Miller WL: Steroidogenic enzymes. Endocr Dev 2008;13:1–18.
24 Voutilainen R, Miller WL: Developmental expression of genes for the stereoidogenic enzymes P450scc (20,22-desmolase), P450c17 (17α-hydroxylase/17,20-lyase), and P450c21 (21-hydroxylase) in the human fetus. J Clin Endocrinol Metab 1986;63:1145–1150.
25 Hillier SG, Whitelaw PF, Smyth CD: Follicular oestrogen synthesis: the 'two-cell, two-gonadotrophin' model revisited. Mol Cell Endocrinol 1994;100:51–54.
26 Devoto L, Kohen P, Gonzalez RR, Castro O, Retamales I, Vega M, Carvallo P, Chrisenson LK, Strauss JF III: Expression of steroidogenic acute regulatory protein in the human corpus luteum throughout the luteal phase. J Clin Endocrinol Metab 2001;86:5633–5639.
27 Morohashi K, Sogawa K, Omura T, Fujii-Kuriyama Y: Gene structure of human cytochrome P-450 (SCC), cholesterol desmolase. J Biochem 1987;101: 879–887.
28 Storbeck KH, Swart P, Swart AC: Cytochrome P450 side-chain cleavage: insights gained from homology modelling. Mol Cell Endocrinol 2007;265/266:65–70.
29 Black SM, Harikrishna JA, Szklarz GD, Miller WL: The mitochondrial environment is required for activity of the cholesterol side-chain cleavage enzyme, cytochrome P450scc. Proc Natl Acad Sci USA 1994;91:7247–7251.

30 Tajima T, Fujieda K, Kouda N, Nakae J, Miller WL: Heterozygous mutation in the cholesterol side chain cleavage enzyme (P450scc) gene in a patient with 46,XY sex reversal and adrenal insufficiency. J Clin Endocr Metab 2001;86:3820–3825.

31 Hiort O, Holterhus PM, Werner R, Marschke C, Hoppe U, Partsch CJ, Riepe FG, Achermann JC, Struve D: Homozygous disruption of P450 side-chain cleavage (CYP11A1) is associated with prematurity, complete 46,XY sex reversal, and severe adrenal failure. J Clin Endocr Metab 2005;90:538–541.

32 Kim CJ, Lin L, Huang N, Quigley CA, AvRuskin TW, Achermann JC, Miller WL: Severe combined adrenal and gonadal deficiency caused by novel mutations in the cholesterol side chain cleavage enzyme, P450scc. J Clin Endocr Metab 2008;93:696–702.

33 Katsumata N, Ohtake M, Hojo T, Ogawa E, Hara T, Sato N, Tanaka T: Compound heterozygous mutations in the cholesterol side-chain cleavage enzyme gene (CYP11A) cause congenital adrenal insufficiency in humans. J Clin Endocr Metab 2002;87:3808–3813.

34 Al Kandari, H, Katsumata, N, Alexander, S, Rasoul, MA: Homozygous mutation of P450 side-chain cleavage enzyme gene (CYP11A1) in 46,XY patient with adrenal insufficiency, complete sex reversal, and agenesis of corpus callosum. J Clin Endocr Metab 2006;91:2821–2826.

35 Rubtsov P, Karmanov M, Sverdlova P, Spirin P, Tiulpakov A: A novel homozygous mutation in CYP11A1 gene is associated with late-onset adrenal insufficiency and hypospadias in a 46,XY patient. J Clin Endocr Metab 2009;94:936–939.

36 Grabensee A, Schaaf K, Mühlenberg R, Hiort O, Holterhus PM, Hauffa BP: Homozygote mutation of the P450 side-chain cleavage enzyme gene (CYP11A1A) in a patient with adrenal failure and sex reversal. Horm Res 2008;70(suppl 1):141.

Prof. Dr. Olaf Hiort
Division of Pediatric Endocrinology and Diabetes
Department of Pediatrics, University of Lübeck
Ratzeburger Allee 160, DE–23538 Lübeck (Germany)
Tel. +49 451 500 2191, Fax +49 451 500 6867, E-Mail hiort@paedia.ukl.mu-luebeck.de

Ghizzoni L, Cappa M, Chrousos G, Loche S, Maghnie M (eds): Pediatric Adrenal Diseases.
Endocr Dev. Basel, Karger, 2011, vol 20, pp 63–79

Clinical and Biochemical Consequences of P450 Oxidoreductase Deficiency

Christa E. Flück · Amit V. Pandey

Department of Pediatrics, Division of Pediatric Endocrinology and Diabetology and Department of Clinical Research, University of Bern, Bern, Switzerland

Abstract

Patients with P450 oxidoreductase (POR) deficiency typically present with adrenal insufficiency, genital anomalies and bony malformations resembling the Antley-Bixler craniosynostosis syndrome. Since our first report in 2004, more than 40 POR mutations have been identified in over 65 patients. POR is the obligate electron donor to all microsomal P450 enzymes, including the steroidogenic enzymes CYP17A1, CYP21A2 and CYP19A1. POR deficiency may cause disordered sexual development manifested as genital undervirilization in 46,XY newborns as well as overvirilization in those who are 46,XX. This may be explained by impaired aromatization of fetal androgens that may cause maternal virilization and low urinary estriol levels during pregnancy. In addition, the alternate 'backdoor' pathway of androgen biosynthesis, which leads to dihydrotestosterone production bypassing androstenedione and testosterone, may also play a role. Functional assays studying the effects of POR mutations on steroidogenesis showed that several POR variants impaired CYP17A1, CYP21A2 and CYP19A1 activities to different degrees, indicating that each POR variant must be studied separately for each potential target P450 enzyme. POR variants may also affect skeletal development and drug metabolism. As most drugs are metabolized by hepatic microsomal P450 enzymes, studies of the impact of POR mutations on drug-metabolizing P450s are particularly important.

Role of Cytochrome P450 Oxidoreductase in the Biochemistry of Electron Transfer

The 82-kDa P450 oxidoreductase (POR) protein was first isolated as a component of the microsomal P450 enzyme system in 1969 [1]. While the human cDNA for POR was cloned in 1989 [2], the POR gene was characterized and sequenced later as part of the human genome project. POR is located on chromosome 7q11.23 and consists of 15 protein-coding exons spanning 32 kb, and a non-coding exon located 38.8 kb upstream [3]. The human POR gene encodes 680 amino acids and shares 94% identity with POR from rat that lacks 3 amino acids at the amino-terminus, and contains 677

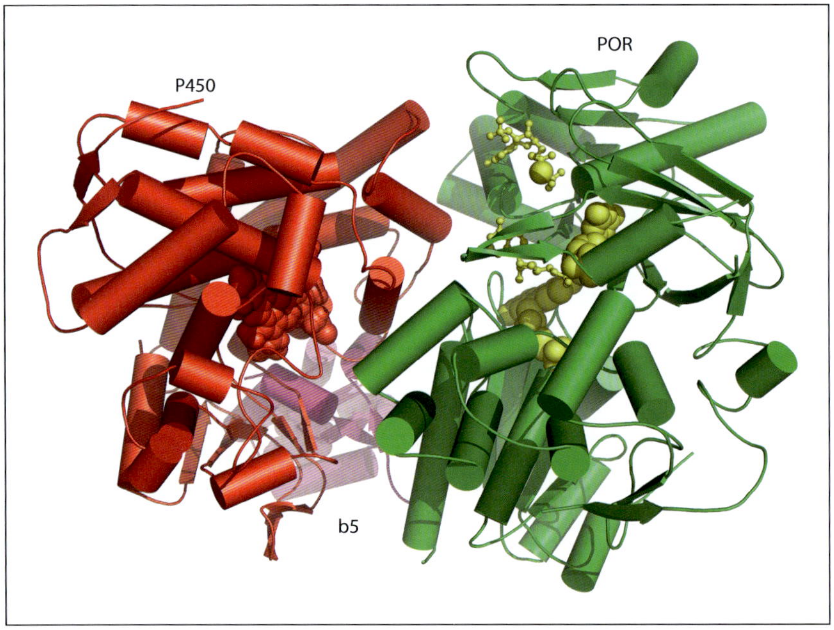

Fig. 1. Electron transfer by POR to microsomal (type II) P450 enzymes such as CYP17A1, CYP21A2 and CYP19A1. NADPH interacts with POR, which is bound to the endoplasmic reticulum, and gives up a pair of electrons, which are received by the FAD moiety. Electron receipt elicits a conformational change, permitting the isoalloxazine rings of the FAD and FMN moieties to move closer together, so that the electrons pass from the FAD to the FMN. Following another conformational change that returns the POR to its original orientation, the FMN domain interacts with the redox-partner binding site of the P450, so that electrons reach the P450 heme iron to achieve catalysis. The interaction of POR and the P450 is coordinated by acidic residues on the surface of the FMN domain of POR, and by basic residues in the redox-partner binding site of the P450. In the case of human CYP17A1, this interaction is facilitated by the allosteric action of cytochrome b_5 and by serine phosphorylation of CYP17A1, which optimize the interaction of the two proteins.

amino acids. The N-terminus of POR consists of a 25- to 30-amino-acid-long hydrophobic sequence which anchors POR toward the cytoplasmic side of the endoplasmic reticulum; this sequence also plays a role in the interaction with cytochrome P450s. Deletion of this N-terminal leads to loss of POR activity. The X-ray crystal structure of a soluble (N-66) form of rat POR lacking the membrane-anchoring region has been determined [4]. POR has two structurally distinct domains, an extended domain that binds reduced nicotinamide adenine dinucleotide phosphate (NADPH) and flavin adenine dinucleotide (FAD) and a domain that binds flavin mononucleotide (FMN); these two domains are separated by a flexible hinge region (fig. 1).

The major function of POR is to transfer electrons from NADPH to microsomal (type II) P450 enzymes (fig. 1). Once the FAD molecule accepts a pair of electrons from NADPH, the hinge flexes to bring the FAD closer to the FMN domain, so that electrons can pass from FAD to FMN. The FMN domain interacts with the redox-

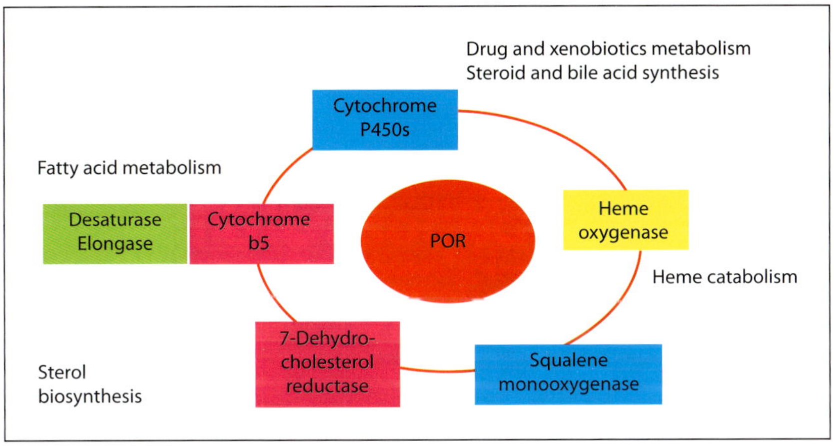

Fig. 2. Overview of the involvement of POR in various biochemical pathways.

partner binding sites of the cytochrome P450 and other electron acceptors to transfer electrons received from the FAD. Unlike mitochondrial ferredoxin reductase, which only has a FAD domain to accept electrons from NADPH and needs ferredoxin as an intermediate to support type I P450 enzymes, POR is able to donate electrons directly to type II P450 enzymes [for review, see 5]. The hinge region is unique to POR and provides structural flexibility to the FMN and FAD regions during electron transfer, both from NADPH and to the P450s and other electron-accepting partners. This structural flexibility permits POR to provide electrons to a wide variety of substrates. For some P450 enzymes, cytochrome b_5 may donate the second electron in the POR catalytic cycle [6], but in the case of CYP17A1 (fig. 1) and some hepatic drug-metabolizing P450 enzymes, cytochrome b_5 acts as an allosteric factor to improve the interaction of POR with P450 [7–9].

POR is the main protein which transfers electrons from NADPH to P450s; therefore, the catalytic activities of all microsomal P450 enzymes depend on POR. The electron transfer from POR to P450s is faster and more efficient than the P450 catalytic reactions, so that in liver and most steroidogenic tissues, the molar ratio of POR to all the microsomal P450s may be as low as 1:20. Only testicular cells contain more POR than CYP17A1 [10], promoting the 17,20-lyase reaction of CYP17A1 far more than the 17α-hydroxylase reaction for the synthesis of C19 precursors of sex steroids [10, 11].

Cytochrome P450s – Partners of POR

Besides microsomal P450s, POR interacts with many more partners for electron transfer (fig. 2). Cytochrome P450 enzymes are heme-containing proteins that catalyze a broad range of oxidative reactions (http://drnelson.uthsc.edu/cytochromeP450c.html).

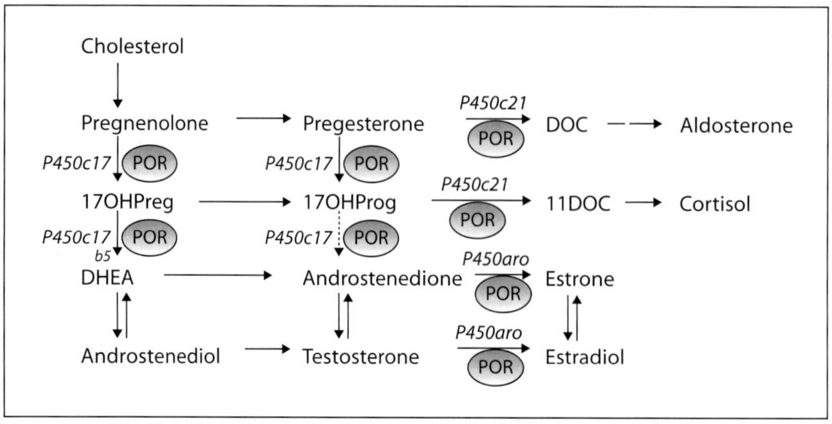

Fig. 3. Steroid biosynthetic pathway. Enzymes requiring POR as electron donor are shown with short names; enzymes that do not require POR are simply depicted as arrows. Note that the human 17,20-lyase activity of CYP17A1 converts 17α-hydroxypregnenolone (17OHPreg) to dehydroepiandrosterone (DHEA) but does not effectively convert 17α-hydroxyprogesterone (17OHProg) to androstenedione. DOC = 11-Deoxycorticosterone; 11DOC = 11-deoxycortisol.

The 57 known human P450 enzymes are grouped in type I P450s, which are found in the mitochondria, and the type II P450s which reside in the endoplasmic reticulum. There are seven type I P450s, which receive electrons from NADPH via the coupled electron chain of ferridoxin reductase and ferridoxin. All type I P450 enzymes are involved in the synthesis of steroids, sterols and bile acids [5]. By contrast, all 50 type II P450s receive electrons from NADPH via the single redox partner POR (fig. 1) [5]. The three steroidogenic type II P450 enzymes are CYP17A1, CYP21A2 and CYP19A1 (fig. 3). Other type II P450 are involved in drug metabolism or in the biosynthetic pathways leading to cholesterol, bile acids and eicosanoids. However, some 'orphan' type II P450s still remain without known function.

Impact of POR Deficiency on Steroidogenesis

Since all steroid-metabolizing reactions catalyzed by CYP17A1, CYP21A2 and CYP19A1 require POR for electron transfer (fig. 3), patients with POR deficiency have a complex pattern of disordered steroidogenesis. Loss of the 17α-hydroxylase and 17,20-lyase activities of CYP17A1 will result in elevated plasma concentrations of deoxycorticosterone (DOC) and corticosterone but decreased levels of cortisol and C19 steroids (dehydroepiandrosterone [DHEA], DHEA sulfate [DHEAS] and androstenedione). These abnormalities may not be noticed at the basal state but are more readily seen following stimulation with adrenocorticotropic hormone (ACTH). Loss of 21-hydroxylase activity results in diminished plasma cortisol (and rarely aldosterone) but elevated 21-deoxycortisol concentrations, especially after ACTH stimulation. However, in contrast to isolated 21-hydroxylase deficiency due to CYP21A2 mutations, POR patients typically have only moderately elevated plasma 17α-hydroxyprogesterone, and rather decreased than increased C19 steroid concentrations due to the combination of CYP21A2 and CYP17A1 deficiencies.

Impairment of CYP17A1 activities by POR will inhibit androgen precursor production in the gonads and lead to disordered development of the external genitalia in the male fetus and to failure of pubertal development and fertility in both sexes, similar to human CYP17A1 mutations. Impairment of CYP19A1 activity due to POR deficiency will result in diminished production of estrogens from androgen precursors in the ovaries of affected individuals. Impairment of aromatase activity will affect the conversion of fetal C19 androgen precursors (DHEA and DHEAS) to estriol and estrone in the feto-placental unit and may cause virilization of the fetus and the mother (manifested by acne, voice changes, and hirsutism). While 46,XX fetuses with aromatase (CYP19A1) gene mutations typically develop severely virilized external genitalia during pregnancy, affected 46,XY patients do not have genital anomalies [12]. So there is an inconsistency in the undervirilization of some males with POR deficiency and the mild virilization of some females with POR deficiency. This apparent paradox may be explained by the 'backdoor pathway' of androgen biosynthesis, which is the main pathway to dihydrotestosterone (DHT) production in tammar wallaby pouch young [13]. In this pathway, DHT is formed from 17α-hydroxyprogesterone or progesterone via 5α- and 3α-reductases, bypassing the conventional intermediates androstenedione and testosterone [13]. From urinary steroid analyses there is evidence both for [14, 15] and against [14, 16] its activity in patients with POR deficiency. However, a role of the backdoor pathway in human androgen production is currently established from patients harboring AKR1C2 (3α-reductase) mutations [Flück CE and Biason-Lauber A: Oral communication, Adrenal Meeting, San Diego 2010].

Human POR Mutations: From Phenotype to Genotype

Peterson et al. [17] reported in 1985 a 46,XY patient with genital ambiguity and an abnormal urinary steroid profile suggesting combined partial deficiencies of the two steroidogenic enzymes CYP17A1 (17α-hydroxylase/17,20-lyase) and CYP21A2 (21-hydroxylase). Subsequently, several similar patients were described with this apparent combination of defects in enzymatic activities of CYP17A1 and CYP21A2 [18–21]. However, in these patients genetic analysis of the CYP17A1 and CYP21A2 genes encoding these enzymes revealed no mutations [21, 22].

Until recently, most disorders of steroidogenesis were caused by mutations in genes encoding steroidogenic enzymes, resulting in diminished or absent enzymatic activities. One exception is congenital lipoid adrenal hyperplasia, caused by mutations in the steroidogenic acute regulatory protein (StAR) which transports cholesterol into mitochondria to initiate steroidogenesis [23]. Only recently, steroid disorders due to genetic cofactor abnormalities such as hexose-6-phosphate dehydrogenase [24] or PAPS synthase [25] have been described. In fact, to explain the complex pattern of partial combined CYP17A1 and CYP21A2 deficiencies, mutations in P450 oxidoreductase (POR), the flavoprotein that supplies electrons to all microsomal P450 enzymes, have been suggested earlier [26]. However, as the POR knockout model revealed embryonic lethality in mice [27, 28], POR was initially excluded as a candidate gene for a human disease.

Nevertheless, in 2004 we reported the first POR mutations in 4 patients who seemed to have steroid abnormalities suggesting combined defects in CYP17A1 and CYP21A2 [29]. Three of these patients had ambiguous genitalia and the Antley-Bixler skeletal malformation syndrome (ABS); 1 patient was an adult with steroid abnormalities, primary amenorrhea and polycystic ovaries [29]. The functional impact of the POR mutations identified in those patients was demonstrated using cytochrome c assays as well as CYP17A1 activity studies. Subsequent studies confirmed the presence of POR mutations in patients with similar patterns of steroid abnormalities with and without ABS [3, 14, 29–36]. To date, approximately 65 patients harboring more than 40 recessive POR mutations have been described in the literature including missense (fig. 4) and frameshift mutations, indels or splice variations (table 1 and www.cypalleles.ki.se). Missense mutations predominate. Among patients of European descent the mutation A287P is the most common mutation; among patients of Japanese heritage the R457H is most common [32]. Interestingly, 12% of the patients reported initially had mutations identified on only one allele, but these patients are phenotypically indistinguishable from those with mutations on both alleles [3, 29, 32, 33]. Meanwhile, genetic studies in a larger cohort of 35 Japanese patients revealed POR mutations on all alleles when extensively searched for [35], making the possibility of manifesting heterozygosity for POR [18] highly unlikely. In this cohort, genotype-phenotype correlation showed that skeletal anomalies, 46,XY disorders of sex development, adrenal dysfunction and pubertal failure were more severe in patients harboring compound heterozygote R457H mutations in combination with an apparent null mutation when compared to homozygote R457H individuals [35].

A study of 842 healthy, unrelated individuals in four ethnic groups (218 African Americans, 260 Caucasian Americans, 178 Chinese Americans, and 185 Mexican Americans) identified a large number of POR sequence variations [37]. These included 140 single nucleotide polymorphisms (SNPs), 13 novel missense variations, and 8 indels. In vitro analysis of the 13 novel missense mutations that were expressed in *Escherichia coli* and assayed for their abilities to support cytochrome c reduction, NADPH oxidation, and the 17α-hydroxylase/17,20-lyase activities of CYP17A1 showed variable impairment of catalytic activities for several mutants, depending on the interacting partner. One missense mutant, A503V, was found on ~28% of all alleles [37].

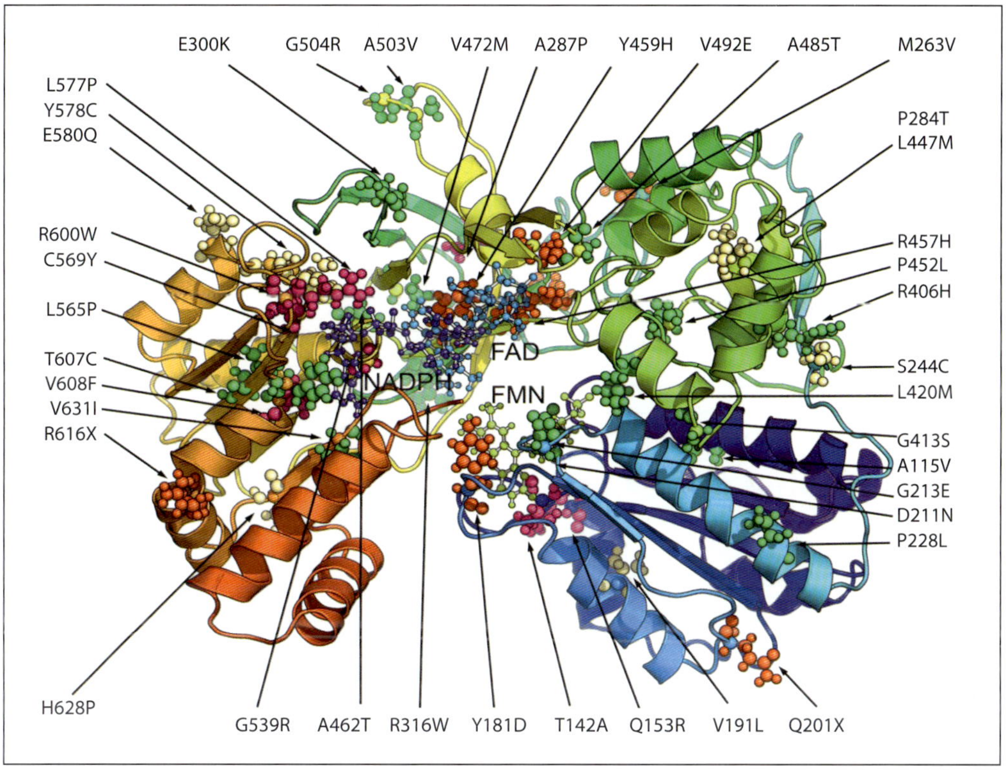

Fig. 4. Model of human POR protein, showing all identified missense mutations. The model has been described [32] and is based on the crystal structure of rat POR lacking 65 amino-terminal residues [4]. The α-carbon backbone is depicted as a narrow ribbon. Ball-and-stick models are used to represent the FAD and FMN moieties (yellow), and the NADPH (cyan). The missense residues identified by sequencing are depicted by charged packed sphere images of different colors, corresponding to their V_{max}/K_m for CYP17A1 activities: red <25%, magenta 25–50%, green >50%.

Structure and Function Testing of POR Variants in vitro and in silico

The activities of most POR mutations have been tested by assays based on CYP17A1 and cytochrome c [29, 32, 36, 37], and for some with all three microsomal P450s involved in steroidogenesis, using recombinant proteins produced in yeast or bacteria [29, 30, 32, 37–39]. In our first report, we tested the impact of POR variants A287P, R457H, V492E, C569Y and V608F on both 17α-hydroxylase and 17,20-lyase activities of CYP17A1 and compared the results with the cytochrome c reduction/NADPH oxidation assays [29]. We found a good correlation between the patients' clinical features and the functional assays based on CYP17A1, but a poor correlation with the assays based on cytochrome c. We made three important observations: First, mutations in the FAD region that destroy the binding of FAD to POR resulted in almost total loss of activity in all assays. Second, the A287P mutation, which is

Table 1. Patients reported with POR deficiency

Patients n	Chromosomal sex 46XX/46XY	OR mutations	Phenotype			Reference (first author)
			ABS features	abnormal genitalia	abnormal steroids	
4	2/2	7/8[1]	3	3	4	Flück, 2004 [29]
3	2/1	6/6	1	2	3	Arlt, 2004 [30]
2	1/1	4/4	2	1	2	Adachi, 2004 [31]
19 (32[2])	6[3]/10[3]	34/38[1]	19 (32)	12[3]	10[3]	Huang, 2005 [32]
10	6/4	19/20[1]	9	9	10	Fukami, 2006 [34]
3	2/1	6/6	0	2	3	Fukami, 2006 [34]
7	2/5	14/14	5	2	7	Homma, 2006 [14]
1	1/0	1/2[1]	1	1	1	Scott, 2007 [3]
12 (35)	19/16	70/70	28	26		Fukami, 2009 [35]
4	3/1	8/8	3	3	4	Sahakitrungruang, 2009 [36]

[1] POR mutations not identified on all alleles.
[2] 19 out of 32 patients with the ABS phenotype had POR mutations.
[3] Karyotype, description of genitalia or steroid profile not known for all patients.

not directly involved in the electron transport but located close to the hinge region (residues 234–283), had reduced CYP17A1 activities but nearly wild-type activity in the cytochrome c assays. Third, mutations C569Y and V608F, which are located in the NADPH-binding domain, had higher Michaelis constants for NADPH in the cytochrome c assay suggesting that these mutants might affect binding of NADPH to POR. Subsequent studies tested the impact of many more missense mutations on CYP17A1 activities (fig. 4) [32, 36, 37].

We have also compared the effects of POR variants on CYP19A1 activity with the effects on CYP17A1 activities (fig. 5) [38]. Mutations R457H and V492E caused a complete loss of CYP19A1 activity (fig. 5a, b), confirming that POR mutations disrupting the electron transfer will severely affect any interacting P450. By contrast, POR mutants A287P, C569Y and V608F, had variable effects on CYP19A1 and CYP17A1 activities (fig. 5a). POR mutations C569Y and V608F, which were identified in a compound heterozygote state in a mildly affected female patient with a polycystic ovary syndrome-like phenotype, had more than 50% of wild-type CYP17A1 activities but less than 50% of wild-type CYP19A1 activity [38]. In addition, those mutants seemed

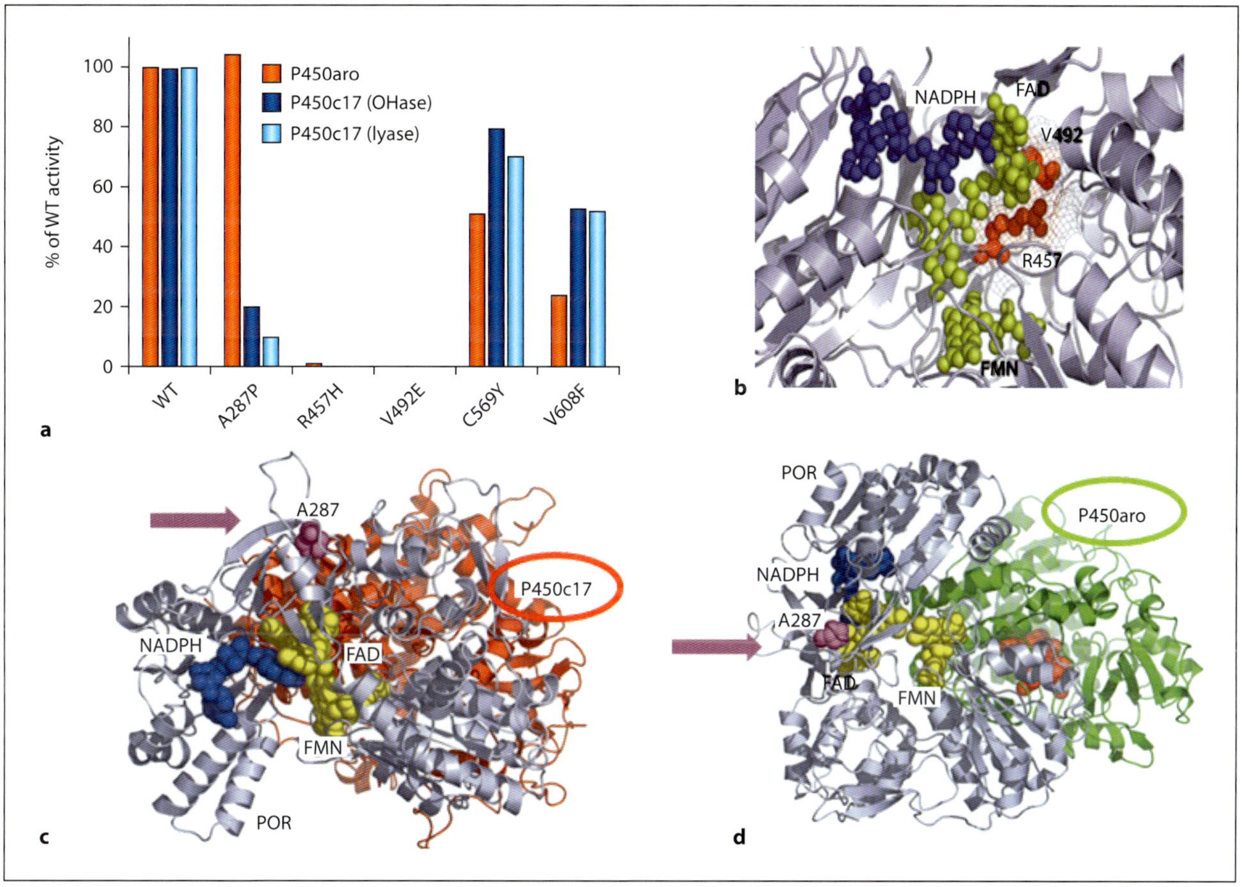

Fig. 5. Differential impact of POR missense mutations on CYP17A1 versus CYP19A1 activities. **a** CYP17A1 (17α-hydroxylase (OHase) and 17,20-lyase (lyase)) and CYP19A1 (P450aro) activities were assessed in vitro when supported by either wild-type (WT) or mutant POR proteins. Proteins for the assays were produced in a 'humanized yeast system' [29, 38]. Catalytic efficiencies were calculated as V_{max}/K_m and are given as percentage of WT. **b** Structure-function analysis of mutant POR proteins R457H and V492E (depicted in red) [38]. Residues R457 and V492 are essential for electron transfer from FAD to FMN (yellow balls). **c**, **d** Docking studies of POR (blue) with partners P450c17 (CYP17A1, red) and P450aro (CYP19A1; green): (**c**) A287 (purple balls) localizes to the surface of POR and lies within the interface of POR with P450c17. (**d**) Docking of POR with P450aro reveals no involvement of A287 residue in their interaction.

to be more sensitive to the amount of NADPH available [38]. Conversely, the A287P mutant did not affect CYP19A1 activity but reduced CYP17A1 activities remarkably.

Molecular modeling and protein docking studies suggested that A287P disrupts the interaction of POR with CYP17A1 (fig. 5c) but will not affect the interaction of POR with CYP19A1 (fig. 5d) [38]. The differential inhibition of wild-type POR and A278P, S244C, H628P and Y181D mutants on enzymatic activity of CYP21A2 was studied by others [39]. Mutation Y181D, which is involved in the FMN binding, impaired both CYP21A2 and CYP17A1 activities similarly, but the A287P mutant while impairing CYP17A1 activities, did not affect CYP21A2 activity. The basis of the selectivity of some POR variants towards certain P450s or other electron acceptors will become clearer as studies with other target proteins are carried out. Thus far, we believe that the impact of specific POR variants that do not lead to obvious changes disrupting the structure (such as loss of FAD or FMN or a truncated protein) cannot be predicted from a single cytochrome c or P450 assay but need to be tested with each P450 and interaction partner specifically.

POR and Skeletal Malformations

Patients with POR deficiency were earlier diagnosed as having ABS with genital anomalies by geneticists. ABS is a skeletal malformation syndrome that was first described in 1975 [40]. ABS is characterized by craniosynostosis, midface hypoplasia, choanal atresia, radiohumeral or radioulnar synostosis, joint contractures, arachnodactyly, and bowing of the femora [41–43]. Many patients with ABS have activating mutations in the gene for fibroblast growth factor receptor 2 (FGFR2) [22, 44–47]. Fibroblast growth factors (FGFs) are mitogens involved in bone growth and development [48]. They bind to four different tyrosine kinase receptors on the cell surface (FGFRs). Dominant gain-of-function mutations in FGFRs (predominantly FGFR2) cause a variety of craniosynostosis syndromes including Pfeiffer, Apert, Jackson-Weiss, and Crouzon syndromes [49–52]. The same FGFR2 mutations can cause different clinical syndromes, suggesting that these syndromes are phenotypic variants of a single genetic disorder.

In contrast to all other craniosynostosis syndromes, about 50% of patients with ABS have genital anomalies [22, 43]; furthermore, ABS patients with genital anomalies do not have FGFR2 mutations [22]. When we studied 32 ABS patients with and without hormonal abnormalities in steroidogenesis, 18 patients had recessive POR mutations, 10 had dominant FGFR2 or FGFR3 mutations but no POR mutations, and only 1 had compound heterozygote POR mutations and a FGFR1 mutation; no mutations in either POR or FGFR were found in 3 [32]. This study indicates that POR and FGFR2 mutations segregate, and that patients with an ABS-like phenotype and anomalies in genital development and steroidogenesis have a distinct disease, namely POR deficiency. However, meanwhile we identified 1 patient with ABS and genital anomalies who has compound heterozygote POR mutations and a mutation in FGFR-like 1, a protein related to FGFR1-3 [53]. Mutant FGFRL1 was found to affect FGF signaling and receptor localization. Thus, digenic disorder in ABS with genital anomalies seems to be present in very rare cases.

The pathogenesis underlying the skeletal malformations in POR deficiency is not known in great detail. Several lines of evidence suggest a possible role for cholesterol biosynthesis and POR in bone development. First, cholesterol biosynthesis requires squalene epoxidase, a non-P450 enzyme, and 14α-demethylase (CYP51), both of which require POR for electron transfer [54, 55]. Second, human disorders of cholesterol biosynthesis such as the Smith-Lemli-Opitz syndrome, which is caused by mutations in the gene for 7-dehydrocholesterol reductase, can cause skeletal anomalies. Third, cholesterol is required for normal activity and signal transduction by hedgehog proteins, which are crucial for the regulation of growth, morphogenesis and bone formation in embryogenesis [56]. In addition, retinoic acid seems involved in the development of skeletal malformations in POR deficiency. Inappropriate levels (excess or deficiency) of retinoic acid can cause a wide range of limb malformations [57]. Retinoic acid is metabolized by microsomal CYP26A1, which depends on POR for electron transfer. Reducing retinoic acid in the diet of pregnant mice carrying POR knockout pups partially ameliorated their phenotype [28] suggesting a role for retinoic acid toxicity in bone malformations of POR deficiency. In line with these findings, gene expression analysis of conditional limb POR knockout mice recently confirmed that POR deletion leads to excess retinoic acid and impaired cholesterol metabolism [58].

Impact of POR on Drug-Metabolizing P450s

The majority of drugs in clinical use are metabolized by hepatic, POR-dependent P450 enzymes [59]. Sequence variations in these P450 enzymes cause variations in drug responses among individuals and among ethnic groups. Because all type II P450s depend on POR for electron supply, POR variants may also contribute to variations in drug metabolism. Before POR mutations were identified in patients with ABS and disordered steroidogenesis, patients with a similar phenotype were reported after in utero exposure to fluconazole, suggesting that this drug might be teratogenic [60–62]. Fluconazole is an inhibitor of lanosterol 14α-demethylase (CYP51) which converts lanosterol to ergosterol in the cholesterol biosynthetic pathway; a genetic defect in CYP51 was postulated but not found [63]. Nevertheless, lymphoblast cells isolated from POR patients have higher lanosterol levels than cells from controls or from patients with ABS due to FGFR2 mutations [63], probably because POR mutations affect CYP51 enzyme activity. In fact, one of these patients exposed to fluconazole in utero [63], was later found to harbor homozygote POR mutations A287P [29]. Fluconazole also inhibits the hepatic drug-metabolizing type II P450s CYP2C9, CYP2C19 and CYP3A4 [64], which may be affected by POR variants.

Further evidence for the importance of POR in hepatic P450 activities comes from a liver-specific knockout of POR in mice [65]. Homozygous deletion of the POR gene (called Cpr in mice) in the liver decreased microsomal cytochrome P450 and heme oxygenase activities, and decreased pentobarbital clearance and total plasma cholesterol. However, unlike global POR knockout mice, which are embryonic lethal [27], liver-specific knockouts are viable, and are normal in gross appearance, growth, and fertility [65].

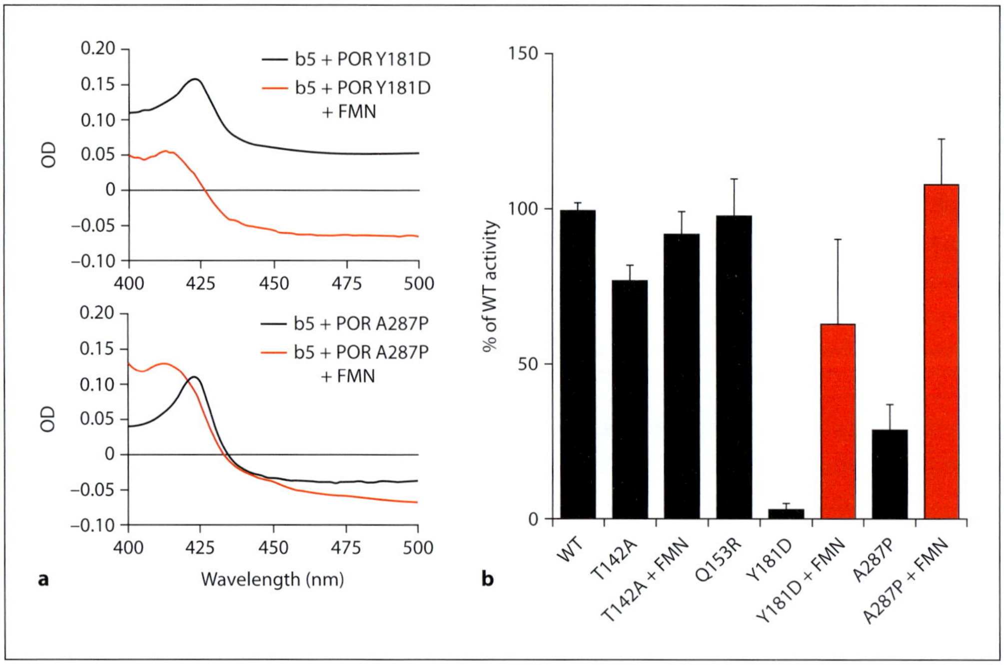

Fig. 6. Restoration of POR-dependent cytochrome b_5 and CYP3A4 activities by external flavin in vitro. **a** Cyt b_5 reduction was tested with mutant POR proteins with and without adding external FMN to the reaction mixture monitoring absorption maxima at 423 nm [68]. **b** CYP3A4 activity was tested with the fluorogenic substrate BOMCC [68]. External FMN rescues the activity of 3A4 supported by POR mutants Y181D and A287P.

Some functional data concerning the impact of POR variants on hepatic drug-metabolizing P450s is now emerging. The activities of 10 P450 enzymes (CYP1A2, 2A6, 2B6, 2C8, 2C9, 2C19, 2D6, 2E1, 3A4/5 and 4A9/11) were examined in pooled liver microsome samples using test drugs that are specific for each enzyme [66]. However, since P450 genetic polymorphisms are quite common, some of the variations reported may be due to variations in the P450 rather than in POR, and some loss of POR and P450 activities might reflect variations in sample handling. The 35 POR mutants identified from patients and from the sequencing of 842 POR genes from normal individuals [37] have been expressed in bacteria and their activities measured in vitro with bacterially-expressed CYP1A2 and CYP2C19, again showing that some POR variants will affect the activity of some P450 enzymes, but not others [67]. We found that POR mutations in the FMN-binding domain (e.g. Y181D) inhibit CYP3A4 activity and cytochrome b_5 reduction (fig. 6) [68]. Computational docking experiments suggested that external FMN could dock in close proximity to the FAD moiety of POR and receive electrons donated by NADPH, forming a 'rescue electron transfer ladder'. Supplementing flavin in our in vitro POR assays, we were able to restore activities of both CYP3A4 and Cyt b_5 enzymes not only for POR Y181D but also for A287P which is not part of the FMN moiety (fig. 6) [68]. However, which specific POR variant with what partner P450 may respond to external FMN, and whether (ribo)flavin may also work as a treatment for affected patients, remains to be seen.

Conclusions and Perspectives

POR deficiency is a novel form of congenital adrenal hyperplasia. However, as POR is the obligate electron donor for all microsomal P450s, POR sequence variations have broader implications and may not only affect steroidogenesis and sexual development, but also skeletal development, drug metabolism and other functions related to type II P450s (fig. 2). Therefore, the clinical spectrum of POR deficiency is quite broad, and ranges from severely handicapped ABS individuals to mildly affected normal-looking adults with compromised fertility. The prevalence of POR deficiency remains to be determined. Several patients previously misdiagnosed for having 21-hydroxylase (CYP21A2) or 17α-hydroxylase (CYP17A1) deficiency have been identified. The diagnosis of POR deficiency may be considered from clinical and hormonal findings, but requires confirmation by genetic analysis. Treatment may include replacement of glucocorticoids and sex steroids (rarely mineralocorticoids) as assessed by low basal or stimulated serum hormone levels for each individual patient. The skeletal malformations of POR deficiency require orthopedic management, and most reported mortality is related to skeletal abnormalities causing respiratory problems (e.g. choanal obstruction). Another important clinical issue remains the effect of POR variants on drug-metabolizing P450 enzymes. Pharmacogenetic testing for relevant POR variants in 842 normal subjects revealed more than 100 SNPs and a common POR variant, A503V, in about 28% of alleles [37]. In vitro testing of several of these mutations found variations in activities depending on substrates and partner P450s, indicating that each POR mutant must be assayed separately with potential target P450s [69, 70]. Computational docking studies and functional assays suggest that the activity loss of some POR variants may be rescued by external flavin. Further evaluation of flavin supplementation in patients with POR deficiency is needed to find out whether metabolic effects of some POR mutations could be restored by excess flavin.

Acknowledgement

This work was support by grants from the Swiss National Science Foundation (320000-116299 to C.E.F. and 3100A0-113719 to A.V.P.).

References

1 Lu AY, Junk KW, Coon MJ: Resolution of the cytochrome P-450-containing ω-hydroxylation system of liver microsomes into three components. J Biol Chem 1969;244:3714–3721.

2 Yamano S, Aoyama T, McBride OW, Hardwick JP, Gelboin HV, Gonzalez FJ: Human NADPH-P450 oxidoreductase: complementary DNA cloning, sequence and vaccinia virus-mediated expression and localization of the CYPOR gene to chromosome 7. Mol Pharmacol 1989;36:83–88.

3 Scott RR, Gomes LG, Huang N, Van Vliet G, Miller WL: Apparent manifesting heterozygosity in P450 oxidoreductase deficiency and its effect on coexisting 21-hydroxylase deficiency. J Clin Endocrinol Metab 2007;92:2318–2322.

4 Wang M, Roberts DL, Paschke R, Shea TM, Masters BSS, Kim JJ: Three-dimensional structure of NADPH-cytochrome P450 reductase: prototype for FMN- and FAD-containing enzymes. Proc Natl Acad Sci USA 1997;94:8411–8416.

5 Miller WL: Regulation of steroidogenesis by electron transfer. Endocrinology 2005;146:2544–2550.

6 Guengerich FP, Johnson WW: Kinetics of ferric cytochrome P450 reduction by NADPH-cytochrome P450 reductase: rapid reduction in the absence of substrate and variations among cytochrome P450 systems. Biochemistry 1997;36:14741–14750.

7 Auchus RJ, Lee TC, Miller WL: Cytochrome b_5 augments the 17,20-lyase activity of human P450c17 without direct electron transfer. J Biol Chem 1998; 273:3158–3165.

8 Yamazaki H, Nakano M, Imai Y, Ueng YF, Guengerich FP, Shimada T: Roles of cytochrome b_5 in the oxidation of testosterone and nifedipine by recombinant cytochrome P450 3A4 and by human liver microsomes. Arch Biochem Biophys 1996;325: 174–182.

9 Loughran PA, Roman LJ, Miller RT, Masters BS: The kinetic and spectral characterization of the *E. coli*-expressed mammalian CYP4A7: cytochrome b_5 effects vary with substrate. Arch Biochem Biophys 2001;385:311–321.

10 Yanagibashi K, Hall PF: Role of electron transport in the regulation of the lyase activity of C21 side-chain cleavage P-450 from porcine adrenal and testicular microsomes. J Biol Chem 1986;261: 8429–8433.

11 Lin D, Black SM, Nagahama Y, Miller WL: Steroid 17α-hydroxylase and 17,20-lyase activities of P450c17: contributions of serine-106 and P450 reductase. Endocrinology 1993;132:2498–2506.

12 Grumbach MM, Auchus RJ: Estrogen: consequences and implications of human mutations in synthesis and action. J Clin Endocrinol Metab 1999;84:4677–4694.

13 Auchus RJ: The backdoor pathway to dihydrotestosterone. Trends Endocrinol Metab 2004;15:432–438.

14 Homma K, Hasegawa T, Nagai T, Adachi M, Horikawa R, Fujiwara I, Tajima T, Takeda R, Fukami M, Ogata T: Urine steroid hormone profile analysis in cytochrome P450 oxidoreductase deficiency: implication for the backdoor pathway to dihydrotestosterone. J Clin Endocrinol Metab 2006;91:2643–2649.

15 Shackleton C, Marcos J, Arlt W, Hauffa BP: Prenatal diagnosis of P450 oxidoreductase deficiency (ORD): a disorder causing low pregnancy estriol, maternal and fetal virilization, and the Antley-Bixler syndrome phenotype. Am J Med Genet 2004;129A:105–112.

16 Hershkovitz E, Parvari R, Wudy SA, Hartmann MF, Gomes LG, Loewental N, Miller WL: Homozygous mutation G539R in the gene for P450 oxidoreductase in a family previously diagnosed as having 17,20-lyase deficiency. J Clin Endocrinol Metab 2008;93: 3584–3588.

17 Peterson RE, Imperato-McGinley J, Gautier T, Shackleton CH: Male pseudohermaphroditism due to multiple defects in steroid-biosynthetic microsomal mixed-function oxidases. A new variant of congenital adrenal hyperplasia. N Engl J Med 1985; 313:1182–1191.

18 Malunowicz E, Romer TE, Szarras-Czapnik M, Mielniczuk Z, Gajewska D: Combined deficiency of 17α-hydroxylase and 21-hydroxylase in an 8-year-old girl. Endokrynol Pol 1987;38:117–124.

19 Augarten A, Pariente C, Gazit E, Chayen R, Goldfarb H, Sack J: Ambiguous genitalia due to partial activity of cytochromes P450c17 and P450c21. J Steroid Biochem Mol Biol 1992;41:37–41.

20 Leiberman E, Hershkovitz E, Lauber-Biason A, Phillip M, Zachmann M: Subnormal cortisol response to adrenocorticotropin in isolated partial 17,20-lyase deficiency. J Pediatr Endocrinol Metab 1997;10:387–390.

21 Adachi M, Tachibana K, Asakura Y, Suwa S, Nishimura G: A male patient presenting with major clinical symptoms of glucocorticoid deficiency and skeletal dysplasia, showing a steroid pattern compatible with 17α-hydroxylase/17,20-lyase deficiency, but without obvious CYP17 gene mutations. Endocr J 1999;46:285–292.

22 Reardon W, Smith A, Honour JW, Hindmarsh P, Das D, Rumsby G, Nelson I, Malcolm S, Ades L, Sillence D, Kumar D, DeLozier-Blanchet C, McKee S, Kelly T, McKeehan WL, Baraitser M, Winter RM: Evidence for digenic inheritance in some cases of Antley-Bixler syndrome? J Med Genet 2000;37:26–32.

23 Miller WL: Congenital lipoid adrenal hyperplasia: the human gene knockout for the steroidogenic acute regulatory protein. J Mol Endocrinol 1997;19: 227–240.

24 Draper N, Walker EA, Bujalska IJ, Tomlinson JW, Chalder SM, Arlt W, Lavery GG, Bedendo O, Ray DW, Laing I, Malunowicz E, White PC, Hewison M, Mason PJ, Connell JM, Shackleton CH, Stewart PM: Mutations in the genes encoding 11β-hydroxysteroid dehydrogenase type 1 and hexose-6-phosphate dehydrogenase interact to cause cortisone reductase deficiency. Nat Genet 2003;34:434–439.

25 Noordam C, Dhir V, McNelis JC, Schlereth F, Hanley NA, Krone N, Smeitink JA, Smeets R, Sweep FC, Claahsen-van der Grinten HL, Arlt W: Inactivating PAPSS2 mutations in a patient with premature pubarche. N Engl J Med 2009;360:2310–2318.

26 Miller WL: Congenital adrenal hyperplasia. N Engl J Med 1986;314:1321–1322.

27 Shen AL, O'Leary KA, Kasper CB: Association of multiple developmental defects and embryonic lethality with loss of microsomal NADPH-cytochrome P450 oxidoreductase. J Biol Chem 2002;277:6536–6541.

28 Otto DM, Henderson CJ, Carrie D, Davey M, Gundersen TE, Blomhoff R, Adams RH, Tickle C, Wolf CR: Identification of novel roles of the cytochrome P450 system in early embryogenesis: effects on vasculogenesis and retinoic acid homeostasis. Mol Cell Biol 2003;23:6103–6116.

29 Flück CE, Tajima T, Pandey AV, Arlt W, Okuhara K, Verge CF, Jabs EW, Mendonca BB, Fujieda K, Miller WL: Mutant P450 oxidoreductase causes disordered steroidogenesis with and without Antley-Bixler syndrome. Nat Genet 2004;36:228–230.

30 Arlt W, Walker EA, Draper N, Ivison HE, Ride JP, Hammer F, Chalder SM, Borucka-Mankiewicz M, Hauffa BP, Malunowicz EM, Stewart PM, Shackleton CH: Congenital adrenal hyperplasia caused by mutant P450 oxidoreductase and human androgen synthesis: analytical study. Lancet 2004;363:2128–2135.

31 Adachi M, Tachibana K, Asakura Y, Yamamoto T, Hanaki K, Oka A: Compound heterozygous mutations of cytochrome P450 oxidoreductase gene (POR) in two patients with Antley-Bixler syndrome. Am J Med Genet 2004;128A:333–339.

32 Huang N, Pandey AV, Agrawal V, Reardon W, Lapunzina PD, Mowat D, Jabs EW, Van Vliet G, Sack J, Flück CE, Miller WL: Diversity and function of mutations in P450 oxidoreductase in patients with Antley-Bixler syndrome and disordered steroidogenesis. Am J Hum Genet 2005:76:729–749.

33 Fukami M, Horikawa R, Nagai T, Tanaka T, Naiki Y, Sato N, Okuyama T, Nakai H, Soneda S, Tachibana K, Matsuo N, Sato S, Homma K, Nishimura G, Hasegawa T, Ogata T: Cytochrome P450 oxidoreductase gene mutations and Antley-Bixler syndrome with abnormal genitalia and/or impaired steroidogenesis: molecular and clinical studies in 10 patients. J Clin Endocrinol Metab 2005;90:414–426.

34 Fukami M, Hasegawa T, Horikawa R, Ohashi T, Nishimura G, Homma K, Ogata T: Cytochrome P450 oxidoreductase deficiency in three patients initially regarded as having 21-hydroxylase deficiency and/or aromatase deficiency: diagnostic value of urine steroid hormone analysis. Pediatr Res 2006;59:276–280.

35 Fukami M, Nishimura G, Homma K, Nagai T, Hanaki K, Uematsu A, Ishii T, Numakura C, Sawada H, Nakacho M, Kowase T, Motomura K, Haruna H, Nakamura M, Ohishi A, Adachi M, Tajima T, Hasegawa Y, Hasegawa T, Horikawa R, Fujieda K, Ogata T: Cytochrome P450 oxidoreductase deficiency: identification and characterization of biallelic mutations and genotype-phenotype correlations in 35 Japanese patients. J Clin Endocrinol Metab 2009;94:1723–1731.

36 Sahakitrungruang T, Huang N, Tee MK, Agrawal V, Russell WE, Crock P, Murphy N, Migeon CJ, Miller WL: Clinical, genetic, and enzymatic characterization of P450 oxidoreductase deficiency in four patients. J Clin Endocrinol Metab 2009;94:4992–5000.

37 Huang N, Agrawal V, Giacomini KM, Miller WL: Genetics of P450 oxidoreductase: sequence variation in 842 individuals of four ethnicities and activities of 15 missense mutations. Proc Natl Acad Sci USA 2008;105:1733–1738.

38 Pandey AV, Kempna P, Hofer G, Mullis PE, Fluck CE: Modulation of human CYP19A1 activity by mutant NADPH P450 oxidoreductase. Mol Endocrinol 2007;21:2579–2595.

39 Dhir V, Ivison HE, Krone N, Shackleton CHL, Doherty AJ, Stewart PM, Arlt W: Differential inhibition of CYP17A1 and CYP21A2 activities by the P450 oxidoreductase mutant A287P. Mol Endocrinol 2007;21:1958–1968.

40 Antley R, Bixler D: Trapezoidocephaly, midfacial hypoplasia and cartilage abnormalities with multiple synostoses and skeletal fractures. Birth Defects Orig Artic Ser 1975;11:397–401.

41 DeLozier CD, Antley RM, Williams R, Green N, Heller RM, Bixler D, Engel E: The syndrome of multisynostotic osteodysgenesis with long-bone fractures. Am J Med Genet 1980;7:391–403.
42 Hassell S, Butler MG: Antley-Bixler syndrome: report of a patient and review of literature. Clin Genet 1994;46:372–376.
43 Crisponi G, Porcu C, Piu ME: Antley-Bixler syndrome: case report and review of the literature. Clin Dysmorphol 1997;6:61–68.
44 Chun K, Siegel-Bartelt J, Chitayat D, Phillips J, Ray PN: FGFR2 mutation associated with clinical manifestations consistent with Antley-Bixler syndrome. Am J Med Genet 1998;77:219–224.
45 Gripp KW, Stolle CA, McDonald-McGinn DM, Markowitz RI, Bartlett SP, Katowitz JA, Muenke M, Zackai EH: Phenotype of the fibroblast growth factor receptor 2 Ser351Cys mutation: Pfeiffer syndrome type III. Am J Med Genet 1998;78:356–360.
46 Pulleyn LJ, Reardon W, Wilkes D, Rutland P, Jones BM, Hayward R, Hall CM, Brueton L, Chun N, Lammer E, Malcolm S, Winter RM: Spectrum of craniosynostosis phenotypes associated with novel mutations at the fibroblast growth factor receptor 2 locus. Eur J Hum Genet 1996;4:283–291.
47 Schaefer F, Anderson C, Can B, Say B: Novel mutation in the FGFR2 gene at the same codon as the Crouzon syndrome mutations in a severe Pfeiffer syndrome type 2 case. Am J Med Genet 1998;75:252–255.
48 Cohen MM Jr: The new bone biology: pathologic, molecular, and clinical correlates. Am J Med Genet A 2006;140:2646–2706.
49 Muenke M, Schell U, Hehr A, Robin NH, Losken HW, Schinzel A, Pulleyn LJ, Rutland P, Reardon W, Malcolm S, Winter RM: A common mutation in the fibroblast growth factor receptor 1 gene in Pfeiffer syndrome. Nat Genet 1994;8:269–274.
50 Reardon W, Winter RM, Rutland P, Pulleyn LJ, Jones BM, Malcolm S: Mutations in the fibroblast growth factor receptor 2 gene cause Crouzon syndrome. Nat Genet 1994;8:98–103.
51 Rutland P, Pulleyn LJ, Reardon W, Baraitser M, Hayward R, Jones B, Malcolm S, Winter RM, Oldridge M, Slaney SF, et al: Identical mutations in the FGFR2 gene cause both Pfeiffer and Crouzon syndrome phenotypes. Nat Genet 1995;9:173–176.
52 Park WJ, Meyers GA, Li X, Theda C, Day D, Orlow SJ, Jones MC, Jabs EW: Novel FGFR2 mutations in Crouzon and Jackson-Weiss syndromes show allelic heterogeneity and phenotypic variability. Hum Mol Genet 1995;4:1229–1233.
53 Rieckmann T, Zhuang L, Fluck CE, Trueb B: Characterization of the first FGFRL1 mutation identified in a craniosynostosis patient. Biochim Biophys Acta 2009;1792:112–121.
54 Ono T, Bloch K: Solubilization and partial characterization of rat liver squalene epoxidase. J Biol Chem 1975;250:1571–1579.
55 Debeljak N, Fink M, Rozman D: Many facets of mammalian lanosterol 14α-demethylase from the evolutionarily conserved cytochrome P450 family CYP51. Arch Biochem Biophys 2003;409:159–171.
56 Gofflot F, Hars C, Illien F, Chevy F, Wolf C, Picard JJ, Roux C: Molecular mechanisms underlying limb anomalies associated with cholesterol deficiency during gestation: implications of hedgehog signaling. Hum Mol Genet 2003;12:1187–1198.
57 Lee GS, Kochhar DM, Collins MD: Retinoid-induced limb malformations. Curr Pharm Des 2004; 10:2657–2699.
58 Schmidt K, Hughes C, Chudek JA, Goodyear SR, Aspden RM, Talbot R, Gundersen TE, Blomhoff R, Henderson C, Wolf CR, Tickle C: Cholesterol metabolism: the main pathway acting downstream of cytochrome P450 oxidoreductase in skeletal development of the limb. Mol Cell Biol 2009;29: 2716–2729.
59 Ingelman-Sundberg M: Pharmacogenetics of cytochrome P450 and its applications in drug therapy: the past, present and future. Trends Pharmacol Sci 2004;25:193–200.
60 Lee BE, Feinberg M, Abraham JJ, Murthy AR: Congenital malformations in an infant born to a woman treated with fluconazole. Pediatr Infect Dis J 1992;11:1062–1064.
61 Pursley TJ, Blomquist IK, Abraham J, Andersen HF, Bartley JA: Fluconazole-induced congenital anomalies in three infants. Clin Infect Dis 1996;22:336–340.
62 Aleck KA, Bartley DL: Multiple malformation syndrome following fluconazole use in pregnancy: report of an additional patient. Am J Med Genet 1997;72:253–256.
63 Kelley RI, Kratz LE, Glaser RL, Netzloff ML, Wolf LM, Jabs EW: Abnormal sterol metabolism in a patient with Antley-Bixler syndrome and ambiguous genitalia. Am J Med Genet 2002;110:95–102.
64 Niwa T, Shiraga T, Takagi A: Effect of antifungal drugs on cytochrome P450 (CYP) 2C9, CYP2C19, and CYP3A4 activities in human liver microsomes. Biol Pharm Bull 2005;28:1805–1808.
65 Gu J, Weng Y, Zhang QY, Cui H, Behr M, Wu L, Yang W, Zhang L, Ding X: Liver-specific deletion of the NADPH-cytochrome P450 reductase gene: impact on plasma cholesterol homeostasis and the function and regulation of microsomal cytochrome P450 and heme oxygenase. J Biol Chem 2003;278: 25895–25901.

66 Hart SN, Wang S, Nakamoto K, Wesselman C, Li Y, Zhong XB: Genetic polymorphisms in cytochrome P450 oxidoreductase influence microsomal P450-catalyzed drug metabolism. Pharmacogenet Genomics 2008;18:11–24.
67 Agrawal V, Huang N, Miller WL: Pharmacogenetics of P450 oxidoreductase. Effect of sequence variants on activities of CYP1A2 and CYP2C19. Pharmacogenet Genomics 2008;18:569–576.
68 Nicolo C, Fluck CE, Mullis PE, Pandey AV: Restoration of mutant cytochrome P450 reductase activity by external flavin. Mol Cell Endocrinol 2010;321:245–252.
69 Pandey AV, Flück CE, Mullis PE: Altered heme catabolism by heme oxygenase-1 caused by mutations in human NADPH cytochrome P450 reductase. Biochem Biophys Res Commun 2010;400:374–378.
70 Flück CE, Mullis PE, Pandey AV: Reduction in hepatic drug metabolising CYP3A4 activities caused by P450 oxidoreductase mutations identified in patients with disordered steroid metabolism. Biochem Biophys Res Commun 2010;401:149–153.

Christa E. Flück, MD
Pediatric Endocrinology and Diabetology, University Children's Hospital Bern
Freiburgstrasse 15, G3 812, CH–3010 Bern (Switzerland)
Tel. +41 31 632 0499, Fax +41 31 632 8424
E-Mail christa.flueck@dkf.unibe.ch

Ghizzoni L, Cappa M, Chrousos G, Loche S, Maghnie M (eds): Pediatric Adrenal Diseases.
Endocr Dev. Basel, Karger, 2011, vol 20, pp 80–87

Molecular Genetics of 21-Hydroxylase Deficiency

Anna Wedell

Department of Molecular Medicine and Surgery, Karolinska Institutet, Centre for Inherited Metabolic Diseases, Karolinska University Hospital, Stockholm, Sweden

Abstract

More than 95% of all cases of congenital adrenal hyperplasia are caused by deficiency of steroid 21-hydroxylase, an enzyme encoded by the *CYP21A2* gene. The severity of the clinical symptoms varies according to the level of residual 21-hydroxylase activity. The *CYP21A2* gene is located in the HLA class III region, as a component of so called RCCX modules containing homologous genes repeated in tandem. Misalignment followed by unequal crossing over as well as gene conversion events result in a high degree of variation in gene copy number as well as gene sequence in this genomic region. The presence of a highly homologous pseudogene, *CYP21A1P*, forms the basis for the relatively high incidence of 21-hydroxylase deficiency as deleterious sequences can be transferred from *CYP21A1P* to *CYP21A2*. Despite the complexity of the locus, safe approaches for genotyping are established, and genotype phenotype relationships have been documented making genotyping a valuable complement to biochemical investigations in the diagnostics of 21-hydroxylase deficiency. This is of particular importance in relation to family investigations and neonatal screening.

Congenital adrenal hyperplasia (CAH) comprises a group of autosomal recessive disorders caused by complete or partial defects in one of several steroidogenic enzymes involved in the synthesis of cortisol from cholesterol in the adrenal glands. Low levels of cortisol stimulate an overproduction of CRH and ACTH, which continuously stimulates the adrenal gland giving rise to hyperplasia of the adrenal tissue. Depending on which steroidogenic enzyme is affected, synthesis of the other classes of adrenocortical steroid hormones is affected in different ways. More than 95% of all cases of CAH are caused by 21-hydroxylase deficiency (21-OHD, OMIM +201910), which in addition to cortisol impairs synthesis of mineralocorticoids. The steroid precursors prior to this enzymatic block (progesterone and 17-hydroxyprogesterone) accumulate and are shunted through the adrenal androgen biosynthetic pathway, leading to overproduction of adrenal androgens [1]. The severity of the clinical symptoms varies according to the level of residual 21-hydroxylase activity. Complete enzyme deficiency is life-threatening due to neonatal hyponatremic dehydration (salt-wasting), and females affected with severe 21-OHD are born with virilization of external genitalia due to prenatal androgen excess; the single most common cause of ambiguous genitalia in the newborn. Milder forms present with varying degrees of hyperandrogenism after the first year of life. Although there is a continuous spectrum of phenotypic manifestations, it has been useful to divide the disease into three forms according to severity: classical salt-wasting, classical simple virilizing, and non-classical 21-OHD. The classical forms are associated with symptoms from birth and occur with an incidence of around 1:10,000 to 1:15,000 live births, whereas the milder non-classical form is much more prevalent (up to 1:500 in various populations).

The *CYP21A2* Gene and the RCCX Module

Steroid 21-OHD is caused by mutations in the *CYP21A2* gene (previously called *CYP21* or *CYP21B*, GeneID 1589), which is located in the HLA class III region in the major histocompatibility (MHC) locus on the short arm of chromosome 6 (band 6p21.3) [2] together with a highly homologous pseudogene, *CYP21A1P* (previously called *CYP21P* or *CYP21A*). These two genes share about 98% homology in their ten exons and about 96% in the introns, but *CYP21A1P* is inactive because of several deleterious mutations [3, 4]. Both genes are arranged in tandem repeat with the *C4A* and *C4B* genes encoding the fourth component of complement [5, 6]. The *C4/CYP21* unit is flanked by a telomeric *RP* gene and a centromeric *TNX* gene, forming what is referred to as RCCX modules *(RP-C4-CYP21-TNX). RP1* encodes a putative nuclear protein similar to DNA helicase, *RP2* is a truncated, non-functional form of *RP1* [7], *TNXB* encodes an extracellular matrix protein, tenascin X, which overlaps the *CYP21A2* gene on the opposite strand, and *TNXA* is a truncated copy of *TNXB* overlapping *CYP21A1P*, also on the opposite DNA strand [8]. Most haplotypes have a bimodular form composed of two sets of four genes arranged in tandem as follows: *RP1 – C4A – CYP21A1P – TNXA – RP2 – C4B – CYP21A2 – TNXB* (fig. 1).

Mutations Causing 21-OHD

As a result of its structure with repeated genes arranged in tandem, the RCCX region (like the MHC locus in general) is predisposed to a wide variety of genetic rearrangements. Misalignment during meiosis followed by genetic recombination events can result in monomodular as well as trimodular RCCX haplotypes [9, 10] (fig. 2). In addition, there is evidence of gene conversion events; a poorly understood mechanism in which genetic material is transferred between homologous genes in a non-

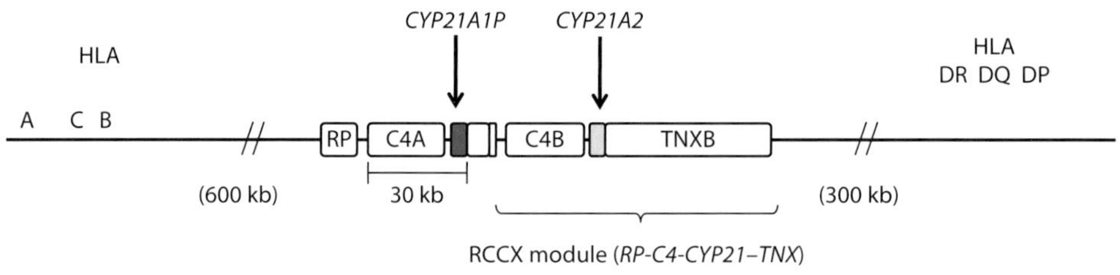

Fig. 1. The *CYP21A2* gene encoding steroid 21-hydroxylase is located in the HLA class III region in the MHC locus on the short arm of chromosome 6 (band 6p21.3) together with a highly homologous pseudogene, *CYP21A1P*. Both genes are arranged in tandem repeat with the *C4A* and *C4B* genes encoding the fourth component of complement. The *C4/CYP21* unit is flanked by a telomeric *RP* gene and a centromeric *TNX* gene, forming what is referred to as RCCX modules *(RP-C4-CYP21-TNX)*. Most haplotypes have a bimodular form composed of two sets of four genes arranged in tandem.

reciprocal fashion. The result is a chromosomal region displaying a high degree of variability both regarding gene copy number and sequence.

The presence of a highly homologous pseudogene containing several inactivating mutations thus forms the background for the relatively high incidence of 21-OHD. As discussed, unequal crossing over can result in gene deletions and duplications and depending on the breakpoints, such deletions and duplications can result in different types of chimaeras between the homologous genes in the RCCX modules. For example, different extents of *CYP21A1P* 5′ sequence can be attached to the 3′ portion of *CYP21A2* forming *CYP21A1P* – *CYP21A2* chimaeras, which will be inactive due to the deleterious pseudogene sequences, or a 5′ portion of the *TNXA* gene can be attached to the 3′ portion of *TNXB* as a result of a gross deletion that includes the whole *CYP21A2* gene. In both cases, the resulting haplotype lacks a functional *CYP21A2* gene. Rare cases have rearrangemens also affecting the function of the *TNXB* gene, resulting in a contiguous gene syndrome including CAH and an Ehlers-Danlos-like phenotype [11].

In addition to unequal crossing over, gene conversion events can result in transfer of sequences from *CYP21A1P* into *CYP21A2*. A wide variety of different gene conversions have been described, comprising variable portions of *CYP21A1P* containing different numbers of mutations. When large portions of *CYP21A2* have been converted into pseudogene sequences, this is commonly referred to as large gene conversions. Individual pseudogene-derived mutations can also be transferred into *CYP21A2* as microconversions [12–15]. Together with deletions removing the complete *CYP21A2* gene (monomodular RCCX haplotypes containing only *CYP21A1P*) and large gene conversions, which together typically account for 25–30% of all disease-causing alleles, nine such recurring pseudogene-derived mutations are responsible for around

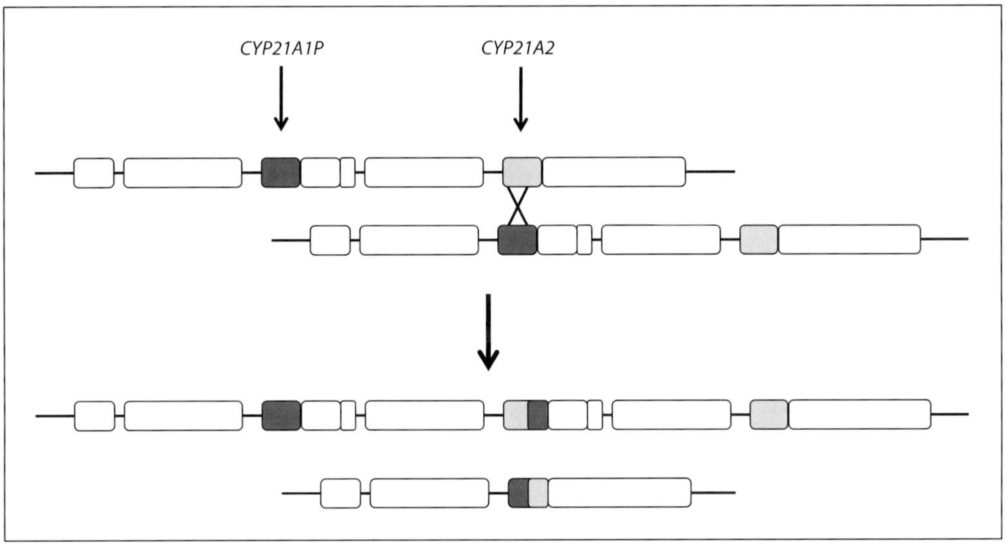

Fig. 2. As a result of its structure with repeated genes arranged in tandem, the RCCX region is predisposed to a wide variety of genetic rearrangements. Misalignment during meiosis followed by genetic recombination can result in monomodular as well as trimodular RCCX haplotypes, and depending on the breakpoints, various forms of chimeric genes can be formed between the homologous genes.

95% of all CAH alleles in most populations, although there are some differences in the relative frequencies of the individual mutations between ethnic groups [1]. These nine common pseudogene-derived mutations are: P30L [c.89C>T], I2 splice [c.290-13A/C>G], del 8 bp E3 [c.329_336delGAGACTAC], I172N [c.515T>A], Cluster E6 [c.707T>A+710T>A+716T>A], V281L [c.841G>T], L307 frameshift [c.920_921insT], Q318X [c.952C>T], and R356W [c.1066C>T] (fig. 3). The P453S [c.1357C>T] mutation is present in a low proportion of *CYP21A1P* alleles, and can thus also spread in a similar fashion. In addition to these pseudogene-derived mutations, almost 100 rare mutations causing 21-OHD have been described (www.cypalleles.ki.se). The vast majority of these has been identified in individual families, or represent recurrent rare alleles specific for particular ethnic groups.

Genotype Phenotype Relationships

In diagnostics of CAH, *CYP21A2* genotyping is a valuable complement to biochemical investigations. The diagnosis can be established by measuring 17-hydroxyprogesterone with or without ACTH stimulation, but this method does not discriminate heterozygotes from the normal population, and it does not always give clear information on disease severity. Genotyping is especially valuable in relation to neonatal screening

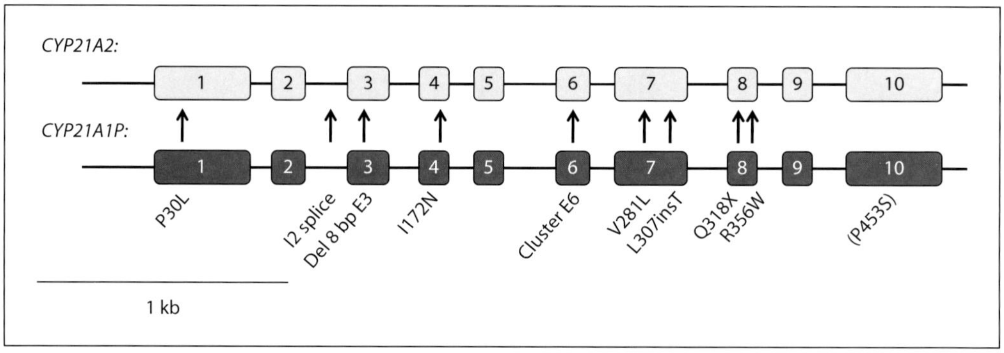

Fig. 3. The *CYP21A2* and *CYP21A1P* are about 98% homologous in their ten exons and 96% in the introns, but *CYP21A1P* is inactive because of several deleterious mutations. These can be transferred to *CYP21A2* through recombination or gene conversion; nine such recurring pseudogene-derived mutations are responsible for the majority of all CAH alleles.

Fig. 4. The common, pseudogene-derived mutations in *CYP21A2,* their corresponding clinical disease presentations, and their residual activities assayed after expression of recombinant enzyme in vitro. SW = Salt-wasting CAH; SV = simple virilizing CAH; NC = non-classical CAH.

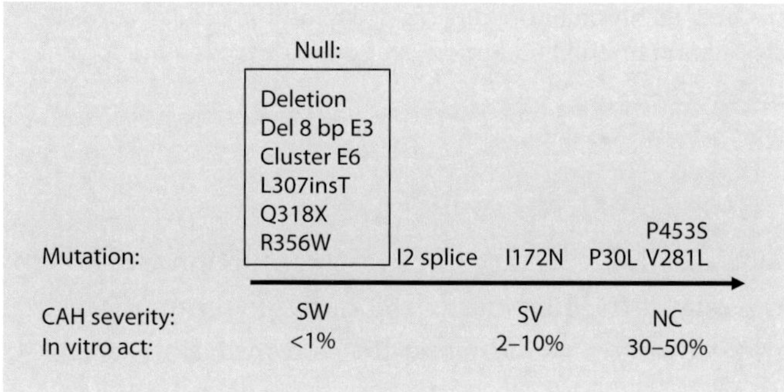

which investigates babies before the onset of symptoms, as these programs have a problem with false positive cases, in particular in premature babies [16]. Genotyping can thus confirm the diagnosis (or carrier state) and at the same time give prognostic information on disease severity. This is especially important in boys detected in the screening, since they do not display genital malformations as a sign of classical disease. In general, there is a good agreement between degree of disease presentation and the underlying combination of mutations (fig. 4), although some exceptions exist especially regarding the degree of hyperandrogenic symptoms [17–22]. These genotype-phenotype relationships have been clarified by clinical investigations on groups of patients representing different genotypes (the genotype group being defined by the mildest of the two mutations present). In vitro enzyme activity assays have also been performed, disclosing varying degrees of residual function in recombinant enzyme carrying different amino acid substitutions [for references, see

1]. These artificial systems do not provide absolute values of activity reflecting the in vivo situation, but they offer a means to make relative comparisons between different mutations. In this way, the severity of rare mutations can be assessed relative to the more common mutations for which more abundant clinical data are available. With some exceptions, the residual activities measured in vitro have been in reasonable agreement with the clinical disease severity. The I2 splice mutation, which is the single most common point mutations causing 21-OHD, activates a cryptic splice site in the splice acceptor site of intron 2, and is associated with very low but measurable activity in vitro.

Molecular Genetic Diagnosis of 21-OHD

A complete characterization of the locus regarding number of RCCX modules and their specific gene content requires Southern blotting with probes for the different genes and using different combinations of restriction endonucleases [23]. To resolve complex rearrangements, family members are preferably included. However, these laborious procedures are not convenient for clinical routine use and this detailed level of information regarding the whole locus is not necessary to establish the clinically relevant genotype. Many different protocols are used in a clinical setting to selectively amplify and genotype *CYP21A2*, avoiding amplification of the non-functional *CYP21A1P* pseudogene. This is facilitated by the small size of the gene; all ten exons cover approximately 3.1 kb of DNA. A convenient approach is to use the 8-bp deletion in exon 3 as a marker for the pseudogene, and perform direct sequencing of all genes amplified using primers discriminating *CYP21A2* from *CYP21A1P* in this position [24]. Multiplex ligation-dependent probe amplification analysis [25] can be performed in parallel using a probe in the same position, giving information on gene copy number. As there are alleles with complex rearrangements containing more than one disease-causing mutation, it is recommended to analyze the complete gene sequence in all cases. If possible, mutations should also be segregated in the family, to verify their presence in separate alleles and to identify de novo mutations which occur at a frequency of approximately 1%.

References

1 White PC, Speiser PW: Congenital adrenal hyperplasia due to 21-hydroxylase deficiency. Endocr Rev 2000;21:245–291.

2 White PC, New MI, Dupont B: HLA-linked congenital adrenal hyperplasia results from a defective gene encoding a cytochrome P-450 specific for steroid 21-hydroxylation. Proc Natl Acad Sci USA 1984;81:7505–7509.

3 White PC, New MI, Dupont B: Structure of human steroid 21-hydroxylase genes. Proc Natl Acad Sci USA 1986;83:5111–5115.

4 Higashi Y, Yoshioka H, Yamane M, Gotoh O, Fujii-Kuriyama Y: Complete nucleotide sequence of two steroid 21-hydroxylase genes tandemly arranged in human chromosome: a pseudogene and a genuine gene. Proc Natl Acad Sci USA 1986;83:2841–2845.
5 White PC, Grossberger D, Onufer BJ, Chaplin DD, New MI, Dupont B, Strominger JL: Two genes encoding steroid 21-hydroxylase are located near the genes encoding the fourth component of complement in man. Proc Natl Acad Sci USA 1985;82:1089–1093.
6 Carroll MC, Campbell RD, Porter RR: Mapping of steroid 21-hydroxylase genes adjacent to complement component C4 genes in HLA, the major histocompatibility complex in man. Proc Natl Acad Sci USA 1985;82:521–525.
7 Shen L, Wu LC, Sanlioglu S, Chen R, Mendoza AR, Dangel AW, Carroll MC, Zipf WB, Yu CY: Structure and genetics of the partially duplicated gene RP located immediately upstream of the complement C4A and the C4B genes in the HLA class III region. Molecular cloning, exon-intron structure, composite retroposon, and breakpoint of gene duplication. J Biol Chem 1994;269:8466–8476.
8 Bristow J, Tee MK, Gitelman SE, Mellon SH, Miller WL: Tenascin-X: a novel extracellular matrix protein encoded by the human XB gene overlapping P450c21B. J Cell Biol 1993;122:265–278.
9 Werkmeister JW, New MI, Dupont B, White PC: Frequent deletion and duplication of the steroid 21-hydroxylase genes. Am J Hum Genet 1986;39:461–469.
10 Blanchong CA, Zhou B, Rupert KL, Chung EK, Jones KN, Sotos JF, Zipf WB, Rennebohm RM, Yu CY: Deficiencies of human complement C4A and C4B and heterozygosity in length variants of RP-C4-CYP21-TNX (RCCX) modules in Caucasians: the load of RCCX genetic diversity on major histocompatibility complex-associated disease. J Exp Med 2000;191:2183–2196.
11 Burch GH, Gong Y, Liu W, Dettman RW, Curry CJ, Smith L, Miller WL, Bristow J: Tenascin-X deficiency is associated with Ehlers-Danlos syndrome. Nat Genet 1997;17:104–108.
12 Speiser PW, New MI, White PC: Molecular genetic analysis of nonclassical steroid 21-hydroxylase deficiency associated with HLA-B14,DR1. N Engl J Med 1988;319:19–23.
13 Higashi Y, Tanae A, Inoue H, Fujii-Kuriyama Y: Evidence for frequent gene conversion in the steroid 21-hydroxylase P-450(C21) gene: implications for steroid 21-hydroxylase deficiency. Am J Hum Genet 1988;42:17–25.
14 Amor M, Parker KL, Globerman H, New MI, White PC: Mutation in the CYP21B gene (Ile-172→Asn) causes steroid 21-hydroxylase deficiency. Proc Natl Acad Sci USA 1988;85:1600–1604.
15 Globerman H, Amor M, Parker KL, New MI, White PC: Nonsense mutation causing steroid 21-hydroxylase deficiency. J Clin Invest 1988;82:139–144.
16 Nordenstrom A, Thilén A, Hagenfeldt L, Larsson A, Wedell A: Genotyping is a valuable diagnostic complement to neonatal screening for congenital adrenal hyperplasia due to steroid 21-hydroxylase deficiency. J Clin Endocrinol Metab 1999;84:1505–1509.
17 Speiser PW, Dupont J, Zhu D, Serrat J, Buegeleisen M, Tusie-Luna M-T, Lesser M, New MI, White PC: Disease expression and molecular genotype in congenital adrenal hyperplasia due to 21-hydroxylase deficiency. J Clin Invest 1992;90:584–595.
18 Wilson RC, Mercado AB, Cheng KC, New MI: Steroid 21-hydroxylase deficiency: genotype may not predict phenotype. J Clin Endocrinol Metab 1995;80:2322–2329.
19 Wedell A, Thilén A, Ritzén EM, Stengler B, Luthman H: Mutational spectrum of the steroid 21-hydroxylase gene in Sweden: implications for genetic diagnosis and association with disease manifestation. J Clin Endocrinol Metab 1994;78:1145–1152.
20 Jääskeläinen J, Levo A, Voutilainen R, Partanen J: Population-wide evaluation of disease manifestation in relation to molecular genotype in steroid 21-hydroxylase (CYP21) deficiency: good correlation in a well defined population. J Clin Endocrinol Metab 1997;82:3293–3297.
21 Krone N, Braun A, Roscher AA, Knorr D, Schwarz HP: Predicting phenotype in steroid 21-hydroxylase deficiency? Comprehensive genotyping in 155 unrelated, well-defined patients from southern Germany. J Clin Endocrinol Metab 2000;85:1059–1065.
22 Stikkelbroeck NMML, Hoefsloot LH, De Wijs IJ, Otten BJ, Hermus ARMM, Sistermans EA: CYP21 gene mutation analysis in 198 patients with 21-hydroxylase deficiency in the Netherlands: six novel mutations and a specific cluster of four mutations. J Clin Endocrinol Metab 2003;88:3852–3859.
23 Morel Y, André J, Uring-Lambert B, Hauptmann G, Bétuel H, Tossi M, Forest MG, David M, Bertrand J, Miller WL: Rearrangements and point mutations of P450c21 genes are distinguished by five restriction endonuclease haplotypes identified by a new probing strategy in 57 families with congenital adrenal hyperplasia. J Clin Invest 1989;83:527–536.

24 Wedell A, Luthman H: Steroid 21-hydroxylase deficiency: two additional mutations in salt-wasting disease and rapid screening of disease-causing mutations. Hum Mol Genet 1993;2:499–504.

25 Schouten JP, McElgunn CJ, Waaijer R, Zwijnenburg D, Diepvens F, Pals G: Relative quantification of 40 nucleic acid sequences by multiplex ligation-dependent probe amplification. Nucleic Acids Res 2002;30:e57.

Prof. Anna Wedell, MD, PhD
Department of Molecular Medicine and Surgery, CMM:02
Karolinska Institutet/Karolinska University Hospital
SE–171 76 Stockholm (Sweden)
Tel. +46 8 517 765 35, Fax +46 8 585 827 60, E-Mail Anna.Wedell@ki.se

Ghizzoni L, Cappa M, Chrousos G, Loche S, Maghnie M (eds): Pediatric Adrenal Diseases.
Endocr Dev. Basel, Karger, 2011, vol 20, pp 88–95

Brain Development and Cognitive, Psychosocial, and Psychiatric Functioning in Classical 21-Hydroxylase Deficiency

Heino F.L. Meyer-Bahlburg

NYS Psychiatric Institute/Department of Psychiatry, Columbia University, New York, N.Y., USA

Abstract

The disturbance of the hypothalamic-pituitary-adrenal axis characteristic of congenital adrenal hyperplasia (CAH) due to 21-hydroxylase deficiency (21-OHD) is likely to affect brain development, yet neuroanatomic work is only beginning. Fetal hyperandrogenemia in 46,XX 21-OHD leads to masculinized brain organization and, consequently, at later stages of development, to masculinized gender-related behavior and cognitive function, including, although relatively uncommonly, gender identity. Genital masculinization as well as its surgical treatment has implications for social stigmatization and sexual functioning. CAH-associated electrolyte crises in infancy and later may result in severe cognitive impairment. Psychiatric disorders are somewhat increased, especially in patients with severe degrees of CAH.

Congenital adrenal hyperplasia (CAH) due to 21-hydroxylase deficiency (21-OHD) involves multiple hormone abnormalities, among them deficiency of cortisol, aldosterone, and epinephrine, excess of corticotropin-releasing hormone, corticotropin, 17α-hydroxyprogesterone, and adrenal androgens (further converted to testosterone and dihydrotestosterone), as well as non-physiologic hormone replacement therapy with glucocorticoid and aldosterone medications [1]. These hormones affect multiple somatic systems that may have psychological implications, among them at all ages the glucose household, during prenatal development the differentiation of the reproductive tract and (presumably) the brain, and during puberty the growth of primary sex characteristics, development of secondary sex characteristics, further differentiation of the brain, and activation of previously organized brain circuits [2]. Best understood in this context is the role of sex steroids. Animal research has documented that the sexual differentiation of the mammalian brain depends primarily on androgens and (except in some primates) estrogens, possibly also progestogens, and appears to take place primarily during hormone-sensitive and hormone-producing periods of development [3], namely the pre- or perinatal period, puberty, and (in females), first pregnancy. Sex-related or sex-dimorphic behaviors affected thereby include juvenile play behavior (e.g., rough and tumble play), sexual attraction (e.g., proximity-seeking, proceptive behavior), copulatory behavior (receptivity, lordosis vs. mounting/intromission), parenting behavior, and orientation in space (spatial perception). Whereas pertinent neuroanatomic work (postmortem or with brain imaging) is increasingly being conducted also in humans, hardly any such work has been reported on patients with CAH. Therefore, this paper will mostly address psychological effects of CAH, in the domains of gender behavior and identity, psychosexual functioning, cognitive functioning, and psychiatric functioning. Most of the studies of these domains have been conducted in 46,XX girls and women born with genital ambiguity due to 21-OHD CAH. Thus, they will be the primary focus of this paper, with some notes added on 46,XX CAH raised male.

Gender Development in 46,XX 21-OHD CAH

Evidence

In humans in general, development of gender – in terms of behavior and identity – depends on genes, sex hormones, brain organization, and psychosocial factors. In 46,XX CAH, the hormone abnormalities are thought to yield a somewhat atypical sexual differentiation of the brain, particularly the high prenatal levels of male sex hormones to which the genetically female brain is exposed, but no positive findings have been reported to date [4]. In addition, the variability in sex-hormone modulation by glucocorticoid replacement therapy may affect brain differentiation and activation postnatally. Moreover, one also expects social and psychological effects on diverse aspects of gender development from such somatic gender atypicalities as genital ambiguity, postnatal virilization (e.g., hirsutism), body built (short stature, stockiness, obesity), and the outcome and side effects of genital surgery.

Masculinization of gender-related behavior (childhood play behavior, aggression, occupational preferences, leisure time activities, maternalism, etc.) in 46,XX CAH has been extensively studied in all age groups from early childhood to mid-adulthood. Several studies from our own group have demonstrated a dose-response relationship (on the group level) between CAH severity, which is associated with increased prenatal androgen production, and behavioral masculinization, along with marked inter-individual variability; we also ruled out a separate contribution of the 21-hydroxylase gene to gender behavior [5]. The findings on gender behavior also extend to sexual orientation [6].

The effects of the prenatal hormonal milieu on the development of gender identity are less clear cut. Our own team was unable to show in CAH girls a dose-response relationship between CAH severity (i.e., prenatal androgens) and dimensionally assessed gender identity [7]. A review of the world literature yielded gender dysphoria and/or patient-initiated gender change in 5.2% of 46,XX CAH raised female [8]; detailed case histories suggest a contribution of prolonged postnatal androgen exposure and of uncorrected or poorly corrected genital ambiguity to such a development [9].

Implications for the Clinician
In regard to gender assignment of strongly masculinized 46,XX newborns with 21-OHD CAH, the existing data do not warrant a routine assignment to the male gender. This conclusion was also endorsed by recent international consensus statements [10]. In cases of erroneous assignment of 46,XX CAH newborns to the male gender, gender reassignment needs to be considered, once the correct diagnosis has been made (see below). Gender-atypical behavior by itself should not be interpreted as an indication of gender dysphoria, but parents (and some patients) may need reassurance. If a patient develops a bisexual or homosexual orientation, she and sometimes her parents may need assistance in adjusting to and accepting this development. Low maternalism may need to be taken into consideration by some CAH women, when they make plans for partnering and family formation.

Genital Ambiguity, Genital Surgery, and Psychosexual Functioning

Evidence
Marked genital ambiguity at birth may have significant long-term psychological consequences. It is likely to affect gender assignment and gender rearing ('sex-typing') and sometimes even puts at risk the acceptance of a child by the parents. At later ages, stigmatization by family members, peers, and romantic partners are reported by many patients [11]. Body image development may also be adversely affected. The genital malformation may preclude vaginal intercourse for the woman, thereby interfere with conception (in addition to other neuroendocrine factors). Moreover, it may adversely affect marriage prospects and social status, dependent on cultural values. Genital surgery, therefore, does not only concern medical goals such as facilitating unobstructed urinary emptying without incontinence or infection, facilitating unobstructed menstrual flow, and supporting reproductive capacity, but also serves psychosocial goals. Among these are conforming the genital appearance to the assigned gender, facilitating the development of a consistent gender identity, reducing the risk of stigmatization, facilitating peno-vaginal intercourse (where desired), and permitting age-appropriate development of romantic/sexual partnering [12].

The optimal timing of genital surgery in disorders of sex development (DSD) in general (including CAH) continues to be vigorously debated. The consensus of medical professionals tends to favor surgery early in development [10], whereas leading DSD activists push for a delay of surgery to the age of consent. The only published two surveys of CAH women on this issue showed that the majority favored infancy and childhood (before puberty) for genital surgery [13, 14]. Therefore, and because of the potential psychosocial risks of atypical genitalia, a general moratorium on genital surgery seems inappropriate. Given the diversity of opinions, a careful preparation of the parents for really informed consent is advised.

Outcome criteria for genital surgery include three categories [12]. Among the medical criteria are acute surgical complications, scarring, later anatomic changes (such as fistulas, clitoral atrophy, penile strictures, urethral dilations, vaginal stenosis, compromised urination, uneven tissue growth during puberty, compromised menstrual flow), reproductive potential, reduction of gonadal cancer risk, and cancer risk of the neovagina. Psychosocial, gender-related outcome criteria include general aesthetic appearance (cosmesis) and gender-confirming appearance, such as gender typicality of the external genital anatomy during childhood, capacity for urination in the gender-typical posture, development of gender-typical genital anatomy during puberty, and the development of gender-typical secondary sex characteristics during puberty. Outcome criteria relevant for sexuality include nerve conductance by pudendal evoked responses (intra- or early postsurgical), tissue sensitivity to warmth, cold, and vibration by genito-sensory analyzer [15], erotic sensitivity, orgasmic capacity, genital pain sensitivity, anatomic potential for peno-vaginal intercourse, anatomic capacity for partner's sexual satisfaction, gender typicality of body image and associated reactions to romantic/erotic situations (e.g., avoidance), and overall satisfaction with one's love life.

Recent studies of the outcome of genital surgery including psychosexual functioning in CAH women vary considerably in the aspects covered and the results shown [13, 15–18]. However, women with classical CAH as a group appear to show reduction of the clitoral sensitivity to warmth, cold, and vibration, of erotic sensitivity, libido and lubrication/sexual arousal, and orgasmic capacity and increase in genital pain. Postsurgically, the anatomic potential for peno-vaginal intercourse is often improved, but also thwarted in many others by remaining tightness of the vaginal introitus or vaginal stenosis. Psychosexual milestones are commonly somewhat delayed and the overall satisfaction with one's love life reduced. The adverse outcomes increase with the severity of the 21-OHD condition, but are highly variable across individuals with the same degree of CAH severity. For instance, our team had shown that even among women who had undergone clitorectomy, orgasmic capacity varied from 'absent' to 'above average'.

Implications for the Clinician

The last two to three decades have brought significant advances towards improving the long-term sexual-function outcome of women with CAH. Women are not seen any more as passive receptacles for men's penises, but are recognized as active sexual beings seeking a satisfactory quality of sexual life. Increasingly, clitoral surgery in cases of mild to moderate clitoromegaly is avoided. In cases with more marked genital masculinization, the replacement of clitorectomy and clitoral recession by clitoral reduction with a strong focus on the preservation of the neurovascular bundle are commonly recommended. Vaginal dilatation is not any more recommended prior to adolescence.

The emotional risks of patients' genital exposure for medical teaching purposes have been widely recognized, and the teaching practice altered. Finally, the increasing availability of CAH support groups has also been helpful for many patients.

More work is still needed to improve the overall outcome. More emphasis needs to be routinely placed on long-term functional outcome of surgery, and follow-up studies should be devised accordingly. Guidelines should be developed for minimum requirements of training and experience for surgeons who perform genital surgery in DSD patients (e.g., the establishment of surgical centers of excellence). In childhood cases with marked genital ambiguity, the parents should be provided with detailed pros and cons that enables them to provide truly informed consent. The surgeon on the DSD team should be complemented by mental-health staff. Such personnel should be able to include sexual counseling and sexual therapy in their services for adolescents and adults with DSD, and such counseling and therapy should focus on relationships rather than mere sex-organ function, and also include affirmative considerations of bisexual and homosexual lifestyles. It is also important to generally increase the availability of CAH and/or DSD support groups, and also to bring about a dissemination of recent advances in surgical techniques and aftercare to resource-poor countries.

Cognitive Development

Evidence

With appropriate control groups of demographically comparable or even genetically related non-CAH women, the majority of women with CAH are not impaired in their overall intelligence, although a recent study yielded an unexplained exception [19]. If there are problems, they are more likely to occur in the more severe forms of CAH, presumably related to the finding that the number of electrolyte crises experienced correlates with the degree of intellectual impairment of 46,XX CAH [20]. Several studies have reported white-matter abnormalities in the brains of both 46,XX and 46,XY CAH. As hypothetical mechanisms have been proposed an impaired myelination process during oligodendroglial differentiation due to the endogenous hormone imbalance or to non-physiologic glucocorticoid hormone replacement therapy, and, alternatively, subclinical cerebral ischemia from microangiographic damage due to aldosterone deficiency [21]. It is still unknown whether these white-matter abnormalities have any functional implications.

Of the specific cognitive abilities, the (androgen-influenced) spatial perception has been studied most, and the majority of studies show an increase in CAH women compared to controls, as was expected on the basis of animal research [22].

Implications for the Clinician

Monitoring of cognitive development by neuropsychological testing and of school performance is advised, especially in CAH children with a history of electrolyte crises.

Psychiatric Development

Evidence

Although the documentation leaves much to be desired, social stigmatization – mostly related to gender-atypical body development – may be experienced (and anticipated) at all ages [11], and body image problems appear to be frequent. Nevertheless, standard quality of life measures tend to show scores for CAH women within normal limits [18], but such measures are usually not devised to capture body-image or sexuality-related problems. Studies on psychiatric problems using questionnaire measures also tend to show outcome within normal limits, especially in children, but two recent studies using structured psychiatric interviews show increased prevalence of attention deficit/hyperactivity (ADHD) and anxiety disorders in children and adolescents [23] and increased depression, anxiety, and substance abuse disorders in adults [24]. Whether such findings have any relation to the recently described reduction in size and alteration of function of CAH patients' amygdala in response to negative facial emotions [25] is not known.

Implications for the Clinician

Optimal medical treatment that achieves good height, weight, and genital appearance, and prevents postnatal virilization is likely to have important benefits in terms of psychosocial adjustment. The more gender-atypical the appearance of a girl or woman, the more important it is to monitor for stigmatization by family members, the peer group, romantic partners, and people at the work place. In addition, monitoring for psychiatric problems is advised, especially in CAH girls and women with the more severe forms of CAH.

Notes on 46,XX CAH Raised Male

46,XX CAH newborns with the more severe degrees of genital masculinization have an increased chance to be misdiagnosed and assigned to the male gender, especially in resource-poor societies; increased mortality due to delayed diagnosis and electrolyte crises has also been documented. Neonatal screening for CAH reduces both of these outcomes. Current consensus is that all newborns with 46,XX CAH should be assigned and raised as girls [10].

In the absence of newborn screening, however, it is expected that misassignment of some 46,XX CAH to the male gender will continue. Data on long-term gender identity outcome in such males shows gender dysphoria or patient-initiated gender change in 12.1% of reported cases with gender identity information [8]. Unfortunately, the quality of long-term follow-up data on functioning in the various psychological domains of interest is insufficient for general conclusions.

If the diagnosis is corrected early, such children should be reassigned to female. Gender reassignment in infancy can proceed without psychological evaluation of the child, if it is based on sound medical diagnosis and genuinely supported by the parents. At later ages, however, much caution is advised. Once the child is mature enough for psychological gender assessment, gender reassignment must be based on a thorough psychological gender evaluation. Such an evaluation should use a multi-informant (child, both parents, possibly other caretakers) and multi-method approach (standardized questionnaires, semistructured interviews, systematic play observations). In cases with ambiguous evidence, when no clear decision can be made, the originally assigned gender should be maintained and any gender-confirming genital surgery delayed. The parents may need repeated explanations and support.

Conclusions

The effects of 21-OHD CAH, especially its more severe variants, on body and brain lead to significant consequences in multiple domains of psychological functioning. Thus, there is a need for mental-health services including assessment, psychoeducation, counseling, and therapy, all preferably rendered in close cooperation with the specialists involved in medical case management [26]. Given the lack of suitably trained personnel, training opportunities for such personnel need to be created. Moreover, professional organizations should establish websites to provide validated information on CAH to patients and their families, and the formation of support groups should also be encouraged.

References

1 White PC, Speiser PW: Congenital adrenal hyperplasia due to 21-hydroxylase deficiency. Endocr Rev 2000;21;245–291.

2 Arnold AP: The organizational-activational hypothesis as the foundation for a unified theory of sexual differentiation of all mammalian tissues. Horm Behav 2009;55:570–578.

3 Schulz KM, Molenda-Figueira HA, Sisk CL: Back to the future: the organizational-activational hypothesis adapted to puberty and adolescence. Horm Behav 2009;55:597–604.

4 Ciumas C, Lindén Hirschberg A, Savic I: High fetal testosterone and sexually dimorphic cerebral networks in females. Cereb Cortex 2009;19:1167–1174.

5 Meyer-Bahlburg HFL, Dolezal C, Baker SW, Ehrhardt AA, New MI: Gender development in women with congenital adrenal hyperplasia as a function of disorder severity. Arch Sex Behav 2006; 35:667–684.

6 Meyer-Bahlburg HFL, Dolezal C, Baker SW, New MI: Sexual orientation in women with classical or non-classical congenital adrenal hyperplasia as function of degree of prenatal androgen excess. Arch Sex Behav 2008;37:85–99.

7 Meyer-Bahlburg HFL, Dolezal C, Baker SW, Carlson AD, Obeid JS, New MI: Prenatal androgenization affects gender-related behavior but not gender identity in 5- to 12-year-old girls with congenital adrenal hyperplasia. Arch Sex Behav 2004;33:97–104.

8 Dessens AB, Slijper FM, Drop SL: Gender dysphoria and gender change in chromosomal females with congenital adrenal hyperplasia. Arch Sex Behav 2005;34:389–397.

9 Meyer-Bahlburg HFL, Gruen RS, New MI, Bell JJ, Morishima A, Shimshi M, Bueno Y, Vargas I, Baker SW: Gender change from female to male in classical CAH. Horm Behav 1996;20:319–332.

10 Speiser PW, Azziz R, Baskin LS, Ghizzoni L, Hensle TW, Merke DP, Meyer-Bahlburg, HFL, Miller WL, Montori VM, Oberfield SE, Ritzén M, White PC: Congenital adrenal hyperplasia due to steroid 21-hydroxylase deficiency: an Endocrine Society Clinical Practice Guideline. J Clin Endocrinol Metab 2010;95:4133–4160.
11 Khuri J, Meyer-Bahlburg HFL, Baker SW, New MI: Stigma and psychosocial trauma in women with classical congenital adrenal hyperplasia. Int J Transgender 2008;10:193.
12 Meyer-Bahlburg HFL: Gender assignment and reassignment in intersexuality: controversies, data, and guidelines for research. Adv Exp Med Biol 2002;511: 199–223.
13 Nordenskjöld A, Holmdahl G, Frisén L, Falhammar H, Filipsson H, Thorén M, Janson PO, Hagenfeldt K: Type of mutation and surgical procedure affect long-term quality of life for women with congenital adrenal hyperplasia. J Clin Endocrinol Metab 2008; 93:380–386.
14 Wisniewski AB, Migeon CJ, Malouf MA, Gearhart JP: Psychosexual outcome in women affected by congenital adrenal hyperplasia due to 21-hydroxylase deficiency. J Urol 2004;171:2497–2501.
15 Crouch NS, Liao LM, Woodhouse CRJ, Conway GS, Creighton SM: Sexual function and genital sensitivity following feminizing genitoplasty for congenital adrenal hyperplasia. J Urol 2008;179:634–638.
16 Meyer-Bahlburg HFL, Baker SW, Dolezal C, Carlson AD, Obeid JS, New MI: Long-term outcome in congenital adrenal hyperplasia: gender and sexuality. Endocrinologist 2003;13:227–233.
17 Gastaud F, Bouvattier C, Duranteau L, Brauner R, Thibaud E, Kutten F, Bougnères P: Impaired sexual and reproductive outcomes in women with classical forms of congenital adrenal hyperplasia. J Clin Endocrinol Metab 2007;92:1391–1396.
18 Frisén L, Nordenström A, Falhammar H, Filipsson H, Holmdahl G, Janson PO, Thorén M, Hagenfeldt K, Möller A, Nordenskjöld A: Gender role behavior, sexuality, and psychosocial adaptation in women with congenital adrenal hyperplasia due to CYP21A2 deficiency. J Clin Endocrinol Metab 2009;94:3432–3439.
19 Johannsen TH, Ripa CPL, Reinisch JM, Schwartz M, Mortensen EL, Main KM: Impaired cognitive function in women with congenital adrenal hyperplasia. J Clin Endocrinol Metab 2006;91:1376–1381.
20 Baker SKW: An exploratory family study of IQ and specific cognitive abilities in classical congenital adrenal hyperplasia: examined by family, genetic, and salt wasting subgroups. Doctoral dissertation. New York, Graduate School of Arts and Sciences, Columbia University, 1990.
21 Bergamaschi R, Livieri C, Uggetti C, Candeloro E, Egitto MG, Pichiecchio A, Cosi V, Bastianello S: Brain white matter impairment in congenital adrenal hyperplasia. Arch Neurol 2006;63:413–416.
22 Mueller SC, Temple V, Oh E, VanRyzin C, Williams A, Cornwell B, Grillon C, Pine DS, Ernst M, Merke DP: Early androgen exposure modulates spatial cognition in congenital adrenal hyperplasia. Psychoneuroendocrinology 2008;33:973–980.
23 Mueller SC, Ng P, Sinaii N, Leschek EW, Green-Golan L, VanRyzin C, Ernst M, Merke DP: Psychiatric characterization of children with genetic causes of hyperandrogenism. Eur J Endocrinol (in press).
24 Meyer-Bahlburg HFL, Gruen R, Baker SW, New MI: Lifetime psychiatric diagnoses in women with congenital adrenal hyperplasia. Int J Transgender 2008; 10:200.
25 Ernst M, Maheu FS, Schroth E, Hardin J, Golan LG, Cameron J, Allen R, Holzer S, Nelson E, Pine DS, Merke DP: Amygdala function in adolescents with congenital adrenal hyperplasia: a model for the study of early steroid abnormalities. Neuropsychologia 2007;45:2104–2113.
26 Meyer-Bahlburg HFL: Treatment guidelines for children with disorders of sex development. Neuropsychiatr Enfance Adolesc 2008;56:345–349.

Heino F.L. Meyer-Bahlburg, Dr. rer. nat.
NYS Psychiatric Institute/Department of Psychiatry
Columbia University, 1051 Riverside Drive, Unit 15
New York, NY 10032 (USA)
Tel. +1 212 543 5299, Fax +1 212 543 6003, E-Mail meyerb@childpsych.columbia.edu

Ghizzoni L, Cappa M, Chrousos G, Loche S, Maghnie M (eds): Pediatric Adrenal Diseases.
Endocr Dev. Basel, Karger, 2011, vol 20, pp 96–105

Long-Term Outcome of Prenatal Dexamethasone Treatment of 21-Hydroxylase Deficiency

Svetlana Lajic · Anna Nordenström · Tatja Hirvikoski

Department of Molecular Medicine and Surgery, Karolinska University Hospital/Karolinska Institutet, Stockholm, Sweden

Abstract

Prenatal treatment of congenital adrenal hyperplasia (CAH) with dexamethasone (DEX) has been in use since the mid-1980s. Its effectiveness for reducing virilization of external genitalia is well established. DEX treatment has to be started in the 6th–7th postmenstrual week and continued until the results of the prenatal diagnosis are available. Hence, the dilemma is that 7 out of 8 fetuses (boys and unaffected girls) are treated unnecessarily. Girls with CAH are treated until term. Accumulating evidence from animal studies and follow-up data has raised concerns regarding the long-term consequences of this controversial treatment. We have previously reported that direct neuropsychological assessment of children exposed to DEX and controls show normal full-scale IQ, learning and long-term memory. However, the children exposed to DEX during the first trimester had an impaired verbal working memory which was significantly associated with low self-perceived scholastic competence. In addition, the children showed increased self-rated social anxiety. The same cohort of children answered questions concerning friends, activities and gender-related behaviors. The results indicate less masculine and more neutral behavior in short-term DEX-exposed boys. These findings indicate that long-term follow-ups of this group of patients are of extreme importance and that future DEX treatment of CAH may be questioned. We therefore encourage additional studies on larger cohorts in order to draw more decisive conclusions about the safety of the treatment. Until then, it is important that the parents are thoroughly informed about the potential risks and uncertainties, as well as the benefits, of this treatment.

Prenatal Diagnosis and Treatment

The rationale for prenatal treatment of congenital adrenal hyperplasia (CAH) is based on the hypothesis that female in utero virilization can be overcome if adequate fetal adrenal suppression is achieved. David and Forest [1] described the first successful

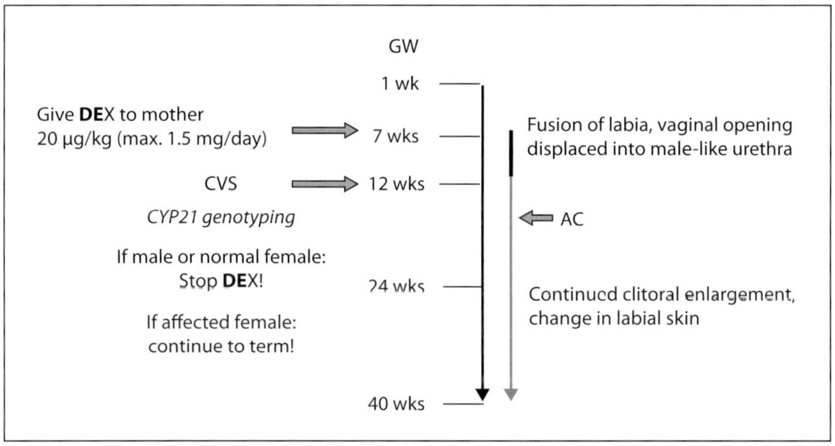

Fig. 1. Protocol for prenatal treatment of CAH. The vertical gray arrow indicates critical periods of androgen action and shows that the clitoris is sensitive to androgens throughout the gestational period. Boys and unaffected girls receive treatment approximately between GW 6 and 12–13 if a CVS is performed or between 6 and 16 if amniocentesis (AC) is performed for prenatal diagnostics. CAH-affected girls are treated to term.

prenatal treatment of a girl born with 21-OHD and thereby defined the outlines of the protocol used today for prenatal treatment of CAH (fig. 1). Dexamethasone (DEX) is the drug of choice because it escapes inactivation by placental 11β-hydroxysteroid dehydrogenase and thus suppresses the fetal adrenal cortex more efficiently than hydrocortisone [1]. Suppression of the fetal adrenal cortex will reduce the levels of adrenal androgens and lead to normal female sex differentiation. It has recently been shown that the pituitary-adrenal axis is indeed in operation at the early stage when DEX treatment is initiated [2]. Expression of the enzyme 3β-hydroxysteroid dehydrogenase type 2, which is important for the synthesis of cortisol, is detected as early as 8–9 weeks postconception. In addition, at this time point the fetal adrenal gland is secreting cortisol under the stimulation of ACTH, mediated via the type 2 melanocortin receptor, and the ACTH-secreting corticotrophs are suppressible by DEX [2].

A woman who has previously given birth to a child with severe CAH can be offered treatment with DEX in the 6th–7th gestational week (GW) of pregnancy. The typical dose used is 20 µg/kg body weight/day, divided into three daily doses. A few weeks later, during weeks 11–12, a chorionic villous sampling (CVS) is done to enable genotyping for 21-OHD. *CYP21A2* genotyping is an invaluable diagnostic tool, and in order to ascertain accurate molecular diagnostics of the fetus, it is important to analyze both parents as well as the proband to ensure a correct interpretation of the segregation of mutations in this complicated locus. If the fetus is a male or a healthy female, treatment is stopped, but if it is a female with CAH, treatment is continued until term (fig. 1). In women treated until term, the efficiency of the treatment is monitored with measurements of maternal serum DHEAS (from GW 7) and estriol (from GW 16). Low levels of DHEAS indicate adequate fetal adrenal suppression while a low level of estriol is a marker of maternal adrenal suppression and compliance. Discontinuation of DEX treatment should be tapered, in both short- and long-term-treated women. Due to the recessive mode of inheritance, only 1 out of 8 fetuses will benefit from DEX treatment and 7 out of 8 will thus be subjected to high doses of glucocorticoids (GCs) unnecessarily during early embryonic life. Today, fetal sex typing (SRY detection) can be performed using PCR amplification of cell-free fetal DNA in maternal blood as early as weeks 6–9 [3–5]. Applying this methodology in the genetic diagnostics of CAH would abolish the need for DEX administration and CVS in women carrying male fetuses and only 3 out of 8 fetuses would then be subjected to unnecessary treatment. Nevertheless, even early sex typing of the fetus still involves the ethical dilemma that a few healthy fetuses will be treated for a short period of time for the sake of the benefit of others.

Short-Term Outcome of Prenatal Treatment

The general agreement is that DEX is efficient in reducing or ameliorating virilization of CAH-affected girls. Treated newborns have consistently been less virilized than their older sisters under the condition that the mother was compliant and that the treatment was initiated in time (mean Prader score 1–2 in treated versus 4 in untreated girls) [6–11]. Most treated girls thus require little or no surgical intervention, which is a great relief for the family and may improve the future quality of life of the child. The effects of excessive androgens on the developing female brain may also be prevented by prenatal treatment. However, side effects must be considered in the pregnant woman as well as in the child, as DEX is a very potent synthetic GC that may interfere with maternal well-being as well as the somatic and psychological development of the growing child. Accumulating evidence tells us that the hormonal and nutritional milieu in utero probably predisposes the child to a range of diseases later in adult life [12, 13]. The fact that most children receive DEX treatment unnecessarily adds to the concerns as well as the possibility that side effects may differ in children treated short term (GW 7–14) versus long term (GW 7–40).

Maternal side effects observed during DEX treatment are those normally expected with GC therapy, i.e. weight gain, cutaneous striae, edema, sleep disturbances and mood fluctuations [6–11]. Most of the maternal side effects disappeared after discontinuation of treatment but, in our experience, approximately 50% of the mothers treated during early pregnancy and all mothers treated until term reported some type of discomfort that could be attributed to DEX [11]. Therefore, it is very important to inform the women in advance about possible side effects in order to achieve better coping and improve compliance.

Accumulating evidence indicates that prenatally administered GCs have long-term negative effects not only in experimental animals but also in humans [for a review, see 14]. It is imperative that prenatal DEX treatment of children at risk for CAH is carried out in centers where proper follow-up of treated cases can be achieved. The concept of fetal programming is a fact and we should be observant not only of fetal growth or child behavior but also of metabolic outcomes in treated cases. Most of the prenatally DEX-exposed children at risk for CAH show a normal pre- and postnatal growth, although several adverse events with failure to thrive have been observed in both short- and full-term-treated cases [11]. Long-term studies on the metabolic and cardiovascular status are still lacking. This should be investigated, as prenatal exposure to GCs programs the fetal HPA axis and may lead to altered susceptibility to metabolic and cardiovascular disease. In sheep, at 2 years of age, offspring repeatedly treated with betamethasone during fetal life had elevated levels of ACTH in response to corticotropin-releasing hormone, but had reduced basal and stimulated cortisol levels [15]. This demonstrates the importance of studying outcome at different points in time over the life course. In primates, prenatal DEX during late gestation resulted in offspring with elevated levels of basal- and stress-stimulated cortisol already at the prepubertal stage [16, 17]. This reflects an increased drive of the HPA axis rather than attenuated feedback, as seen in rodents. In addition, the animals exhibited hypertension, impaired glucose tolerance and hyperinsulinemia [17]. Little is known about the long-term impact of antenatal GC on human HPA axis function, but a recent report indicates that 30 years after exposure to betamethasone, the exposed individuals (particularly women) exhibited insulin resistance [18].

Long-Term Behavioral Outcome of Prenatally Treated Children at Risk for CAH

GCs are necessary for normal CNS development during fetal life, affecting such processes as cell proliferation as well as neuronal growth and differentiation. However, an excess of GCs may have negative effects on the sensitive developing brain. At the onset of the prenatal treatment (GW 6–7), the basic features of the CNS can already be identified and the neuroblasts have begun to migrate to their permanent locations where the differentiation occurs [19]. Consequently, these processes may be affected by the DEX exposure. The effects of synthetic GCs on the developing brain have mostly been studied in children treated with synthetic GCs perinatally due to a risk of preterm delivery [20]. GCs enhance the maturation of the fetal lung, thus reducing respiratory distress, but can also lead to negative effects on both cognition and psychological well-being *(ibid)*. However, the effects may be different in prenatal treatment of CAH, which, in most cases, occurs in early pregnancy (CAH-affected girls are treated to term) and using lower doses of GC than in the treatment of respiratory distress.

There are only a few long-term follow-up studies of behavior in prenatally treated children at risk for CAH [21–24] (for a summary, see table 1). The studies consist of children of different age groups and partly use different assessment methods, which impedes comparisons of the results.

Table 1. Summary of the long-term follow-up studies on prenatal DEX treatment of children at risk for CAH

	Participants		Results from different assessment methods		
	DEX-exposed children	control group	parental questionnaires	children's self-rating scales and interviews	neuropsychological tests
Trautman et al., 1995 [25]	n = 26 (3 CAH-affected)	n = 14 untreated children at risk for CAH (3 CAH-affected)	No differences in general development (*R-DPDQ/MCDI*) or temperament (*ITQ/TTQ/BSQ*); DEX-treated children had higher scores on shyness and emotionality, lower on sociability (*EAS*), as well as more internalizing and total problems (*CBCL*)	–	–
	age reported for the total study group (n = 40): 6 months to 5.5 years, mean age = 2.5 (±1.3)				
Meyer-Bahlburg et al., 2004 [23]	n = 36 0–15 months	n = 15 0–15 months	no differences in general development (*KIDS/R-DPDQ*)	–	–
	n = 89 15 months to 6 years	n = 126 15 months to 6 years	no differences in general development (*CDI/R-DPDQ*)	–	–
	n = 44 6–12 years	n = 162 6–12 years	no differences on school performance (*CBCL School scale)*	–	–
	total n = 174 (48 CAH-affected: 31 girls, 17 boys)	total n = 313 (195 CAH-affected: 100 girls, 95 boys)			

Table 1. Continued

	Participants		Results from different assessment methods		
	DEX-exposed children	control group	parental questionnaires	children's self-rating scales and interviews	neuropsychological tests
Hirvikoski et al., 2007 [22]	n = 26 7–17 years mean = 10.95 (±2.33) (9 CAH-affected: 4 girls, 5 boys)	n = 35 7–17 years mean = 10.38 (±2.33), healthy controls	no differences on school performance (*CBCL School scale)*	CAH-unaffected children reported poorer scholastic competence (*SPPC*) and increased social anxiety (*SASC-R*)	No differences in IQ (*WISC-III*), handedness, learning or memory (*NEPSY*); CAH-unaffected children had poorer working memory (*WISC-III*)
Hirvikoski et al., 2008 [21]	same cohort as Hirvikoski et al., 2007 [22]		no differences on behavioural problems (*CBCL*); no differences on shyness (*SPAI-C-P*); DEX-treated children had higher scores on sociability (*EAS*)	–	–
Hirvikoski et al., unpubl. data	n = 26 7–17 years mean = 10.95 (±2.33) (9 CAH-affected: 4 girls, 5 boys)	n = 35 7–17 years mean = 10.38 (±2.33) healthy controls		CAH-unaffected boys showed more neutral behaviors than controls (KI-GRB)	–

Note: R-DPDQ = Revised Denver Prescreening Developmental Questionnaire; MCDI = Minnesota Child Development Inventory; CBCL = Child Behaviour Checklist; ITQ =Infant Temperament Questionnaire; TTQ = Toddler Temperament Questionnaire; BSQ = Behavioral Style Questionnaire; EAS = EAS Temperament Survey for Children; KIDS = Kent Infant Development Scale; CDI = Child Development Inventory; WISC-III = Wechsler Intelligence Scales for Children; NEPSY = Developmental Neuropsychological Assessment; SPPC = Self-Perception Profile for Children; SASC-R = The Social Anxiety Scale for Children–Revised; SPAI-C-P = Social Phobia and Anxiety Inventory for Children – Parent Report; KI-GRB = The Karolinska Inventory of Gender Role Behaviour. For descriptions of the instruments as well as references, see the original studies.

Temperament, Psychological Well-Being and Behavioral Problems

A pilot study, using parental questionnaires as the method of assessment indicated that prenatally DEX-exposed preschool children showed more shyness and emotionality, as well as less sociability [24]. The mothers also reported more internalizing problems on a questionnaire measuring behavioral problems in children who were between 2 and 5½ years of age *(ibid)*. In the Swedish cohort, consisting of school-age children, no differences were observed in the parental rating of behavioral problems or temperament except that the DEX-treated children were rated as slightly more sociable than controls [21]. In the comparison of three groups (CAH-affected, DEX-treated; CAH-unaffected, DEX-treated; and controls), this difference no longer reached statistical significance due to small sample size. Likewise, in parental ratings of social anxiety, no between-group differences could be identified. However, in the children's self-ratings, the CAH-unaffected short-term-treated children reported more social anxiety compared to the control group [22] (fig. 2).

Cognition and School Performance

In the American cohort, using parental ratings, no differences were observed in the general development of children up to 6 years of age [23]. In addition, no differences could be detected in parent-rated school performance in children from 6 to 12 years of age, as compared to controls *(ibid)*. This result was replicated in the Swedish cohort using the same assessment methods [22]. However, in the children's self-ratings, the CAH-unaffected short-term-treated children reported poorer scholastic competence. This result was significantly correlated with a poorer working memory capacity observed in the CAH-unaffected group, assessed by direct neuropsychological testing (fig. 3). This may not be surprising since working memory is a key executive function which is important for many abilities needed in school work. The working memory consists in the ability to maintain information for a mental operation during a short period of time [25] and is thus required for such tasks as reading comprehension, mental arithmetic and ability to follow instructions. Like other complex cognitive functions and behaviors, working memory emerges from a neural network and not a single brain structure. Working memory functions are predominantly associated with effective connectivity in the prefrontal cortex, as well as the catecholaminergic and dopaminergic prefrontal subcortical networks and their connectivity with cerebellar regions [26]. It should be noted that the observed between-group difference in neuropsychological tests was specific and not general. In the assessment of general cognitive abilities such as IQ, no between-group differences were observed (fig. 4). Likewise, there were no differences in measures of learning and long-term memory.

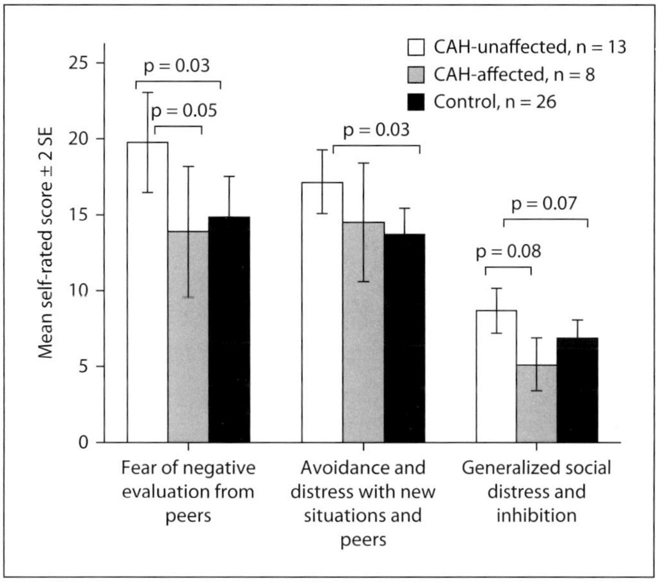

Fig. 2. CAH-unaffected, short-term-treated children reported more social anxiety than controls. SASC-R = The Social Anxiety Scale for Children.

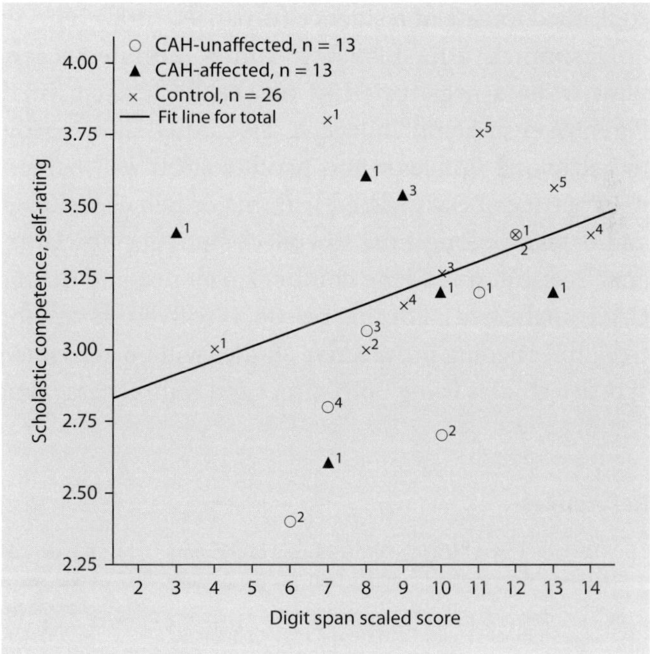

Fig. 3. Self-reported scholastic competence was significantly correlated with a poorer working memory capacity observed in the CAH-unaffected group. Only children ≥8 years old completed the self-rating questionnaires; the numbers indicate subject counts.

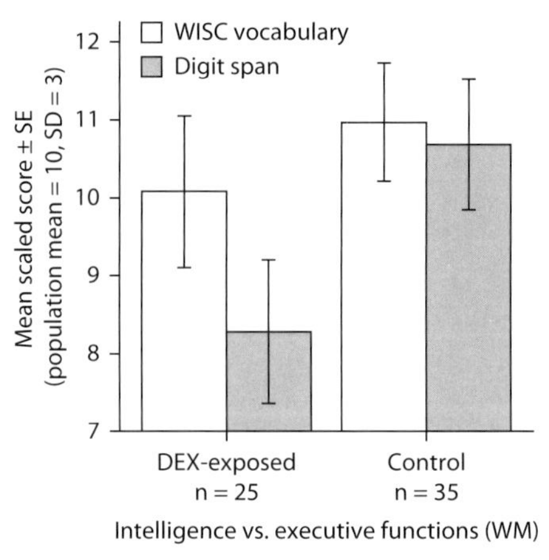

Fig. 4. Results show a significantly poorer performance in verbal working memory (measured with digit span from the WISC battery) in the DEX-treated group, while there are no between-group differences in general verbal reasoning ability (measured with the vocabulary from WISC). WISC = Wechsler Intelligence Scale for Children.

Gender Role Behavior

There are no published studies investigating gender role behavior in prenatally DEX-treated children at risk for CAH, although preliminary results from the Swedish cohort suggest that there may be a shift from masculine to neutral behaviors in DEX-treated, CAH-unaffected boys. No between-group differences were observed in feminine behaviors. In girls, no differences were identified in any gender role behavior when controlled for site of residence (urban vs. rural) [Hirvikoski et al., unpubl. data].

In summary, the published studies indicate no general effect on cognitive functions while a negative effect on specific (executive) functions has been observed. Parental ratings and children's self-ratings show partly disparate results with regard to behavioral problems and psychological well-being. However, the treatment does not appear to be associated with major behavioral problems. CAH is a rare disorder and it takes a long time to collect study groups large enough to enable data to be analyzed in terms of the children's age, sex, and diagnostic status (CAH-affected vs. CAH-unaffected). For this reason, it will also take a long time before the first results from the ongoing prospective studies will be available [27]. Meanwhile, larger retrospective studies using both direct and indirect assessment methods are needed.

References

1 David M, Forest MG: Prenatal treatment of congenital adrenal hyperplasia resulting from 21-hydroxylase deficiency. J Pediatr 1984;105:799–803.

2 Goto M, Piper Hanley K, Marcos J, et al: In humans, early cortisol biosynthesis provides a mechanism to safeguard female sexual development. J Clin Invest 2006;116:953–960.

3 Rijnders RJ, van der Schoot CE, Bossers B, et al: Fetal sex determination from maternal plasma in pregnancies at risk for congenital adrenal hyperplasia. Obstet Gynecol 2001;98:374–378.

4 Morel Y, Tardy V, Costa JM, et al: 21-Hydroxylase deficiency: new strategies emerging from molecular studies. Ann Endocrinol (Paris) 2003;64:456–470.

5 Bischoff FZ, Lewis DE, Simpson JL: Cell-free fetal DNA in maternal blood: kinetics, source and structure. Hum Reprod Update 2005;11:59–67.
6 Forest MG, David M, Morel Y: Prenatal diagnosis and treatment of 21-hydroxylase deficiency. J Steroid Biochem Mol Biol 1993;45:75–82.
7 Forest MG, Dörr HG: Prenatal therapy in congenital adrenal hyperplasia due to 21-hydroxylase deficiency: retrospective follow-up study of 253 treated pregnancies in 215 families. Endocrinologist 2003; 13:252–259.
8 Mercado AB, Wilson RC, Cheng KC, et al: Prenatal treatment and diagnosis of congenital adrenal hyperplasia owing to steroid 21-hydroxylase deficiency. J Clin Endocrinol Metab 1995;80:2014–2020.
9 Carlson AD, Obeid JS, Kanellopoulou N, et al: Congenital adrenal hyperplasia: update on prenatal diagnosis and treatment. J Steroid Biochem Mol Biol 1999;69:19–29.
10 New MI, Carlson A, Obeid J, et al: Prenatal diagnosis for congenital adrenal hyperplasia in 532 pregnancies. J Clin Endocrinol Metab 2001;86: 5651–5657.
11 Lajic S, Wedell A, Bui TH, et al: Long-term somatic follow-up of prenatally treated children with congenital adrenal hyperplasia. J Clin Endocrinol Metab 1998;83:3872–3880.
12 Barker DJ, Eriksson JG, Forsén T, Osmond C: Fetal origins of adult disease: strength of effects and biological basis. Int J Epidemiol 2002;31:1235–1239.
13 Seckl JR, Holmes MC: Mechanisms of disease: glucocorticoids, their placental metabolism and fetal 'programming' of adult pathophysiology. Nat Clin Pract Endocrinol Metab 2007:3:479–488.
14 Seckl JR: Prenatal glucocorticoids and long-term programming. Eur J Endocrinol 2004;151(suppl 3): 49–62.
15 Sloboda DM, Timothy JMM, Li S, et al: Prenatal betamethasone exposure results in pituitary-adrenal hyporesponsiveness in adult sheep. Am J Physiol 2007;292:E61–E70.
16 Uno H, Eisele S, Sakai A, et al: Neurotoxicity of glucocorticoids in the primate brain. Horm Behav 1994;28:336–348.
17 De Vries A, Holmes MC, Heijnis A, et al: Prenatal dexamethasone exposure induces changes in nonhuman primate offspring cardiometabolic and hypothalamic-pituitary-adrenal axis function. J Clin Invest 2007;117:1058–1067.
18 Dalziel SR, Walker NK, Parag V, et al: Cardiovascular risk factors after antenatal exposure to betamethasone: 30-year follow-up of a randomised controlled trial. Lancet 2005;365:1856–1862.
19 Lajic S, Nordenstrom A, Hirvikoski T: Long-term outcome of prenatal treatment of congenital adrenal hyperplasia. Endocr Dev 2008;13:82–98.
20 Baud O, Sola A: Corticosteroids in perinatal medicine: how to improve outcomes without affecting the developing brain? Semin Fetal Neonatal Med 2007;12:273–279.
21 Hirvikoski T, Nordenström A, Lindholm T, et al: Long-term follow-up of prenatally treated children at risk for congenital adrenal hyperplasia: does dexamethasone cause behavioural problems? Eur J Endocrinol 2008;159:309–316.
22 Hirvikoski T, Nordenström A, Lindholm T, et al: Cognitive functions in children at risk for congenital adrenal hyperplasia treated prenatally with dexamethasone. J Clin Endocrinol Metab 2007;92: 542–548.
23 Meyer-Bahlburg HF, Dolezal C, Baker SW, et al: Cognitive and motor development of children with and without congenital adrenal hyperplasia after early-prenatal dexamethasone. J Clin Endocrinol Metab 2004;89:610–614.
24 Trautman PD, Meyer-Bahlburg HF, Postelnek J, New MI: Effects of early prenatal dexamethasone on the cognitive and behavioral development of young children: results of a pilot study. Psychoneuroendocrinology 1995;20:439–449.
25 Baddeley A: Working Memory. Oxford, Clarendon Press, 1986.
26 Arnsten AF: Fundamentals of attention-deficit/hyperactivity disorder: circuits and pathways. J Clin Psychiatry 2006;67(suppl 8):7–12.
27 Lajic S, Nordenstrom A, Ritzen EM, Wedell A:: Prenatal treatment of congenital adrenal hyperplasia. Eur J Endocrinol 2004;151(suppl 3):63–69.

Svetlana Lajic, MD, PhD
Department of Molecular Medicine and Surgery
Centre for Molecular Medicine, L8:02, Karolinska University Hospital
SE–171 76 Stockholm (Sweden)
Tel. +46 8 5177 3922, Fax +46 8 5177 3620, E-Mail Svetlana.Lajic@ki.se

Ghizzoni L, Cappa M, Chrousos G, Loche S, Maghnie M (eds): Pediatric Adrenal Diseases.
Endocr Dev. Basel, Karger, 2011, vol 20, pp 106–115

Role of Genetic Variation in Regulation of Aldosterone Biosynthesis

Samantha Alvarez-Madrazo[a] · John M. Connell[b] · E. Marie Freel[a]

[a]BHF Glasgow Cardiovascular Research Centre, University of Glasgow, Glasgow, and [b]School of Medicine, University of Dundee, Dundee, UK

Abstract

Aldosterone biosynthesis is not only altered in rare mendelian disorders. Recent evidence suggests that common polymorphisms in the genes mediating the final stages of aldosterone and cortisol production (*CYP11B1* and *CYP11B2* respectively) are also associated with milder alterations in adrenal corticosteroid biosynthesis. These abnormalities consist of a decrease in adrenal 11β-hydroxylase activity and a subtle, life-long excess of aldosterone secretion which may lead to long-term cardiovascular risks. An interaction between the *CYP11B1* and *CYP11B2* genes may exist but is yet to be elucidated. This article describes the studies which highlight the importance of adrenal steroid synthesis in the development of hypertension and cardiovascular dysfunction as well as the role of common polymorphisms in adrenal synthetic genes in altering corticosteroid biosynthesis.

Genetic Regulation of Aldosterone Biosynthesis; Understanding Population Cardiovascular Risk

In the last decade, aldosterone has assumed central stage as a cardiovascular hormone important not only for normal electrolyte and blood pressure homeostasis, but also in the pathophysiology of syndromes of cardiovascular dysfunction. For these reasons, it is important to understand better the way in which population variability in aldosterone levels is controlled. The purpose of this brief review is to provide a summary of the most recent data on the genetic control of aldosterone production.

Population Cardiovascular Dysfunction and Aldosterone

The first case of primary aldosteronism (PA) leading to hypertension and hypokalaemia caused by an adrenal adenoma was reported by Jerome Conn in 1955, and thus was termed 'Conn's syndrome' [1]. Although Conn felt that PA would prove to be a frequent cause of hypertension, evidence in the subsequent two decades suggested that gross examples of PA were relatively infrequent and probably limited to a very small proportion of the hypertensive population. However, following the introduction of the aldosterone/renin ratio (ARR) in the 1980s and beyond, it became apparent that excess aldosterone in relation to the level of renin, with evidence of lack of suppressibilty of aldosterone by sodium loading is a relatively common phenomenon occurring in up to 11% of the hypertensive population, and around 20% of patients with resistant hypertension [2]. Less than 45% of these subjects have evidence of an autonomous adrenal adenoma, so that the majority have so-called 'idiopathic hyperaldosteronism (bilateral adrenal hyperplasia)'. Whether these subjects truly have PA is a matter of semantics and it is likely that the term 'primary aldosteronism' incorporates a number of pathophysiological syndromes; nonetheless, patients identified as having an elevated ARR and abnormal response of aldosterone production to sodium loading clearly have an inappropriate level of aldosterone in comparison to its main trophin angiotensin II. This, undoubtedly, has pathophysiological consequences that extend beyond hypertension.

In this regard, there is very good evidence that patients with PA are exposed to substantially greater cardiovascular risk than subjects with an equivalent degree of blood pressure elevation. Thus, the rates of stroke, myocardial infarction, and cardiac dysrhythmia are substantially increased [3]. These events probably reflect activation of mineralocorticoid receptors in a range of target tissues. Data from animal studies suggest that that the adverse tissue effects of aldosterone are dependent on prevailing sodium intake – where this is limited, aldosterone excess seems less harmful [4]. Although similar data from human studies are less extensive, there is some evidence that there is a significant interaction between aldosterone and salt intake to determine end organ damage, such as renal microalbuminuria excretion [5].

The above detail focuses on true aldosterone excess. However, the absolute importance of the hormone likely lies in its contribution to cardiovascular dysfunction in a broader context. For example, it is now clear that aldosterone levels, often within the so-called normal range, contribute to outcome following myocardial infarction. In addition, recent evidence from the Framingham Offspring Study demonstrates that plasma aldosterone levels predict the subsequent development of hypertension, and the rise in blood pressure over a period of 5 years [6]. Again, it is important to note that levels of the hormone in this circumstance were not elevated, consistent with the idea that population variation in secretion of aldosterone was an important determinant of the risk of hypertension. Data from our own group has corroborated this finding; we showed that there was a strongly positive correlation between aldosterone levels in a middle-aged cohort (60–70 years) and prevailing blood pressure levels; again, the aldosterone concentrations were not significantly elevated [7]. However, as there was a 10 mm Hg difference in systolic blood pressure between the subjects with the lowest aldosterone levels compared with the highest, this suggests that the population risk of having a long-term tendency to produce higher than average aldosterone levels is significant. These data, therefore, indicate that an understanding of the factors that control population variation in aldosterone levels is important in providing an insight into cardiovascular risk.

Population Variation in Aldosterone Levels

Day-to-day control of aldosterone is determined principally by the renin-angiotensin system and plasma potassium levels. The response of aldosterone to angiotensin II is also set by prevailing sodium intake, so that a low salt diet is associated with an increased aldosterone response to trophic stimulus. In the study of the relation of aldosterone levels and blood pressure in older subjects noted above, we also found that there was a highly significant inverse relationship between the hormone in later life and birth weight [7]. This would be consistent with an influence of programming in foetal life; in turn, this could reflect an epigenetic effect, or an influence of shared genetic determination. However, the distribution of aldosterone within the general population is continuous, consistent with an underlying genetic influence in its control.

Before focussing on specific genes, it is necessary to ask if a phenotype is heritable. In this regard, there are excellent data to confirm that aldosterone levels in human populations are strongly heritable, consistent with an underlying genetic influence. This applies to both plasma levels of the hormone and its urinary excretion [8].

In considering the possible genes that might contribute to the control of aldosterone, it is reasonable to look at rare monogenic syndromes associated with altered production of the hormone, and rodent models of disease where levels are altered. Thus, there are two hypertensive syndromes where aldosterone production is directly affected by single gene disorders: 11β-hydroxylase deficiency and glucocorticoid remediable aldosteronism. In both of these, the defects lie in the genes encoding aldosterone synthase *(CYP11B2)* or 11β-hydroxylase *(CYP11B1)* and the resulting phenotype is mineralocorticoid hypertension. In the Dahl salt-sensitive rat, mutations in both *CYP11B2* and *CYP11B1* lead to hypertension and altered aldosterone secretion in response to its usual trophins [8]. Finally, in an engineered model where the mouse *CYP11B2* gene is attached to the human 3′ region of the growth hormone gene, *CYP11B2* expression during high salt diet is associated with modestly raised aldosterone levels and high blood pressure [9] Thus, there are substantial grounds for asking if polymorphic variation at the *CYP11B* locus is associated with population variation in aldosterone levels or risk of hypertension.

CYP11B1 and *CYP11B2*

CYP11B2 and *CYP11B1* are highly homologous genes that lie on chromosome 8 in man; they share >95% homology in nucleotide sequence and in primary protein

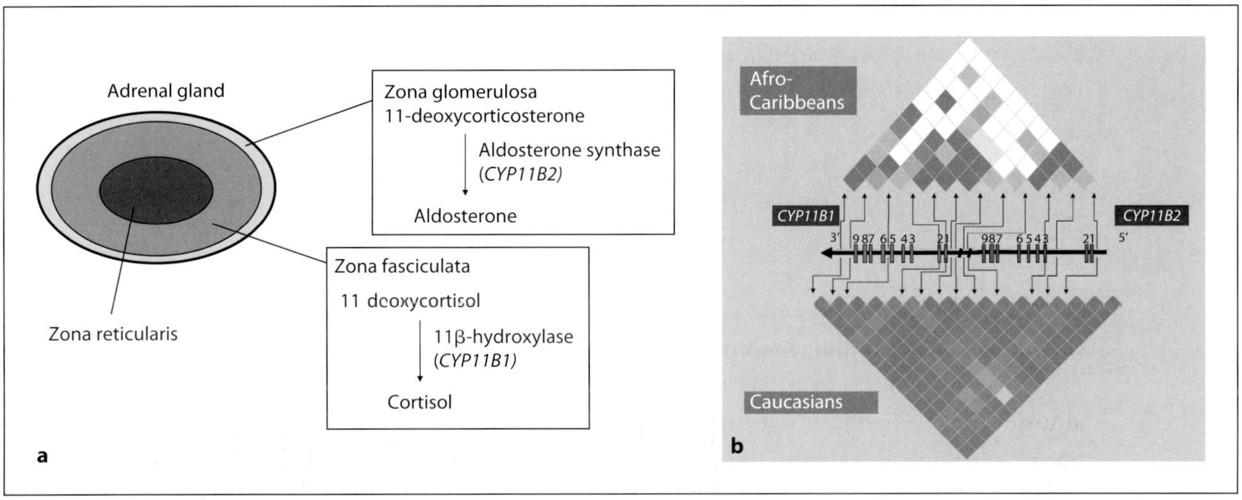

Fig. 1. **a** Adrenal gland zones and last steps in aldosterone and cortisol production. **b** Comparison of LD patterns in British and Afro-Caribbean subjects (a darker shade indicates higher LD in the locus).

structure [10]. *CYP11B2* encodes aldosterone synthase; it is expressed in zona glomerulosa of the adrenal cortex under the control of angiotensin II and potassium. *CYP11B1* encodes 11β-hydroxylase in zona fasciculata under the control of ACTH (fig. 1a). Both enzymes catalyse 11-hydroxylation of substrate (deoxycorticosterone and deoxycortisol); aldosterone synthase also has the capacity to convert corticosterone to 18-hydroxycorticosterone and subsequently aldosterone.

Initial molecular studies of this locus concentrated on a single nucleotide polymorphism in the 5′ regulatory region of *CYP11B2*, which disrupts a putative transcription binding site (-344C/T). We, and others, have shown that T allele carriers demonstrate increased aldosterone excretion compared with C carriers [11, 12]. More recently, we reported that a variant in close linkage disequilibrium (LD) with this variant (intron-2 conversion) was also associated with higher aldosterone levels [13]. However, reporter gene studies have failed to show a direct effect of the -344 C/T variant on gene expression, and the mechanism of this effect remains uncertain [14]. Other studies have examined the possible regulation of aldosterone using a genome-wide association approach but have failed to identify clear loci that might influence levels [15, 16].

A much more robust biochemical phenotype associated with variation *at CYP11B2* is an increase in basal and stimulated levels of the 11-deoxysteroids, deoxycorticosterone (DOC) and 11-deoxycortisol, as well as an increased 11-deoxycortisol/cortisol ratio [17]. We have interpreted this phenotype as evidence of a minor reduction in efficiency in adrenal 11-hydroxylation performed by the enzyme expressed in zona fasciculata as a gene product of *CYP11B1*. Interestingly, this biochemical pattern has previously been described in cohorts of patients with essential hypertension, with no clear explanation of its significance. However, we and other groups have produced convincing data that shows that this biochemical signature is closely associated with markers around the proximal part of the *CYP11B1* gene [18, 19]. In particular, in the MRC BRIGHT population, we found that polymorphic variants in the 5′ untranslated region (UTR) *of CYP11B1* (-1889 T and -1859 G) were associated with an increased level of 11-deoxysteroid metabolites, consistent with the above reduced efficiency [20].

The pattern of a minor apparent reduced efficiency in zona fasciculata 11β-hydroxylation raises the possibility that the regulation of expression of the gene might be altered. We have explored this by examining the regulation of contrasting reporter gene constructs that include the 5′ UTR variants described above in cultured adrenal cells [20]. The constructs containing the variants associated with reduced in vivo conversion of deoxycortisol to cortisol were also associated with reduced expression of a reporter gene in response to ACTH. Thus, it seems likely that the biochemical phenotype is a consequence of altered availability of the 11β-hydroxylase enzyme which is a rate-limiting step. In preliminary confirmatory studies, we have shown that the efficiency of transcription of the allele containing these 5′ UTR polymorphisms is reduced by approximately 35% in human adrenal tissue compared with the contrasting allele [21]. It is of relevance that the haplotype block that includes these variants in *CYP11B1* also includes the -344T and intron-2 conversion variants in *CYP11B2* that are associated with increased aldosterone production; the same studies of mRNA availability have shown that there is a modest (~18%) increase from this allele of *CYP11B2*.

These data are internally consistent and suggest that there are common polymorphic variants in *CYP11B1* that lead to reduced gene transcription, although the precise mechanism that links sequence variation with altered availability of mRNA remains to be determined; some of the variants described disrupt potential binding sites for putative transcription factors, but other mechanisms may be relevant. For example, Smithies' group [9] developed a rodent model of aldosterone excess by altering the 3′ UTR region of *CYP11B2* (by replacing the wild-type sequence with that of the bovine growth hormone gene); this resulted in a modest increase in mRNA availability during high sodium intake, with a consequent increase in aldosterone levels and raised blood pressure. Although there are no data in relation to variability in the human *CYP11B1* or *CYP11B2* 3′ UTR regions that might have similar effects on mRNA availability (either directly or via changes in binding of regulatory miRNA), this remains a possible mechanism to account for altered enzyme activity.

CYP11B Locus and Hypertension

A large number of studies have examined the role of *CYP11B2* in essential hypertension; most have focussed on the -344 C/T polymorphism in the 5′ UTR of the gene. We and others reported that the T allele was associated with high blood pressure [8], however many of these studies were small and underpowered case-control investigations. In a more focussed investigation we showed that the T allele was over-represented in hypertensive patients with a raised ARR compared to those with a normal ratio, suggesting that the genetic predisposition was limited to hypertensive subjects with relative aldosterone excess [22]. Nevertheless, other studies have failed to confirm these findings or have implicated the alternative allele in association with hypertension [23]. These disparities may reflect type 2 statistical error; alternatively, there is significant disparity in the pattern of LD across the locus depending on ethnic origin, and it may be that ethnic variation accounts for some of the inconsistencies noted. For example, we have found that the pattern of LD contrasts sharply when populations of strictly UK (Caucasian ancestry) and Jamaican origin (Afro-Caribbean ancestry) are compared (fig. 1b). Nevertheless, despite these disparities, when a meta-analysis is performed, evidence to support a modest effect of the T allele in predisposition to hypertension (17% increase in relative risk) is found [24].

An alternative approach to candidate gene studies is to perform large-scale genome-wide association studies; in the last 2 years, several major studies of hypertension involving many thousands of cases and controls have been performed. None of these have found evidence of association of tag SNPs that might identify the *CYP11B* locus with hypertension [15, 16]. However, this approach may lack the power to identify relatively small effects; the populations comprising these studies are very heterogeneous, making identification of a locus where LD may be variable difficult. It is of interest that one potential locus that has emerged from these studies is 17-hydroxylase, an important enzyme in steroidogenesis which is also implicated in a rare adrenal hypertensive disorder, highlighting the potential of genes expressed in the adrenal cortex to contribute to a common complex phenotype. However, it is also worth noting that these studies have not identified other key loci that are clearly important in blood pressure regulation (for example, those implicated in Bartter's and Gitelman's syndromes). Thus, there is a need for careful large-scale case-control studies of the *CYP11B* locus in homogeneous populations with accurate phenotyping. Although the current balance of evidence supports the notion that *CYP11B1* and *CYP11B2* are implicated in hypertension due to the strong pattern of LD across the locus it is not yet possible to define whether one or both genes contribute to this effect, although the data above on the impact of *CYP11B1* variants on 11-hydroxylation efficiency, and *CYP11B2* on aldosterone levels suggest that both may contribute.

Role of *CYP11B1* and *CYP11B2* in Genesis of Hypertension

Both genes are implicated in the rare monogenic syndromes described above that either raise or lower blood pressure in association with activation of the mineralocorticoid receptor. However, the above evidence suggests that more common polymorphic variants in the regulatory regions of the genes are associated with long-term (albeit minor) biochemical changes and consequent development of hypertension, probably in later adult life. In this regard, it is noteworthy that the Framingham investigators reported that aldosterone levels in young/middle-aged adults positively predicted the rise in blood pressure and development of hypertension over a 5-year period; in this study, the hormone levels were distributed across the normal range [6]. We subsequently reported that there was also a positive association between aldosterone levels in adult subjects aged between 60 and 70 years and blood pressure [7]. Both of these studies suggest that the regulation of aldosterone in the normal population contributes to the determination of the set level of blood pressure. As we have also noted, aldosterone concentrations are heritable and variation in *CYP11B2* contributes to this, a plausible explanation for some of the association between the gene and risk of hypertension is a direct effect on synthesis of the hormone. We have recently found evidence that the expression levels of *CYP11B2* in the adrenal cortex are related to polymorphic variants, suggesting that the availability of the enzyme aldosterone synthase, which is rate-limiting, may be the functional explanation for this association [21].

The link between variation in *CYP11B1* and risk of hypertension may be less immediately clear. In the rare monogenic syndrome of 11β-hydroxylase deficiency, the near absolute lack of enzyme activity results in an inability to synthesis cortisol and an increase in ACTH drive to the adrenal; in this circumstance, excess deoxycorticosterone (DOC) is produced. This hormone can transactivate the mineralocorticoid receptor, leading to sodium retention and hypertension. As a consequence, renin levels and production of aldosterone are suppressed and, as a by-product of increased ACTH drive, there is significant over-production of adrenal androgens.

The polymorphic variation that we and others have described in *CYP11B1* is associated with a pattern of steroid synthesis that is consistent with a minor reduction in efficiency of 11β-hydroxylation activity; this is a markedly different biochemical phenotype from that in the rare monogenic disorder. However, it is associated with a sustained and lifelong increase in production of deoxycortisol, and a raised ratio of that steroid to its product, cortisol. A similar increase in the DOC/corticosterone ratio would be expected. Intriguingly, this pattern of steroid synthesis has been reported in subjects with essential hypertension in studies where no genetic stratification has been performed. Thus, it may be that polymorphic variation in regulatory regions of *CYP11B1* that leads to reduced expression of the gene both in vitro and in vivo, results in a minor fall in capacity for 11β-hydroxylation in the zona fasciculata that leads to the biochemical phenotype described in the studies of essential hypertension.

How might this minor biochemical abnormality result in risk of development of high blood pressure? First, the phenotype is associated with a small increase in DOC production. As noted, this steroid can exert an effect on the mineralocorticoid receptor, although it is not clear if the relatively trivial increase in levels would be capable of exerting any substantial effect. Secondly, the minor loss of efficiency in production of cortisol (essentially within the range of physiological activity) should be associated with a compensatory increase in ACTH drive to the adrenal. Indeed, we have already shown that this is the case, so that the genotype associated with raised deoxycortisol/cortisol ratio is also associated with higher morning levels of ACTH [25]. Additionally, in a separate study of subjects with significant hypertension, we showed that subjects with this genotype demonstrated a correlation between aldosterone and ACTH-dependent steroids, consistent with the notion that the same trophin (ACTH) was contributing to the regulation of aldosterone, cortisol and adrenal androgen production; no such correlation was seen in the contrasting genotype [26]. We have hypothesised elsewhere that this chronic (lifelong) and very minor increase in ACTH drive to the adrenal may increase the secretion of aldosterone, either by a direct effect on the secretory pathway, a hyperplastic effect on zona glomerulosa, or an influence on expression of *CYP11B2* [8]. At present these remain speculative mechanisms, but are amenable to further investigation.

In summary, interest in regulation adrenal steroid synthesis has traditionally focussed on rare inborn errors of steroid biosynthesis. We believe that there is now good evidence that common polymorphic variation in the same genes is associated with altered steroid biosynthesis across the lifespan, and consequent risk of cardiovascular dysfunction. The present evidence suggests that both *CYP11B1* and *CYP11B2* are involved in the phenotypes of a minor reduction in efficiency of 11-hydroxylation activity and modest excess of aldosterone secretion; a single haplotype includes variants that are associated with both biochemical phenotypes and risk of hypertension. Whether there is a genuine interaction between the two genes remains to be fully determined, but we propose that there may be a digenic effect (similar to that noted in the Dahl salt-sensitive hypertensive rat) that predisposes to high blood pressure later in life.

The implications of the above studies may lie more in the recognition of the importance of adrenal steroid synthesis in the development of cardiovascular dysfunction and the role of mineralocorticoids in hypertension and its outcomes than in identification of any adrenal locus that might lend itself to direct intervention. Mineralocorticoid receptor antagonists are important drugs in the treatment of heart failure, but we propose that they have substantial potential to treat high blood pressure and effectively prevent its deleterious outcomes. Finally, the role of aldosterone synthase in regulating production of the hormone identifies this as a potential novel therapeutic target for intervention.

References

1 Conn JW: Primary aldosteronism: a new clinical syndrome. J Lab Clin Med 1955;45:3–17.
2 Gaddam KK, Pimenta E, Husain S, Calhoun DA: Aldosterone and cardiovascular disease. Curr Probl Cardiol 2009;34:51–84.
3 Milliez P, Girerd X, Plouin PF, Blacher J, Safar ME, Mourad JJ: Evidence for an increased rate of cardiovascular events in patients with primary aldosteronism. J Am Coll Cardiol 2005;45:1243–1248.
4 Brilla CG, Weber KT: Mineralocorticoid excess, dietary sodium, and myocardial fibrosis. J Lab Clin Med 1992;120:893–901.
5 Pimenta E, Gaddam KK, Pratt-Ubunama MN, Nishizaka MK, Aban I, Oparil S, Calhoun DA: Relation of dietary salt and aldosterone to urinary protein excretion in subjects with resistant hypertension. Hypertension 2008;51:339–344.
6 Vasan RS, Evans JC, Larson MG, Wilson PW, Meigs JB, Rifai N, Benjamin EJ, Levy D: Serum aldosterone and the incidence of hypertension in nonhypertensive persons. N Engl J Med 2004;351:33–41.
7 Reynolds RM, Walker BR, Phillips DI, Dennison EM, Fraser R, MacKenzie SM, Davies E, Connell JM: Programming of hypertension: associations of plasma aldosterone in adult men and women with birthweight, cortisol, and blood pressure. Hypertension 2009;53:932–936.
8 Connell JM, MacKenzie SM, Freel EM, Fraser R, Davies E: A lifetime of aldosterone excess: long-term consequences of altered regulation of aldosterone production for cardiovascular function. Endocr Rev 2008;29:133–154.
9 Makhanova N, Hagaman J, Kim HS, Smithies O: Salt-sensitive blood pressure in mice with increased expression of aldosterone synthase. Hypertension 2008;51:134–140.
10 Mornet E, Dupont J, Vitek A, White PC: Characterization of two genes encoding human steroid 11β-hydroxylase (P-450(11)β). J Biol Chem 1989;264:20961–20967.
11 Davies E, Holloway CD, Ingram MC, Inglis GC, Friel EC, Morrison C, Anderson NH, Fraser R, Connell JM: Aldosterone excretion rate and blood pressure in essential hypertension are related to polymorphic differences in the aldosterone synthase gene CYP11B2. Hypertension 1999;33:703–707.
12 Paillard F, Chansel D, Brand E, Benetos A, Thomas F, Czekalski S, Ardaillou R, Soubrier F: Genotype-phenotype relationships for the renin-angiotensin-aldosterone system in a normal population. Hypertension 1999;34:423–429.
13 Alvarez-Madrazo S, Padmanabhan S, Mayosi BM, Watkins H, Avery P, Wallace AM, Fraser R, Davies E, Keavney B, Connell JM: Familial and phenotypic associations of the aldosterone renin ratio. J Clin Endocrinol Metab 2009;94:4324–4333.
14 Bassett MH, Zhang Y, Clyne C, White PC, Rainey WE: Differential regulation of aldosterone synthase and 11β-hydroxylase transcription by steroidogenic factor-1. J Mol Endocrinol 2002;28:125–135.
15 Levy D, Ehret GB, Rice K, Verwoert GC, Launer LJ, Deghan A, et al: Genome-wide association study of blood pressure and hypertension. Nat Genet 2009; 41:677–687.
16 Newton-Cheh C, Johnson T, Gateva V, Tobin M, Bochud M, Coin L, et al: Genome-wide association study identifies eight loci associated with blood pressure. Nat Genet 2009;41:666–676.
17 Davies E, Holloway CD, Ingram MC, Friel EC, Inglis GC, Swan L, Hillis WS, Fraser R, Connell JM: An influence of variation in the aldosterone synthase gene (CYP11B2) on corticosteroid responses to ACTH in normal human subjects. Clin Endocrinol (Oxf) 2001;54:813–817.
18 Ganapathipillai S, Laval G, Hoffmann IS, Castejon AM, Nicod J, Dick B, Frey FJ, Frey BM, Cubeddu LX, Ferrari P: CYP11B2-CYP11B1 haplotypes associated with decreased 11β-hydroxylase activity. J Clin Endocrinol Metab 2005;90:1220–1225.
19 Keavney B, Mayosi B, Gaukrodger N, Imrie H, Baker M, Fraser R, Ingram M, Watkins H, Farrall M, Davies E, Connell JM: Genetic variation at the locus encompassing 11β-hydroxylase and aldosterone synthase accounts for heritability in cortisol precursor (11-deoxycortisol) urinary metabolite excretion. J Clin Endocrinol Metab 2005;90:1072–1077.
20 Barr M, MacKenzie SM, Friel EC, Holloway CD, Wilkinson DM, Brain NJ, Ingram MC, Fraser R, Brown M, Samani NJ, Caulfield M, Munroe PB, Farrall M, Webster J, Clayton D, Dominiczak AF, Connell JM, Davies E: Polymorphic variation in the 11β-hydroxylase gene associates with reduced 11-hydroxylase efficiency. Hypertension 2007;49: 113–119.
21 MacKenzie S, Stewart PM, Plouin PF, Fraser R, Connell JM, Davies E: Differential transcription of 11β-hydroxylase and aldosterone synthase in human adrenocortical tissue. Endocr Abstr 2009;19: P302.

22 Lim PO, Macdonald TM, Holloway C, Friel E, Anderson NH, Dow E, Jung RT, Davies E, Fraser R, Connell JM: Variation at the aldosterone synthase (CYP11B2) locus contributes to hypertension in subjects with a raised aldosterone-to-renin ratio. J Clin Endocrinol Metabolism 2002;87:4398–4402.
23 Pojoga L, Gautier S, Blanc H, Guyene TT, Poirier O, Cambien F, Benetos A: Genetic determination of plasma aldosterone levels in essential hypertension. Am J Hypertens 1998;11:856–860.
24 Sookoian S, Gianotti TF, Gonzalez CD, Pirola CJ: Association of the C-344T aldosterone synthase gene variant with essential hypertension: a meta-analysis. J Hypertens 2007;25:5–13.
25 Freel EM, Ingram M, Wallace AM, White A, Fraser R, Davies E, Connell JM: Effect of variation in CYP11B1 and CYP11B2 on corticosteroid phenotype and hypothalamic-pituitary-adrenal axis activity in hypertensive and normotensive subjects. Clin Endocrinol (Oxf) 2008;68:700–706.
26 Freel EM, Ingram M, Friel EC, Fraser R, Brown M, Samani NJ, Caulfield M, Munroe P, Farrall M, Webster J, Clayton D, Dominczak A, Davies E, Connell JM: Phenotypic consequences of variation across the aldosterone synthase and 11β-hydroxylase locus in a hypertensive cohort: data from the MRC BRIGHT Study. Clin Endocrinol (Oxf) 2007;67: 832–838.

Dr. E. Marie Freel
BHF Glasgow Cardiovascular Research Centre, University of Glasgow
126 University Place, Glasgow G12 8TA (UK)
Tel. +44 141 330 3412, Fax +44 141 330 1689, E-Mail E.Freel@clinmed.gla.ac.uk

Ghizzoni L, Cappa M, Chrousos G, Loche S, Maghnie M (eds): Pediatric Adrenal Diseases.
Endocr Dev. Basel, Karger, 2011, vol 20, pp 116–126

Circadian CLOCK-Mediated Regulation of Target-Tissue Sensitivity to Glucocorticoids: Implications for Cardiometabolic Diseases

Tomoshige Kino[a] · George P. Chrousos[b]

[a]Unit on Molecular Hormone Action, Program in Reproductive and Adult Endocrinology, Eunice Kennedy Shriver National Institute of Child Health and Human Development, National Institutes of Health, Bethesda, Md., USA, and [b]1st Department of Pediatrics, Athens University Medical School, Athens, Greece

Abstract

Glucocorticoids, the end-products of the hypothalamic-pituitary-adrenal (HPA) axis, influence the functions of virtually all organs and tissues through the nuclear glucocorticoid receptor (GR). Circulating levels of glucocorticoids fluctuate naturally in a circadian fashion under the strong influence of the hypothalamic suprachiasmatic nucleus (SCN) circadian CLOCK system, and regulate the transcriptional activity of the GR in the brain and peripheral target tissues. We recently reported that the basic helix-loop-helix transcription factor Clock, which is a histone acetyltransferase and a central component of the self-oscillating transcription factor loop that generates circadian rhythms, represses GR transcriptional activity by acetylating lysine residues within the 'lysine cluster' located in the hinge region of the receptor. This Clock-mediated repression of GR transcriptional activity oscillates in inverse phase to the HPA axis, acting as a target tissue counter-regulatory mechanism to the diurnally fluctuating circulating glucocorticoids. Interestingly, mild evening elevations of cortisol, as occurs in chronic stress situations, and frequent uncoupling of the SCN CLOCK-directed HPA axis from the daily oscillation of target tissue sensitivity to glucocorticoids, as happens in trans-time zone travel and night shift work, produce functional hypercortisolism and, hence, multiple components of the metabolic syndrome with resultant cardiovascular complications.

Glucocorticoids, steroid hormones secreted by the adrenal gland cortices, have strong and diverse actions on every aspect of human physiology [1]. These hormones act as end-effectors of the hypothalamic-pituitary-adrenal (HPA) axis, which, together with the locus ceruleus/autonomic nervous system, plays a central role in both baseline homeostasis and the adaptive response to stress [2]. Glucocorticoids are necessary for proper functioning of virtually all organs and tissues, including the central nervous system (CNS), and the respiratory, cardiovascular, immune, and musculoskeletal

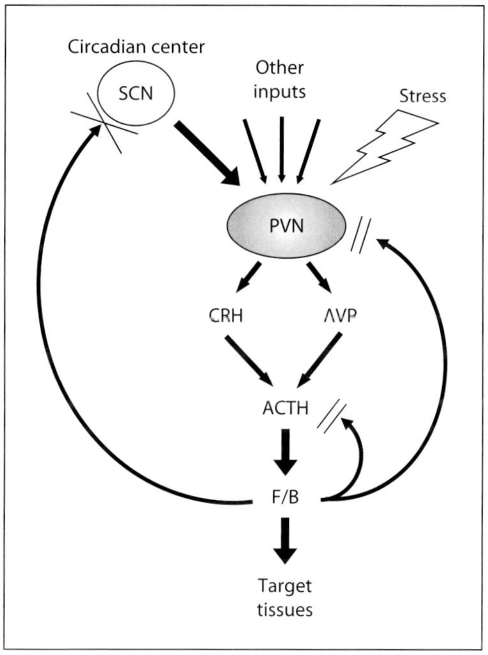

Fig. 1. Organization of the HPA axis. The PVN center of the HPA axis receives circadian inputs from the SCN and stress-related and other regulatory inputs from multiple brain and peripheral sites. CRH and AVP are secreted into the hypophyseal portal system and transported to the anterior pituitary gland where they synergistically stimulate ACTH secretion. ACTH released into the systemic circulation stimulates the *zona fasciculata* of the adrenal cortices to secrete cortisol or corticosterone in humans and rodents, respectively. The 'master' circadian pacemaker of the SCN, thus, produces a circadian rhythm of glucocorticoid concentrations in the blood. Importantly, this site of the brain does not have GR and, hence, is not influenced by circulating glucocorticoids [modified from 16]. ACTH = Adrenocorticotropic hormone, AVP = arginine vasopressin, CRH = corticotropin-releasing hormone, F/B = cortisol/corticosterone, PVN = paraventricular nucleus, SCN = suprachiasmatic nucleus.

systems [1]. Upon exposure to stress, glucocorticoids are secreted into the systemic circulation and alter target tissue activities, dramatically influencing the function of the CNS, shifting intermediary metabolism towards catabolism and modulating immune function [1–3].

The secretion of glucocorticoids is tightly regulated; indeed, the circadian, negative feedback and stress-related activities of the HPA axis are integrated by higher brain centers (fig. 1) [2, 4]. Circulating glucocorticoid levels, i.e. cortisol in humans and corticosterone in rodents, are under the strong circadian influence of the suprachiasmatic nucleus (SCN) of the hypothalamus [4]. In humans, cortisol levels have a diurnal rhythm curve in which the zenith is reached in the early morning and the nadir at midnight, with the purpose of helping adjust the body's activities to the regular periodicity of day/night changes. The overall sensitivity of tissues to glucocorticoids is also regulated by various physiologic and pathologic processes [5, 6]. For example, glucocorticoid action in cells is specifically adjusted during the different phases of the cell cycle [7, 8], while adrenomedullary cells exposed to very high concentrations of glucocorticoids directly diffusing from the adjacent adrenal cortex are resistant to them [9, 10]. In addition, several autoimmune/allergic/inflammatory disorders, the metabolic syndrome, septic conditions and even infection with the human immunodeficiency virus type-1, have been associated with alterations in the responsiveness of specific organs and tissues to glucocorticoids [5, 6, 11]. Underlying mechanisms for the alterations of local glucocorticoid actions, however, have not been well elucidated as yet. In this brief review we summarize recent findings demonstrating circadian regulation of the glucocorticoid signaling system by the biological CLOCK and discuss the potential clinical implications of this regulation.

Glucocorticoid Signaling System

The diverse actions of glucocorticoids at their target tissues are mediated by the glucocorticoid receptor (GR) [2, 12]. This receptor, also called 'nuclear receptor superfamily 3, group C, member 1 (NR3C1)', is expressed virtually in all the organs and tissues of the human body, and belongs to the steroid/sterol/thyroid/retinoid/orphan nuclear receptor superfamily, which consists of over 130 members expressed from the early metazoans to humans [5, 6]. The human GR gene, located in the short arm of chromosome 5 (5q31.3), is composed of 10 exons, and encodes two protein molecules GRα and GRβ through alternative use of specific exon 9 α and β [13].

GRα is the ubiquitously expressed classic receptor that binds to and mediates most of the known actions of glucocorticoids, while GRβ, although also expressed widely, cannot bind glucocorticoids and its physiologic actions have not been fully elucidated as yet [13]. It recently became evident that the GRα variant mRNA is translated from at least 8 initiation sites into multiple amino-terminal GRα isoforms, termed A through D (A, B, C1–C3 and D1–D3), with distinct specific transcriptional activities on glucocorticoid-responsive genes [5, 14]. These GR molecules are also differentially expressed in several different cell lines and tissues [14]. Given that GRα and GRβ share a common mRNA domain that contains the same translation initiation sites [13], the GRβ variant mRNA is also most likely translated from the same initiation sites to a similar host of 8 β isoforms [5].

The human GRα consists of 777 amino acids and has 3 major distinct functional domains, the N-terminal or immunogenic (NTD), the DNA-binding (DBD) and the ligand-binding (LBD) domains (fig. 2a) [15]. GRα has also a hinge region (HD), located between the DBD and LBD and spanning amino acids 481–520, which plays a role in the affinity of the DBD to DNA recognition sequences and in the determination of the numbers of nucleotides spacing tandem response elements [16]. The LBD of GRα consists of 12 α-helices and 4 β-sheets, among which helices 3, 4, 11 and 12 form the glucocorticoid-binding pocket [12]. GRα is located primarily in the cytoplasm in the absence of glucocorticoid ligand, as part of hetero-oligomeric complexes containing heat shock proteins (HSPs) 90, 70, 50, 20 and, possibly, other proteins as well [5, 15]. After binding to its agonist ligand, GRα undergoes conformational changes, dissociates from the HSPs, homo- or heterodimerizes, and translocates as a dimer and/or monomer into the nucleus through the nuclear pore, via an active ATP-dependent process mediated by its nuclear localization signals (NL)-1 and -2 [6, 17]. NL-1 is located in the junction of DBD and HD, while NL-2 spans the entire LBD [17].

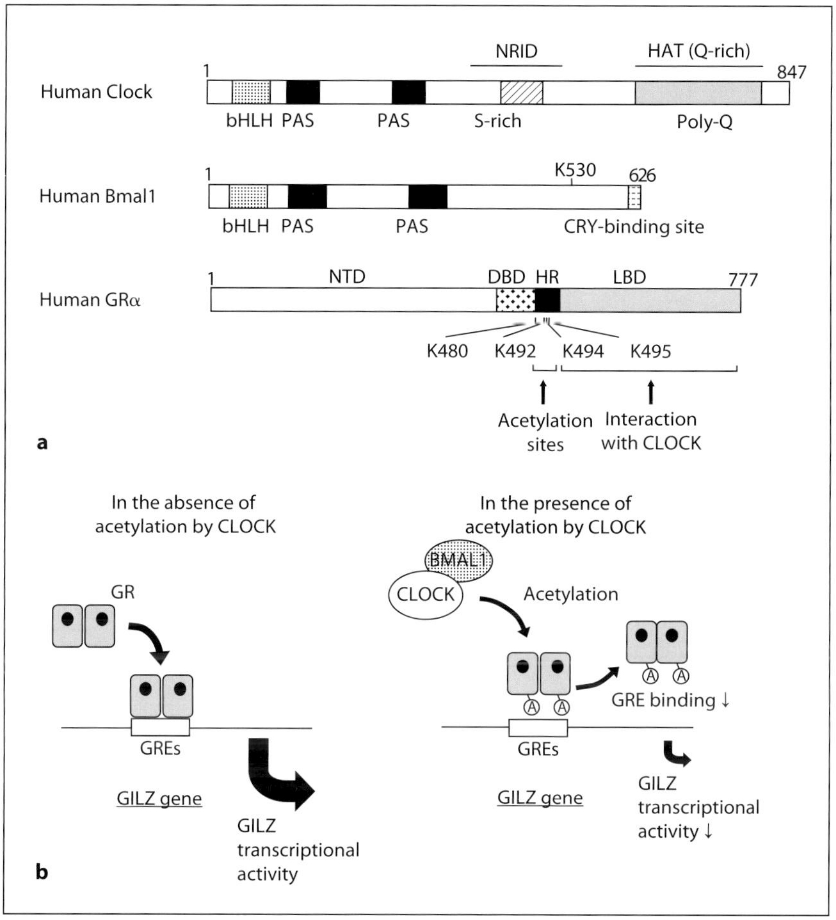

Fig. 2. Linearized structures of the human Clock, Bmal1 and GR molecules and the regulation GR transcriptional activity by Clock. **a** Human Clock and Bmal1 are 847 and 626 amino acid proteins, respectively, that contain one basic helix-loop-helix (bHLH) and two Per-Amt-Sim (PAS) domains in their N-terminal portion. Clock has also one nuclear receptor interacting domain (NRID) in its middle portion and a histone acetyltransferase domain (HAT) in its C-terminal area, while Bmal1 has a CRY-binding site in its C-terminal end. A serine-rich (S-rich) and a glutamine-rich (Q-rich) region, the latter containing a poly-glutamine (poly-Q) stretch, are also shown. The GR has three main functional domains, the amino-terminal, DNA-binding and ligand-binding domains and a lysine cluster in its hinge region. **b** Clock physically interacts with the ligand-binding domain of the GR through the domain enclosed in its C-terminal part and suppresses GR-induced transcriptional activity by acetylating via its intrinsic HAT activity a lysine cluster located in the GR hinge region. Acetylation by Clock reduces the affinity of the GR to its cognate DNA sequences, the GREs. A = Acetylation, Bmal1 = brain-muscle-arnt-like protein 1,DBD = DNA-binding domain, GR = glucocorticoid receptor, GRE = glucocorticoid response element, HR = hinge region, K = lysine residue, LBD = ligand-binding domain, NTD = N-terminal domain [modified from 16, 34, 47].

Inside the nucleus, the ligand-activated GRα directly interacts as a homo- or heterodimer with specific DNA sequences, the glucocorticoid response elements (GREs), in the promoter regions of target genes, or as a monomer or dimer with other transcription factors via protein-protein interactions, indirectly influencing the activity of the latter on their own target genes [5, 6]. GR contains two transactivation domains, activation function (AF)-1 and -2, located at its NTD and LBD, respectively, through which it interacts with many proteins and protein complexes, such as the nuclear receptor coactivator [p160, p300/CREB-binding protein (CBP) and p300/CBP-associated factor (p/CAF)] complexes and the SWI/SNF and vitamin D receptor-interacting protein/thyroid hormone receptor-associated protein (DRIP/TRAP) chromatin-remodeling complexes, eventually influencing the activity of RNA polymerase II and its ancillary factors, ultimately altering the transcription rates of glucocorticoid-responsive genes [5, 15, 18].

GRα also interacts with the nuclear receptor corepressor (NCoR) and its homolog silencing mediator of retinoic acid and thyroid hormone receptor (SMRT), which are macromolecular docking platforms for nuclear receptors and many transcription factors, repressing the transcriptional activity of the GR by attracting histone deacetylase/Sin3 complexes and other repressing agents [18]. The p160 type coactivators and the NCoR/SMRT type corepressors establish equilibrium in their interaction with the GR to respectively facilitate or hinder its transcriptional activity [19].

In addition to transactivation or transrepression of the glucocorticoid-responsive genes explained above, GRα modulates other signal transduction cascades through mutual protein-protein interactions with specific transcription factors, by influencing the ability of these factors to stimulate or inhibit the transcription rates of their respective target genes [1, 5]. This activity may be more important than the GRE-mediated one, granted that mice harboring a mutant GRα, which is active in terms of protein-protein interactions but inactive in terms of dimerization and transactivation via DNA GREs, survive and procreate, in contrast to mice with a deletion of the entire GR gene that die immediately after birth from severe respiratory distress syndrome [20, 21]. The former mouse model and additional in vitro results indicate that GR interacts with and influences other transcription factors primarily as a monomer [20, 22].

The protein-protein interactions of GRα with other transcription factors may take place on promoters that do not contain GREs, as well as on promoters that have both GRE(s) and responsive element(s) of transcription factors that interact with GRα [23]. Suppression of transactivation of other transcription factors through protein-protein interactions may be particularly important in the regulation of immune function and suppression of inflammation by glucocorticoids [20, 22]. A substantial part of the effects of glucocorticoids on the immune system may be explained by the interaction between GRα with nuclear factor-κB (NF-κB), activator protein-1 (AP-1), signal transducers and activators of transcription (STATs) and, probably, interferon regulatory factors [5, 6, 24].

In addition to coregulators and other transcription factors that modulate GR-induced transcriptional activity, several distinct signaling pathways influence the transcriptional activity of the GR via posttranslational modifications of the receptor protein [5]. These include methylation, acetylation, nitrosylation, sumoylation and ubiquitination, as well as phosphorylation, which has been studied best. Indeed, several kinases, such as the cell-cycle-related kinases, mitogen-activated kinases (MAPKs) and the glycogen synthase kinases, phosphorylate specific serine or threonine residues of the GR, while energy-sensing AMP-activated protein kinase indirectly phosphorylates the GR through p38 MAPK [13, 25, 26]. Interestingly, the majority of these residues are located in the AF-1 domain of the human GR NTD, thus, phosphorylation of some or all of these amino acids modulates GR-induced transcriptional activity through alteration of coregulator attraction to the promoter region of glucocorticoid-responsive genes, possibly by changing their affinity and overall interaction with the AF-1 domain of the GR [25, 27].

The Circadian CLOCK and Its Interactions with the Glucocorticoid Signaling System

We recently found that the circadian rhythm transcription factor Clock acetylates the GR at a multiple lysine cluster located in its hinge region, and represses GR-induced transcriptional activity of several glucocorticoid-responsive genes (fig. 2) [16]. Clock, the 'circadian locomotor output cycle kaput', and its heterodimer partner 'brain-muscle-arnt-like protein 1' (Bmal1) belong to the basic helix-loop-helix (bHLH)-PER-ARNT-SIM (PAS) superfamily of transcription factors [28, 29], and play an essential role in the formation of the circadian oscillation rhythm of both hypothalamic SCN and peripheral CLOCK systems that function as internal circadian timekeepers [30]. The CLOCK system located in the SCN of the hypothalamus acts as the 'master' oscillator and generator of the body's circadian rhythm [31, 32], while the peripheral CLOCK system virtually distributed in all organs and tissues, including the CNS outside the SCN, acts generally as a 'slave' CLOCK under the influence of the SCN. Human Clock and Bmal1 are 847 and 626 amino acid proteins, respectively, and contain one bHLH and two PAS domains in their N-terminal portion (fig. 2a). In addition, Clock has one nuclear receptor interacting domain (NRID) in its middle portion and the histone acetyltransferase domain (HAT) in the C-terminal area, while Bmal1 has a CRY-binding site in its C-terminal end [33, 34].

The Clock/Bmal1 heterodimer binds to the E-box response elements located in the promoter region of other essential CLOCK genes, such as *Periods (Per1, Per2* and *Per3)* and *Cryptochromes (Cry1* and *Cry2)* and stimulates their transcription. Accumulated proteins Pers and Crys then form a complex with caseine kinase 1ε and δ, are phosphorylated, translocate into the cell nucleus and repress the transcriptional activity of the Clock/Bmal1 heterodimer by inhibiting its binding to the E-box response elements located in their own promoters, thus ultimately forming a negative feedback transcriptional loop that maintains an oscillation of their gene expression with a period of approximately every 24 h [30, 35, 36]. The HAT activity of Clock is essential for creating diurnal oscillation of this transcription factor loop, possibly by acetylating histones associated with its responsive genes, as well as by acetylating its partner Bmal1, a change that increases the attraction of Cry, facilitating thus the repressive effect of this molecule [34, 37].

In addition to the regulation of this principal transcriptional loop, Clock/Bmal1 stimulates expression of other clock-related proteins, such as Rev-erbα, retinoic acid receptor-related orphan receptor α (RORα), Dec1, Dec2 and albumin gene D site-binding protein (Dbp), which form an auxiliary loop that stabilizes the main regulatory loop composed of Clock/Bmal1, Pers and Crys. Importantly, the transcription factors of the main and auxiliary loops control numerous 'downstream' clock-responsive genes to influence a variety of biological activities [29, 30, 38], so that the circadian CLOCK system can directly adjust body's fundamental activities, such as feeding, thermoregulation, energy expenditure and intermediary metabolism, synchronizing the organism with the day/night changes of the environment linked to the rotation of the planet.

The glucocorticoid signaling system is also a target of the circadian CLOCK system [4]. In addition to the diurnal secretion of circulating glucocorticoids controlled by the master CLOCK, peripheral CLOCK negatively regulates GR-induced transcriptional activity through acetylation of the GR [16]. Indeed, Clock physically interacts with the LBD of the GR through its NRID and acetylates lysine residues located at amino acid positions 480, 492, 494 and 495 of the human GR in its hinge region. The acetylation of GR further suppresses binding of this receptor to GREs, explaining the mechanism of its repressive effect [16] (fig. 2b). It is likely that acetylation-mediated repression of GR transcriptional activity by Clock functions as a local counter-regulatory mechanism to diurnally oscillating circulating glucocorticoids [4]. Since phosphorylation-mediated transcriptional regulation of GR, the most well-examined epigenetic regulation of GR activity, influences GR transcriptional activity in a gene- and tissue-specific fashion [25], it is quite probable that acetylation of the GR also acts gene-, and possibly, tissue-specifically.

Pathophysiologic Implications of the Circadian CLOCK-Mediated GR Signaling System

This presumed gene- and tissue-specific negative regulation of GR transcriptional activity by Clock indicates that diurnal oscillating of gene expression may be necessary for some glucocorticoid-responsive genes to maintain their proper biologic actions. It is known that subjects under chronic stress, who have a flattened cortisol circadian rhythm because of increased hormone levels at night, frequently develop symptoms and signs of chronic hypercortisolism, i.e. components of the metabolic syndrome, such as central obesity, dyslipidemia, insulin resistance and hypertension,

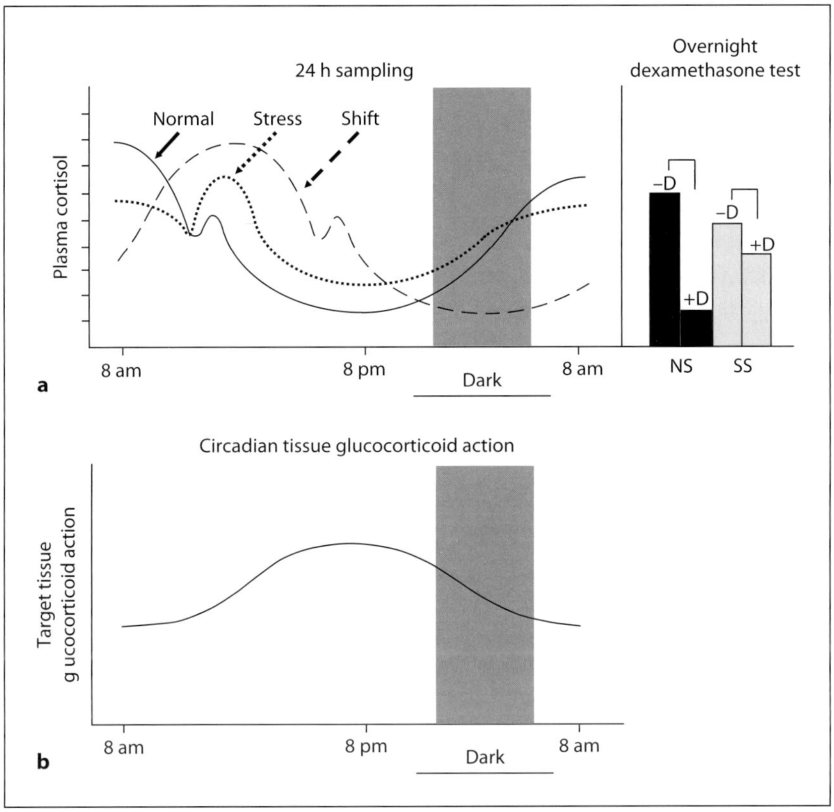

Fig. 3. A heuristic scheme of the circadian secretion of cortisol in normal, chronically stressed and phase-shifted humans (top left panel) and the responses of chronically stressed individuals to midnight dexamethasone administration (top right panel). The corresponding circadian change of target tissue sensitivity to glucocorticoids (bottom panel). It is obvious that even mild evening cortisol elevations, as those seen in chronically stressed individuals, will exert disproportionately increased glucocorticoid effects because of the natural circadian target tissue sensitivity increase at this time of the day. Frequent abrupt phase shifts in circulating cortisol levels due to night-shift work or trans-time zone travel causes uncoupling of circulating cortisol levels and tissue glucocorticoid sensitivity of target tissues, influenced by the peripheral CLOCK, increase the overall glucocorticoid effect and, hence, the risk for metabolic syndrome and associated cardiovascular complications. CS = Chronically stressed individuals, D = midnight dexamethasone administration, NS = non-stressed individuals [modified from 48].

and, hence, are at an increased risk of developing cardiovascular complications [39, 40] (fig. 3). This evidence indicates that proper gene- and tissue-specific diurnal regulation of glucocorticoid hormone action is essential for a healthy life. Interestingly, a fraction (~10%) of the energy-controlling genes are also under circadian regulation in a tissue-specific fashion [41–44].

Uncoupling of the cortisol circadian rhythm, under the influence of the SCN CLOCK system, and the circadian peripheral sensitivity to glucocorticoids, under the influence of the peripheral CLOCK, also leads to development of components of the metabolic syndrome and its cardiovascular complications, as it occurs with chronic hypercortisolism (fig. 3) [4]. For example, rotating shift workers, whose circadian system is repetitively reset by night-time activity/day-time sleep, and persons exposed to frequent jet lag, because of traveling across time zones, are at an increased risk for cardiometabolic disease [45, 46]. Thus, the HPA axis and the SCN and peripheral circadian CLOCK systems are in close cooperation in the regulation of mental and physical activities of human beings, possibly by regulating a substantial number of shared genes.

Concluding Remarks

During the last decade, significant progress has been made in the field of chronobiology and the molecular functions of CLOCK system, which helps adjust the body's daily activities to the predictable regular day/night changes of the environment. The adaptive response to unpredicted, acute changes in the environment, on the other hand, is serviced on demand by the stress system including the HPA axis. Interestingly, the SCN master CLOCK, which creates this diurnal oscillation of circulating glucocorticoids, at the same time moderates their peripheral actions via the peripheral ubiquitous CLOCK, which suppresses GR action by acetylation.

Acknowledgements

Literary work of this article was funded partly by the Intramural Research Program of the Eunice Kennedy Shriver National Institute of Child Health and Human Development, National Institutes of Health, Bethesda, Md., USA and the University of Athens, Athens, Greece.

References

1 Kino T, Chrousos GP: Glucocorticoid effects on gene expression; in Steckler T, Kalin NH, Reul JM (eds): Handbook on Stress and the Brain. Amsterdam, Elsevier, 2005, pp 295–312.
2 Chrousos GP: The hypothalamic-pituitary-adrenal axis and immune-mediated inflammation. N Engl J Med 1995;332:1351–1362.
3 Chrousos GP: Glucocorticoid therapy; in Felig P, Frohman LA (eds): Endocrinology and Metabolism. New York, McGraw-Hill, 2001, pp 609–632.
4 Nader N, Chrousos GP, Kino T: Interactions of the circadian CLOCK system and the HPA axis. Trends Endocrinol Metab 2010;21:277–286.
5 Chrousos GP, Kino T: Intracellular glucocorticoid signaling: a formerly simple system turns stochastic. Sci STKE 2005;304:pe48.
6 Kino T, De Martino MU, Charmandari E, Mirani M, Chrousos GP: Tissue glucocorticoid resistance/hypersensitivity syndromes. J Steroid Biochem Mol Biol 2003;85:457–467.
7 Abel GA, Wochnik GM, Ruegg J, Rouyer A, Holsboer F, Rein T: Activity of the GR in G2 and mitosis. Mol Endocrinol 2001;16:1352–1366.
8 Cidlowski JA, Michaels GA: Alteration in glucocorticoid binding site number during the cell cycle in HeLa cells. Nature 1977;266:643–645.

9 Wurtman RJ: Stress and the adrenocortical control of epinephrine synthesis. Metabolism 2002;51(suppl 1):11–14.
10 Ehrhart-Bornstein M, Bornstein SR: Cross-talk between adrenal medulla and adrenal cortex in stress. Ann NY Acad Sci 2008;1148:112–117.
11 Chrousos GP, Kino T: Glucocorticoid action networks and complex psychiatric and/or somatic disorders. Stress 2007;10:213–219.
12 Nicolaides NC, Galata Z, Kino T, Chrousos GP, Charmandari E: The human glucocorticoid receptor: molecular basis of biologic function. Steroids 2010;75:1–12.
13 Kino T, Su YA, Chrousos GP: Human glucocorticoid receptor isoform β: recent understanding of its potential implications in physiology and pathophysiology. Cell Mol Life Sci 2009;66:3435–3448.
14 Lu NZ, Cidlowski JA: Translational regulatory mechanisms generate N-terminal glucocorticoid receptor isoforms with unique transcriptional target genes. Mol Cell 2005;18:331–342.
15 Kino T, Chrousos GP: Glucocorticoid and mineralocorticoid receptors and associated diseases. Essays Biochem 2004;40:137–155.
16 Nader N, Chrousos GP, Kino T: Circadian rhythm transcription factor CLOCK regulates the transcriptional activity of the glucocorticoid receptor by acetylating its hinge region lysine cluster: potential physiological implications FASEB J 2009;23:1572–1583.
17 Savory JG, Hsu B, Laquian IR, Giffin W, Reich T, Hache RJ, Lefebvre YA: Discrimination between NL1- and NL2-mediated nuclear localization of the glucocorticoid receptor Mol Cell Biol 1999;19:1025–1037.l
18 Rosenfeld MG, Lunyak VV, Glass CK: Sensors and signals: a coactivator/corepressor/epigenetic code for integrating signal-dependent programs of transcriptional response. Genes Dev 2006;20:1405–1428.
19 Wang Q, Blackford JA Jr, Song LN, Huang Y, Cho S, Simons SS Jr: Equilibrium interactions of corepressors and coactivators with agonist and antagonist complexes of glucocorticoid receptors. Mol Endocrinol 2004;18:1376–1395.
20 Reichardt HM, Kaestner KH, Tuckermann J, Kretz O, Wessely O, Bock R, Gass P, Schmid W, Herrlich P, Angel P, Schutz G: DNA binding of the glucocorticoid receptor is not essential for survival. Cell 1998; 93:531–541.
21 Cole TJ, Blendy JA, Monaghan AP, Krieglstein K, Schmid W, Aguzzi A, Fantuzzi G, Hummler E, Unsicker K, Schutz G: Targeted disruption of the glucocorticoid receptor gene blocks adrenergic chromaffin cell development and severely retards lung maturation. Genes Dev 1995;9:1608–1621.
22 Reichardt HM, Tuckermann JP, Gottlicher M, Vujic M, Weih F, Angel P, Herrlich P, Schutz G: Repression of inflammatory responses in the absence of DNA binding by the glucocorticoid receptor. EMBO J 2001;20:7168–7173.
23 Miner JN, Yamamoto KR: Regulatory crosstalk at composite response elements. Trends Biochem Sci 1991;16:423–426.
24 Ogawa S, Lozach J, Benner C, Pascual G, Tangirala RK, Westin S, Hoffmann A, Subramaniam S, David M, Rosenfeld MG, Glass CK: Molecular determinants of crosstalk between nuclear receptors and toll-like receptors. Cell 2005;122:707–721.
25 Kino T, Ichijo T, Amin ND, Kesavapany S, Wang Y, Kim N, Rao S, Player A, Zheng YL, Garabedian MJ, Kawasaki E, Pant HC, Chrousos GP: Cyclin-dependent kinase-5 differentially regulates the transcriptional activity of the glucocorticoid receptor through phosphorylation: clinical implications for the nervous system response to glucocorticoids and stress. Mol Endocrinol 2007;21:552–1568.
26 Nader N, Ng SM, Lambrou GI, Pervanidou P, Wang YH, Chrousos P, Kino T: AMPK regulates metabolic actions of glucocorticoids by phosphorylating the glucocorticoid receptor through p38 MAPK. Mol Endocrinol 2010 (in press).
27 Kino T: Tissue glucocorticoid sensitivity: beyond stochastic regulation on the diverse actions of glucocorticoids. Horm Metab Res 2007;39:420–424.
28 Cermakian N, Sassone-Corsi P Multilevel regulation of the circadian clock. Nat Rev Mol Cell Biol 2000;1:59–67.
29 Ko CH, Takahashi JS: Molecular components of the mammalian circadian clock. Hum Mol Genet 2006; 15:R271–R277.
30 Takahashi JS, Hong HK, Ko CH, McDearmon EL: The genetics of mammalian circadian order and disorder: implications for physiology and disease. Nat Rev Genet 2008;29:764–775.
31 Ralph MR, Foster RG, Davis FC, Menaker M: Transplanted suprachiasmatic nucleus determines circadian period. Science 1990;247.975–978.
32 Rusak B, Zucker I: Neural regulation of circadian rhythms. Physiol Rev 1979;59:449–526.
33 Grimaldi B, Nakahata Y, Sahar S, Kaluzova M, Gauthier D, Pham K, Patel N, Hirayama J, Sassone-Corsi P: Chromatin remodeling and circadian control: master regulator CLOCK is an enzyme. Cold Spring Harb Symp Quant Biol 2007;72:105–112.
34 Hirayama J, Sahar S, Grimaldi B, Tamaru T, Takamatsu K, Nakahata Y, Sassone-Corsi P: CLOCK-mediated acetylation of BMAL1 controls circadian function. Nature 2007;450:1086–1090.

35 Kondratov RV, Kondratova AA, Lee C, Gorbacheva VY, Chernov MV, Antoch MP: Post-translational regulation of circadian transcriptional CLOCK (NPAS2)/BMAL1 complex by cryptochromes. Cell Cycle 2006;5:890–895.

36 Kiyohara YB, Tagao S, Tamanini F, Morita A, Sugisawa Y, Yasuda M, Yamanaka I, Ueda HR, van der Horst GT, Kondo T, Yagita K: The BMAL1 C terminus regulates the circadian transcription feedback loop. Proc Natl Acad Sci USA 2006;103:10074–10079.

37 Doi M, Hirayama J, Sassone-Corsi P: Circadian regulator CLOCK is a histone acetyltransferase. Cell 2006;125:497–508.

38 Ripperger JA, Schibler U: Rhythmic CLOCK-BMAL1 binding to multiple E-box motifs drives circadian Dbp transcription and chromatin transitions. Nat Genet 2006;38:369–374.

39 Chrousos GP, Kino T Glucocorticoid action networks and complex psychiatric and/or somatic disorders, Stress 2007;10:213–219.

40 Chrousos GP, Kino T: Glucocorticoid signaling in the cell. Expanding clinical implications to complex human behavioral and somatic disorders. Ann NY Acad Sci 2009;1179:153–166.

41 Yang X, Downes M, Yu RT, Bookout AL, He W, Straume M, Mangelsdorf DJ, Evans RM: Nuclear receptor expression links the circadian clock to metabolism. Cell 2006;126:801–810.

42 Staels B: When the Clock stops ticking, metabolic syndrome explodes. Nat Med 2006;12:54–55.

43 Panda S, Antoch MP, Miller BH, Su AI, Schook AB, Straume M, Schultz PG, Kay SA, Takahashi JS, Hogenesch JB: Coordinated transcription of key pathways in the mouse by the circadian clock. Cell 2002;109:307–320.

44 Storch KF, Lipan O, Leykin I, Viswanathan N, Davis FC, Wong WH, Weitz CJ: Extensive and divergent circadian gene expression in liver and heart. Nature 2002;417:78–83.

45 Ekstrand K, Bostrom PA, Arborelius M, Nilsson JA, Lindell SE: Cardiovascular risk factors in commercial flight aircrew officers compared with those in the general population. Angiology 1996;47:1089–1094.

46 Scheer FA, Hilton MF, Mantzoros CS, Shea SA: Adverse metabolic and cardiovascular consequences of circadian misalignment. Proc Natl Acad Sci USA 2009;106:4453–4458.

47 Hirayama J, Sahar S, Grimaldi B, Tamaru T, Takamatsu K, Nakahata Y, Sassone-Corsi P: CLOCK-mediated acetylation of BMAL1 controls circadian function. Nature 2007;450:1086–1090.

48 Chrousos GP: The role of stress and the hypothalamic-pituitary-adrenal axis in the pathogenesis of the metabolic syndrome: neuroendocrine and target tissue-related causes. Int J Obes Relat Metab Disord 2000;24(suppl 2):S50–S55.

George P. Chrousos, MD, MACP, MACE, FRCP
1st Department of Pediatrics, Athens University Medical School
Aghia Sohia Children's Hospital, GR–11527 Athens (Greece)
Tel. +30 210 779 4023, Fax +30 210 7759167
E-Mail chrousge@med.uoa.gr

Ghizzoni L, Cappa M, Chrousos G, Loche S, Maghnie M (eds): Pediatric Adrenal Diseases.
Endocr Dev. Basel, Karger, 2011, vol 20, pp 127–136

Glucocorticoid Resistance

E.F.C. van Rossum · E.L.T. van den Akker

Erasmus MC, Rotterdam, The Netherlands

Abstract

Changes in glucocorticoid (GC) receptor sensitivity can be categorized in three different types. First, generalized GC resistance syndrome is a hereditary disease. Patients present with signs and symptoms of increased androgen and/or mineralocorticoid action, combined with biochemical hypercortisolism, but lack of cushingoid features. Second, at a tissue level, transient changes in GC sensitivity are present during disease. Transient changes in GC sensitivity of leukocytes during infectious diseases like sepsis have been found, but also acquired forms or GC resistance occur, in particular in some types of neoplasms, major depression, AIDS, and several autoimmune diseases. Third, at the level of the general population, the diversity in GC sensitivity has a wide interindividual variation which in part can be explained by genetic variation of the GC receptor gene. Several single nucleotide polymorphisms of the gene have been associated with changes in GC sensitivity and its clinical phenotype. In this chapter, four genetic variants are described of which two (rs6198 and rs6189/6190) are associated with a relative GC resistance and two (rs1695 and rs41423247) are associated with a relative GC hypersensitivity.

In clinical practice, cortisol serum concentration is used as a measure of a person's exposure to glucocorticoid (GC) effects. However, the GC effect is also determined by many other factors like fraction bound to cortisol-binding globulin, metabolism and inactivation by 11β-hydroxysteroid dehydrogenase. Furthermore, at a tissue level, GCs exert their effect through the GC receptor. The GC effect at the tissue level is influenced by GC receptor sensitivity.

Changes in GC receptor sensitivity can be categorized in three types: (1) generalized GC resistance; (2) transient tissue-specific changes in GC sensitivity during disease, and (3) differences of GC sensitivity within the normal range of the general population.

Generalized Glucocorticoid Resistance Syndrome

As shown in figure 1a, the production of GCs is under the regulation of the hypothalamus. In response to signals from the central nervous system the secretion of

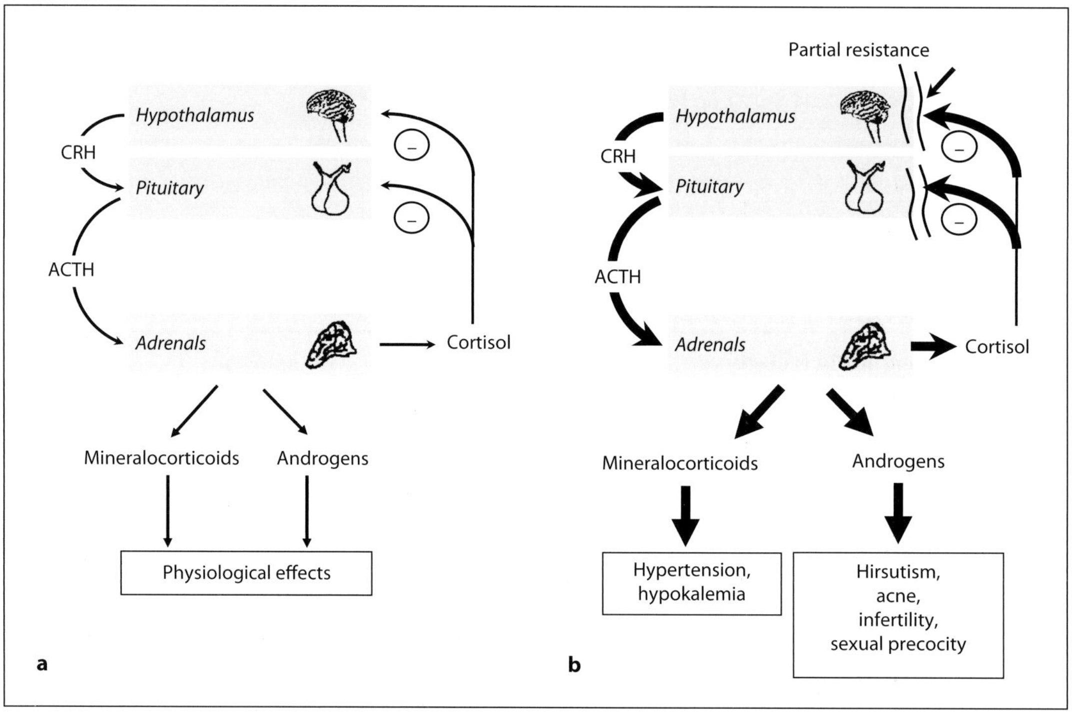

Fig. 1. GCs under hypothalamus regulation: (**a**) normal HPA axis and (**b**) HPA axis in cortisol resistance.

corticotropin-releasing hormone (CRH) and vasopressin is stimulated. These hormones stimulate the pituitary to secrete corticotropin (ACTH) to the circulation, stimulating the adrenal glands to secrete GCs. To control the activity of the HPA axis, negative feedback action by GCs is crucial. If this feedback regulation, mediated by the GC receptor, is disturbed, ACTH levels increase, and the adrenals are stimulated to produce supraphysiological levels of cortisol (fig. 1b). However, due to GC receptor resistance, the effect of cortisol on target tissue is impaired. Since ACTH levels are increased, the adrenal production of mineralocorticoid activity (corticosterone and deoxycorticosterone) and androgens (dehydroepiandrosterone (DHEA), DHEA sulfate and androstenedione) are elevated (fig. 1b) [1, 2].

The first case with GC resistance was reported in 1976 by Vingerhoeds et al. [3]. This patient suffered from biochemical hypercortisolism without clinical signs of Cushing's disease. The ligand-binding affinity of the GC receptor was diminished in this patient, his son and nephew. Later the *GR* gene of the index patient was sequenced and showed a homozygous mutation at position 2054, yielding a valine for aspartic acid substitution at amino acid residue 641 (fig. 2) [4]. The syndrome has an autosomal recessive or dominant mode of inheritance. In the last 20 years, other patients with mutations in the *GR* gene leading to the syndrome of generalized GC resistance

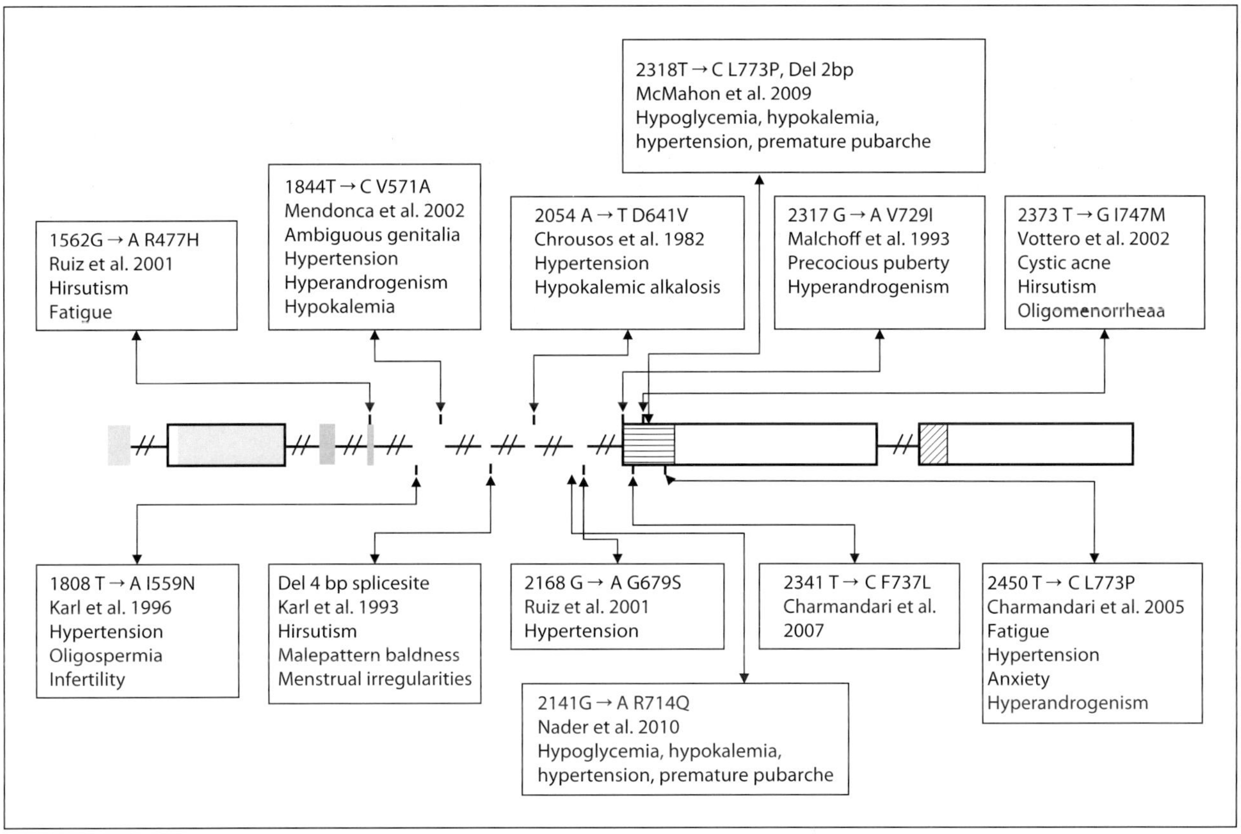

Fig. 2. Mutations in the glucocorticoid receptor gene.

have been described (fig. 2) [1, 5, 6]. All patients were reported to have a *partial* GC resistance. *Total* GC resistance was thought to be incompatible with life. However, in November 2009 an Australian boy was born alive with a homozygous 2-basepair deletion in the ligand-binding domain of the GC receptor [5]. On the first day of life he presented with hypoglycemia, and in the following course with hypertension and fatigue on feeding. He had elevated cortisol levels with no suppression after a massive dose of 12 mg of dexamethasone (DEX). No suppression of lymphocyte interleukin (IL)-6 on DEX was found. Luciferase reporter plasmid transfection assay showed no transactivation activity by DEX. After recovery of a severe sepsis he was reported to be alive at 5 months of age.

Clinical Features

Familial generalized GC resistance is a rare disease but its incidence might be underestimated due to non-specific signs and symptoms. The main characteristics are hypercortisolism without signs of Cushing's syndrome, hypertension, hypokalemia

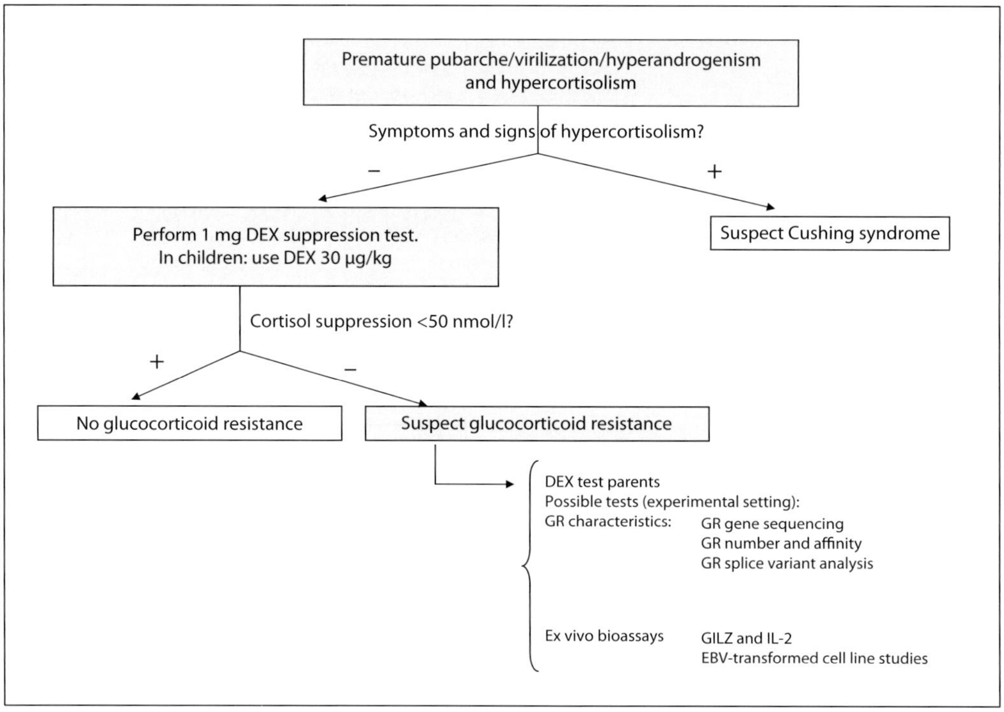

Fig. 3. Diagnostic pathway for patients with suspected cortisol resistance.

(mineralocorticoid excess) and with hirsutism, acne, male pattern baldness (androgen excess) [1, 2]. The clinical presentation in children can be a premature pubarche, increased growth velocity, and advanced bone age as a result of increased androgens. The 1-mg DEX suppression test is used as a diagnostic tool. For children, the DEX dose should be adjusted to their lower body surface compared to adults. Further characterization of the molecular mechanism can be performed in an experimental setting (fig. 3).

Pathogenesis

The effects of cortisol on target genes are the result of a cascade of events. First, the ligand passively diffuses through the cell membrane and binds to the GR. This receptor is present in the cytoplasm of virtually all human cells. In its unbound form the GR is surrounded by chaperone proteins, which keep it inactivated. When the ligand binds to the receptor, its conformation changes, the heat-shock proteins dissociate from the receptor, and the GR translocates to the nucleus [7]. Within the nucleus there are roughly two major modes of GR action. First, GR can affect transcription initiation through binding to positive or negative GC-responsive elements of the target gene, resulting in respectively stimulation or inhibition of transcription. The GR can also act as a transcription factor and indirectly stimulate or inhibit genes by co-factor binding.

In generalized GC resistance, a number of *GR* gene mutations have been reported [1, 5, 6]. These are predominantly located in the ligand-binding domain, but some have been identified in the DNA-binding domain [8]. These mutations lead to a variety of alterations in the GR signaling pathway, e.g. decreased transactivating or transrepressional capacity [9], disturbances in ligand binding [10], decreased GR expression [11], a delay in translocation to the nucleus, changes in interaction with co-activators, alternative splicing or a combination of these changes in GR functioning [9, 11–18]. In some cases, no mutations in the *GR* gene were found and the mechanism leading to the GC resistance is unknown [19, 20]. Several mechanisms leading to GC resistance, suggested in the literature, are: altered phosphorylation status, hormone-induced conformation changes of the GR and nuclear transformation, thermolability of the GR, and enhanced expression of 90-kDa heat-shock protein (hsp90), a chaperone protein [7].

Clinical Investigations and Diagnostic Criteria

What possibilities for diagnostic tests do we have? Figure 3 shows a practical scheme showing clinical and biochemical tests, which can be used for the diagnosis of GC resistance. Plasma ACTH and serum cortisol concentrations are increased. In contrast to Cushing's syndrome, the diurnal rhythm is maintained, although cortisol levels are on average higher. Important for the diagnosis is insufficient suppression of cortisol (normal <50 nmol/l) after a 1-mg overnight DEX suppression test. Urinary cortisol excretion is increased. Serum concentrations of adrenal androgens (dehydroepiandrosterone (DHEA), DHEA sulfate, and androstenedione) and of ACTH-dependent mineralocorticoids (deoxycorticosterone and corticosterone) are also increased. Imaging may show slightly enlarged adrenal glands.

The standard tests used to evaluate hypercortisolism do not discriminate between GC resistance and Cushing's syndrome. A simple test is measurement of bone mineral density (BMD), which is normal or even increased in patients with GC resistance. In contrast, in patients with Cushing's disease, BMD is usually decreased. The increased BMD in GC resistance may be explained by a combination of diminished cortisol effects on bone due to the defect GR, and increased adrenal androgen production, as a result of elevated ACTH levels. No routine diagnostic laboratory test is available. Present tests are labor-intensive and only performed in research laboratories [12]. A fast alternative way to make the diagnosis of hereditary generalized GC resistance more likely is to perform 1-mg DEX suppression tests in family members of the patient [2].

Treatment of Glucocorticoid Resistance

Morning ACTH levels can be suppressed by a low dose of DEX taken around midnight. This leads to a reduction of adrenal overproduction of mineralocorticoids and androgens. Essential in the treatment is that it should be adjusted to the individual signs and symptoms of the patient. Titration to a dose of DEX, which normalizes androgens, blood pressure and serum potassium, seems the optimal therapy. After normalization of mineralocorticoids and androgens the DEX dose to maintain this suppression can be carefully titrated down. To minimize the risks of an effect of DEX in addition to the remaining endogenous cortisol, it is important to slowly decrease the dose of DEX as low as possible. A yearly measurement of BMD is recommended to monitor effects of too high doses of DEX in combination with the endogenous cortisol production. In general, treatment in adults can be started with a dose of about 1 mg DEX at night, in children with a dose of 0.25–0.5 mg/day. This dose can be slowly reduced to a level resulting in normal adrenal hormone levels. To treat hypertension, aldosterone antagonists are recommended, since these also have other effects. In particular, their potassium-sparing and anti-androgenic effects are beneficial for GC-resistant patients. Thiazide or loop diuretics should not be used to control blood pressure because of their potassium-losing effects.

Transient Tissue-Specific Changes in GC Sensitivity

Little is known about transient GC resistance in humans as a result of disease. Disease-related transient changes in GC resistance are tissue-specific which makes them extremely difficult to study in humans. Some in vitro studies reported transient resistance in human leukocytes in response to cytokines [21]. Transient changes in GC sensitivity of leukocytes during infectious diseases like sepsis have been found [22, 23], but also acquired forms of GC resistance occur, in particular in some types of neoplasms. Examples are pituitary tumors (Nelson's syndrome/Cushing's disease), ectopic ACTH-producing tumors, as well as hematological malignancies. In a wide variety of other diseases, both local or systemic, and temporary or chronic forms of GC resistance have been shown, e.g. major depression, AIDS, and several autoimmune diseases [24]. Also in asthma, GC resistance is a well-known clinical problem. Potential mechanisms involved in steroid-resistant asthma are e.g. increased expression of the dominant negative GRβ splice variant, local GC resistance by diminished binding affinity in inflammatory cells induced by certain cytokines (e.g. IL-2, IL-4, IL-13) [25]. Also, impaired nuclear localization and leukemia inhibitory factor, a cytokine decreasing GR expression in animal studies, may both be contributing factors in the development of transient GC resistance.

General Population

In the general population, the diversity in GC sensitivity has a wide interindividual variation but is quite stable within one individual [26]. Part of the interindividual variation can be explained by genetic variation of the GC receptor gene. Several single nucleotide polymorphisms of the gene have been associated with changes in GC sensitivity [27].

The A3669G (rs6198) polymorphism is present in ~35% of persons in the general population. This polymorphism is located at the 3′ UTR end of exon 9β and results in increased stability of GRβ mRNA and increased GRβ protein expression. Increased GRβ expression is thought to have a negative effect on GRα transcriptional activity, resulting in GC insensitivity. This polymorphism is associated with decreased obesity in women [28], decreased total cholesterol levels with increased high-density lipoprotein in men [28] and symptoms of binge eating [29]. This polymorphism is present in a haplotype of the *GR* gene with and without ER22/23EK. When analyzed in a haplotype with only A3669G carriagc (without ER22/23EK), it is associated with mild GC resistance with respect to the transrepressive effects of GCs [30], which are important for immunosuppression. It is associated with an increased risk of autoimmune disease rheumatoid arthritis [31] and a decreased risk of microbial colonization of the nose [32]. Both may be explained by a more active immune system due to this partial GC resistance. The presence of a more active immune status in persons carrying the GR-9β haplotype is further supported by findings of chronic pro-inflammatory status such as elevated levels of IL-6 and C-reactive protein levels measured with high sensitivity (hs-CRP) in a general elderly population [33]. The hs-CRP assay measures CRP levels in the reference range, and it should be noted that although the observed elevation is substantial and significant, it is within the normal range but may lead to effects in the long term. Homozygous carriers for the GR-9β haplotype had an almost three times increased risk of cardiovascular disease (CVD) after 11 years of follow-up in an elderly general population [33]. This GC receptor gene haplotype seems to take many years to sort an effect in the life-long process of atherosclerosis. Cortisol excess and stress are known risk factors for ischemic heart disease, however the exact mechanism of cortisol influencing CVD risk is unknown. These findings suggest that one of the multiple ways that genetically determined cortisol sensitivity is involved in the pathogenesis of CVD is through inflammation. In the field of atherosclerosis and CVD, this observation is important for understanding the mechanism and for identification of a subgroup at risk.

The ER22/23EK polymorphism (rs6189 and rs6190) is present in ~7% of persons. ER22/23EK carriage is associated with mild GC resistance, a healthier metabolic profile with lower total cholesterol, a beneficial body composition with a higher amount of muscle mass and less fat mass, a reduced tendency to develop insulin resistance, and a lower risk of type 2 diabetes mellitus and CVD [34, 35]. In addition, this variant has been related to a decreased risk of dementia [36] and longevity [37].

The mechanism concerning the observed relative GC resistance in ER22/23EK polymorphism carriers has been studied at the molecular level. The ER22/23EK polymorphism had a higher expression of the GR translational variant GR-A [38]. Previously, Yudt and Cidlowski [39] reported that the GC receptor has several translational variants. The first variant, GR-A, results from translation of the GR mRNA from the first AUG codon. The second variant, GR-B, results from translation of the GR mRNA from the second AUG codon. The GR-B protein is transcriptionally more active than the GR-A protein. Higher expression of GR-A was found in ER22/23EK polymorphism carriers leading to decreased expression of the more transcriptionally active GR-B protein, possibly by altering the secondary structure at the mRNA level which probably results in ribosomal leakage and yields more of the GR-A translation variant. The higher intracellular concentrations of the less active variant of the GR may represent the underlying mechanism of the relative GC resistance found in ER22/23EK polymorphism carriers [38].

Polymorphism N363S (rs1695) is present in ~8% of persons. N363S carriage is associated with mild GC hypersensitivity yielding increased body mass index, increased cholesterol levels [35, 40], coronary artery disease [41, 42] and decreased BMD [35]. The Bcl1 polymorphism (rs41423247) is present in ~45% of persons. Bcl1 carriage is associated with mild GC hypersensitivity resulting in more body fat, less lean mass, increased cholesterol [35], increased insulin resistance [43, 44], as well as lower risk on autoimmune disease: Graves' ophthalmopathy [45].

Conclusion

Changes in GC receptor sensitivity can be categorized in three different types. First, generalized GC resistance syndrome is a rare disease, and its incidence might be underestimated at present because it is not be easy for clinicians to diagnose. Hallmarks in patients that warrant further investigation for this diagnosis are signs and symptoms of increased androgen and/or mineralocorticoid action, combined with biochemical hypercortisolism, but lack of cushingoid features. Second, at a tissue level, transient changes in GC sensitivity are present during disease. Future development of tissue-specific treatment is needed to address this. Third, at the level of the general population, differences in GC sensitivity within the normal range are present. Future development of diagnostic tests is needed to identify individual GC sensitivity. The ability to identify individual GC sensitivity can play a role in risk prediction for diseases (like CVD, metabolic syndrome and autoimmune diseases), and it can play a role in predicting response to GC treatment (GC resistance or side effects).

References

1 Charmandari E, Kino T, Ichijo T, Chrousos GP: Generalized glucocorticoid resistance: clinical aspects, molecular mechanisms, and implications of a rare genetic disorder. J Clin Endocrinol Metab 2008;93:1563–1572.

2 Van Rossum EF, Lamberts SW: Glucocorticoid resistance syndrome: a diagnostic and therapeutic approach. Best Pract Res Clin Endocrinol Metab 2006;20:611–626.

3 Vingerhoeds AC, Thijssen JH, Schwarz F: Spontaneous hypercortisolism without Cushing's syndrome. J Clin Endocrinol Metab 1976;43:1128–1133.

4 Hurley DM, Accili D, Stratakis CA, et al: Point mutation causing a single amino acid substitution in the hormone binding domain of the glucocorticoid receptor in familial glucocorticoid resistance. J Clin Invest 1991;87:680–686.

5 McMahon SK, Pretorius CJ, Ungerer JP, et al: Neonatal complete generalized glucocorticoid resistance and growth hormone deficiency caused by a novel homozygous mutation in helix 12 of the ligand-binding domain of the glucocorticoid receptor gene (NR3C1). J Clin Endocrinol Metab 2010;95: 297–302.
6 Nader N, Bachrach BE, Hurt DE, et al: A novel point mutation in helix 10 of the human glucocorticoid receptor causes generalized glucocorticoid resistance by disrupting the structure of the ligand-binding domain. J Clin Endocrinol Metab 2010;95: 2281–2285.
7 Gross KL, Lu NZ, Cidlowski JA: Molecular mechanisms regulating glucocorticoid sensitivity and resistance. Mol Cell Endocrinol 2009;300:7–16.
8 Charmandari E, Kino T, Souvatzoglou E, et al: Natural glucocorticoid receptor mutants causing generalized glucocorticoid resistance: molecular genotype, genetic transmission, and clinical phenotype. J Clin Endocrinol Metab 2004;89:1939–1949.
9 Charmandari E, Kino T, Ichijo T, et al: Functional characterization of the natural human glucocorticoid receptor (hGR) mutants hGRαR477H and hGRαG679S associated with generalized glucocorticoid resistance. J Clin Endocrinol Metab 2006;91: 1535–1543.
10 Hurley DM, Accili D, Stratakis CA, et al: Point mutation causing a single amino acid substitution in the hormone-binding domain of the glucocorticoid receptor in familial glucocorticoid resistance. J Clin Invest 1991;87:680–686.
11 Karl M, Lamberts SW, Koper JW, et al: Cushing's disease preceded by generalized glucocorticoid resistance: clinical consequences of a novel, dominant-negative glucocorticoid receptor mutation. Proc Assoc Am Physicians 1996;108:296–307.
12 Russcher H, Smit P, van Rossum EF, et al: Strategies for the characterization of disorders in cortisol sensitivity. J Clin Endocrinol Metab 2006;91:694–701.
13 Karl M, Lamberts SW, Detera-Wadleigh SD, et al: Familial glucocorticoid resistance caused by a splice site deletion in the human glucocorticoid receptor gene. J Clin Endocrinol Metab 1993;76:683–689.
14 Ruiz M, Lind U, Gåfvels M, et al: Characterization of two novel mutations in the glucocorticoid receptor gene in patients with primary cortisol resistance. Clin Endocrinol (Oxf) 2001;55:363–371.
15 Brufsky AM, Malchoff DM, Javier EC, et al: A glucocorticoid receptor mutation in a subject with primary cortisol resistance. Trans Assoc Am Physicians 1990;103:53–63.
16 Malchoff DM, Brufsky A, Reardon G, et al: A mutation of the glucocorticoid receptor in primary cortisol resistance. J Clin Invest 1993;91:1918–1925.
17 Vottero A, Kino T, Combe H, et al: A novel, C-terminal dominant negative mutation of the GR causes familial glucocorticoid resistance through abnormal interactions with p160 steroid receptor coactivators. J Clin Endocrinol Metab 2002;87:2658–2667.
18 Charmandari E, Raji A, Kino T, et al: A novel point mutation in the ligand-binding domain (LBD) of the human glucocorticoid receptor (hGR) causing generalized glucocorticoid resistance: the importance of the C-terminus of hGR LBD in conferring transactivational activity. J Clin Endocrinol Metab 2005;90:3696–3705.
19 Bronnegard M, Werner S, Gustafsson JA: Primary cortisol resistance associated with a thermolabile glucocorticoid receptor in a patient with fatigue as the only symptom. J Clin Invest 1986;78:1270–1278.
20 Huizenga NA, de Lange P, Koper JW, et al: Five patients with biochemical and/or clinical generalized glucocorticoid resistance without alterations in the glucocorticoid receptor gene. J Clin Endocrinol Metab 2000;85:2076–2081.
21 Strickland I, Kisich K, Hauk PJ, et al: High constitutive glucocorticoid receptor β in human neutrophils enables them to reduce their spontaneous rate of cell death in response to corticosteroids. J Exp Med 2001;193:585–593.
22 Molijn GJ, Koper JW, van Uffelen CJ, et al: Temperature-induced down-regulation of the glucocorticoid receptor in peripheral blood mononuclear leucocyte in patients with sepsis or septic shock. Clin Endocrinol (Oxf) 1995;43:197–203.
23 Van den Akker EL, Koper JW, Joosten K, et al: Glucocorticoid receptor mRNA levels are selectively decreased in neutrophils of children with sepsis. Intensive Care Med 2009;35:1247–1254.
24 Barnes PJ, Adcock IM: Glucocorticoid resistance in inflammatory diseases. Lancet 2009;373:1905–1917.
25 Hamid QA, Wenzel SE, Hauk PJ, et al: Increased glucocorticoid receptor β in airway cells of glucocorticoid-insensitive asthma. Am J Respir Crit Care Med 1999;159:1600–1604.
26 Huizenga NA, Koper JW, de Lange P, et al: Interperson variability but intraperson stability of baseline plasma cortisol concentrations, and its relation to feedback sensitivity of the hypothalamo-pituitary-adrenal axis to a low dose of dexamethasone in elderly individuals. J Clin Endocrinol Metab 1998;83:47–54.
27 Manenschijn L, van den Akker EL, Lamberts SW, van Rossum EF: Clinical features associated with glucocorticoid receptor polymorphisms. An overview. Ann NY Acad Sci 2009;1179:179–198.

28 Syed AA, Irving JA, Redfern CP, et al: Association of glucocorticoid receptor polymorphism A3669G in exon 9β with reduced central adiposity in women. Obesity (Silver Spring) 2006;14:759–764.
29 Cellini E, Castellini G, Ricca V, et al: Glucocorticoid receptor gene polymorphisms in Italian patients with eating disorders and obesity. Psychiatr Genet 2010 (in press).
30 Van den Akker EL, Russcher H, van Rossum EF, et al: Glucocorticoid receptor polymorphism affects transrepression but not transactivation. J Clin Endocrinol Metab 2006;91:2800–2803.
31 Derijk RH, Schaaf MJ, Turner G, et al: A human glucocorticoid receptor gene variant that increases the stability of the glucocorticoid receptor β-isoform mRNA is associated with rheumatoid arthritis. J Rheumatol 2001;28:2383–2388.
32 Van den Akker EL, Nouwen JL, Melles DC, et al: *Staphylococcus aureus* nasal carriage is associated with glucocorticoid receptor gene polymorphisms. J Infect Dis 2006;194:814–818.
33 Van den Akker EL, Koper JW, van Rossum EF, et al: Glucocorticoid receptor gene and risk of cardiovascular disease. Arch Intern Med 2008;168:33–39.
34 Finken MJ, Meulenbelt I, Dekker FW, et al, Dutch POPS-19 Collaborative Study Group: The 23K variant of the R23K polymorphism in the glucocorticoid receptor gene protects against postnatal growth failure and insulin resistance after preterm birth. J Clin Endocrinol Metab 2007;92:4777–4782.
35 Van Rossum EF, Lamberts SW: Polymorphisms in the glucocorticoid receptor gene and their associations with metabolic parameters and body composition. Recent Prog Horm Res 2004;59:333–357.
36 Van Rossum EF, de Jong FJ, Koper JW, et al: Glucocorticoid receptor variant and risk of dementia and white matter lesions. Neurobiol Aging 2008;29:716–723.
37 Van Rossum EF, Feelders RA, van den Beld AW, et al: Association of the ER22/23EK polymorphism in the glucocorticoid receptor gene with survival and C-reactive protein levels in elderly men. Am J Med 2004;117:158–162.
38 Russcher H, van Rossum EF, de Jong FH, et al: Increased expression of the glucocorticoid receptor-A translational isoform as a result of the ER22/23EK polymorphism. Mol Endocrinol 2005;19:1687–1696.
39 Yudt MR, Cidlowski JA: Molecular identification and characterization of a and b forms of the glucocorticoid receptor. Mol Endocrinol 2001;15:1093–1103.
40 Di Blasio AM, van Rossum EF, Maestrini S, et al: The relation between two polymorphisms in the glucocorticoid receptor gene and body mass index, blood pressure and cholesterol in obese patients. Clin Endocrinol (Oxf) 2003;59:68–74.
41 Lin RC, Wang XL, Morris BJ: Association of coronary artery disease with glucocorticoid receptor N363S variant. Hypertension 2003;41:404–407.
42 Rosmond R: Glucocorticoid receptor gene and coronary artery disease: right idea, wrong gene variant? Hypertension 2003;42:e3–e4.
43 Syed AA, Halpin CG, Irving JA, et al: A common intron-2 polymorphism of the glucocorticoid receptor gene is associated with insulin resistance in men. Clin Endocrinol (Oxf) 2008;68:879–884.
44 Rosmond R, Holm G: A 5-year follow-up study of 3 polymorphisms in the human glucocorticoid receptor gene in relation to obesity, hypertension, and diabetes. J Cardiometab Syndr 2008;3:132–135.
45 Boyle B, Korányi K, Patocs A, et al: Polymorphisms of the glucocorticoid receptor gene in Graves' ophthalmopathy. Br J Ophthalmol 2008;92:131–134.

Erica L.T. van den Akker, MD, PhD
Erasmus MC-Sophia, Pediatric Endocrinology
Office sp 1536, Dr. Molewaterplein 60
NL–3015 GJ Rotterdam (The Netherlands)
Tel. +31 10 703 6779, Fax +31 10 703 6811, E-Mail e.l.t.vandenakker@erasmusmc.nl

Ghizzoni L, Cappa M, Chrousos G, Loche S, Maghnie M (eds): Pediatric Adrenal Diseases.
Endocr Dev. Basel, Karger, 2011, vol 20, pp 137–148

Mineralocorticoid Receptor Gene Variants as Determinants of HPA Axis Regulation and Behavior

Roel H. DeRijk[a] · E. Ron de Kloet[a] · Frans G. Zitman[b] · Nienke van Leeuwen[a]

[a]Department of Medical Pharmacology, Leiden Amsterdam Center for Drug Research, Leiden University, and [b]Department of Psychiatry, Leiden University Medical Center, Leiden, The Netherlands

Abstract

We tested if common mineralocorticoid receptor (MR) gene variants contribute to the variability in neuroendocrine control and behavioral reactivity as observed in humans. For that purpose we screened for genetic variability and tested functionality of the identified human MR gene variants in vitro. Four haplotypes were tested for transactivational capacity in vitro and showed profound significant differences when stimulated with cortisol. The MR gene variants were associated with basal levels of cortisol, cortisol levels after dexamethasone administration and with stress-induced hypothalamic-pituitary-adrenal axis and autonomic reactivity. In an elderly cohort, one of the functional MR gene variants, MR-I180V, associated with higher feelings of depression. Moreover, we found an association with neuroticism in a second cohort consisting of depressed patients. In conclusion, we report here new findings on common functional human MR gene variants which reveal a hitherto unknown role of these variants in neuroticism conferring vulnerability to stress-related mental disorders, such as depression and posttraumatic stress syndrome.

Adaptive coping with internal and external challenges depends on optimal regulation of the stress system [1]. Initially, the neuropeptide corticotropin-releasing hormone (CRH) organizes the stress response by activating autonomic output and the hypothalamic-pituitary-adrenal (HPA) axis together with immediate behavioral responses (fig. 1). Secondly, a later response involves regulation by corticosteroids of those activated central systems including neuroendocrine reactivity and behavioral responses. This regulation exerted by the corticosteroids controls the initial behavioral response as well as the management of later adaptive phases.

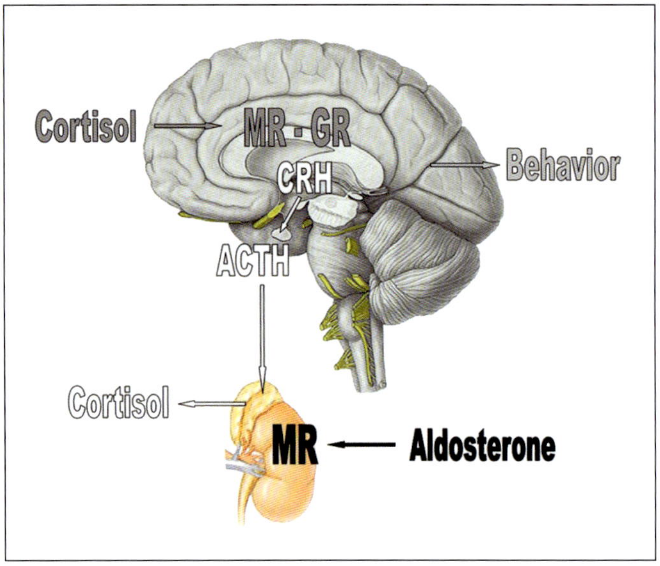

Fig. 1. Hypothalamic-pituitary adrenal (HPA) axis. Corticotropin-releasing hormone (CRH), as released from the paraventricular nucleus (PVN) in the hypothalamic area, stimulates the release of adrenocorticotrophin hormone (ACTH) from the anterior pituitary gland. ACTH stimulates the production and release of corticosteroids from the adrenal cortex, in humans predominantly cortisol. Corticosteroids exert feedback on the regulation of the HPA axis at several levels, including the anterior pituitary corticotrophs and hypothalamus. The hormone also feeds back on limbic structures such as the hippocampus and amygdale affecting emotional responses and learning and memory performance with indirect consequences for HPA axis function. Both high-affinity mineralocorticoid receptors (MR) and the low-affinity glucocorticoid receptors (GR) are central in corticosteroid-mediated feedback on the HPA axis, emotional responses and cognitive processes. These effects of cortisol mediated by both the MR and GR proceed in a context-dependent manner. In addition to the binding of cortisol and 100-fold less aldosterone to MR in the brain, in the kidney cortisol is enzymatically degraded and accordingly the human MR binds predominantly aldosterone, thereby regulating electrolyte balance.

In the periphery, corticosteroids affect a manifold of systems including metabolism, the cardiovascular system and the immune system. Importantly, corticosteroids restrain potentially harmful responses, this action mediated by the relative low affinity glucocorticoid receptor (GR) responding to stress levels of the hormone. Clearly, these actions of corticosteroids are protective, helping the organism to cope with a challenge.

In the brain, corticosteroids act on limbic circuits where they coordinate cognitive and emotional components of fear-motivated behavior. Here two different corticosteroid receptor types mediate the effects of the corticosteroids. The mineralocorticoid receptor (MR), with high affinity for corticosteroids, is involved in appraisal of novel situations and selection of appropriate coping responses. Moreover, via the MR, in concert with other stress mediators, the initial stress reaction is enhanced. Via GR the hormone facilitates the storage of the experience in the memory in preparation of future encounters.

Corticosteroids also feedback on the HPA axis itself, regulating synthesis and release of its hormones. The HPA axis is activated via hypothalamic CRH and arginine vasopressin, inducing the release of pituitary adrenocorticotropin-releasing hormone (ACTH) which initiates the production and release of glucocorticoids from the zona fasciculata from the adrenal cortex. At the level of the pituitary, the GR mediates negative feedback shutting down stress-induced ACTH release. However, feedback on limbic brain regions involved in HPA axis regulation and behavioral responses is far more complex.

In consideration of the ubiquitous effects of corticosteroids, especially during challenges when systems have to perform under strain, optimal HPA axis regulation is crucial. Therefore, an improper HPA axis regulation, both hyper- as well as hypoactivity, is considered a vulnerability factor predisposing for disease.

Corticosteroid Receptors

Most cells express GRs while in the brain, particularly high levels are observed in the hippocampus, a structure which is part of the limbic system. In the periphery, MRs are restricted to epithelial cells in sweat glands, distal colon, kidney and salivary glands mediating aldosterone effects on the regulation of salt homeostasis [2]. In addition, brain MRs are found in peri- and circumventricular structures involved in maintenance of electrolyte balance such as the organum vasculosum lamina terminalis, the medial amygdala and other periventricular brain regions. Moreover, very high levels of MRs are found in the neurons of the hippocampus and other limbic structures, where this receptor is not selective for aldosterone, but also displays high affinity for other steroids such as cortisol and corticosterone as well as deoxycortisol, aldosterone and progesterone. The MR is actually a promiscuous receptor if not protected by 11β-steroid dehydrogenase type 2, which converts cortisol to the inactive cortisone [3].

The classical model for corticosteroid function is based on translocation of the activated receptor from the cytoplasm to the nucleus where it binds responsive elements in regulatory regions of target genes [4]. General transcription factors such as NF-κB and AP-1 interact with the GRs (but probably not with MRs) resulting in mutual modulations. Besides the classical direct genomic actions of corticosteroid receptors, recently fast non-genomic actions of corticosterone on glutamate release and excitatory neurotransmission in brain have been discovered that are likely mediated by membrane variants of both GRs and MRs. These putative membrane MRs and GRs appear identical to their nuclear congener [5, 6]. Importantly, at least in the hippocampal CA1 cell field MR membrane receptors seem to have a lower affinity/efficacy suggesting that such receptors require stress levels of corticosteroids.

Genetic Association Studies

Extensive data from animal models indicate an important role of MRs in neuroendocrine control and behavior [7]. We aimed to determine the role for MRs in normal health and disease in humans. For that purpose we made use of common genetic variability in the MR gene. We hypothesized that natural common genetic variability would confer a risk or protective factor (vulnerability or resilience, respectively) for stress-related disorders. Associations with genetic variants would indicate: (1) involvement of the MR in the tested process, (2) individual vulnerability (or resilience) for a certain disease, and (3) MRs as a new drug target.

Genome-wide association (GWAS) studies are increasingly being employed to test involvement of a gene. Basically, genetic variability (e.g. single nucleotide polymorphism (SNPs), deletions/insertions or copy number variants) are tested for association with a (clinical) phenotype, in an unbiased open approach. However, GWAS have several limitations. First, it is often unclear if the 'genetic tag' is functional, capable of changing the tested phenotype. Second, GWAS requires large cohorts to increase power. Unfortunately, large cohorts almost inevitably suffer from lack of essential data concerning the phenotype. This includes lack of detail of early trauma, drug use or medical history, in addition to a very heterogeneous patient population [8]. Moreover, individual additional genetic variability is not taken into account, while this could have large effects. Finally, extensive gene-environment (G × E) interaction exists which may conceal true associations by inducing additional variability [8].

In contrast to the open GWS approach, which can lead to the identification of new previously not anticipated genes, a candidate gene approach is applied to test involvement of a gene of choice. Based on extensive in vitro animal and human research, functional genetic variants of one's favorite gene are tested in large association or experimental cohort studies. Because this is not an open approach, an a priori hypothesis can be tested using specific test designs. Preferentially, in these studies, focus and much detail is put on those specific functions in which the factor of interest is thought to play a role. This focused and precise testing will greatly increase the power to detect effects.

We focus on the human MR and aim to show involvement in neuroendocrine regulation and behavior, with consequences for health. We describe new human MR gene variants, show functionality in vitro and present data on neuroendocrine regulation. In addition, associations are presented with neuroticism, a behavioral trait predisposing for depression and anxiety disorders.

MR Gene Variants

The gene for the GR is located on chromosome 5q31-32, and the MR on chromosome 4q31.1. The gene structure of both receptors is comparable, with the coding region of

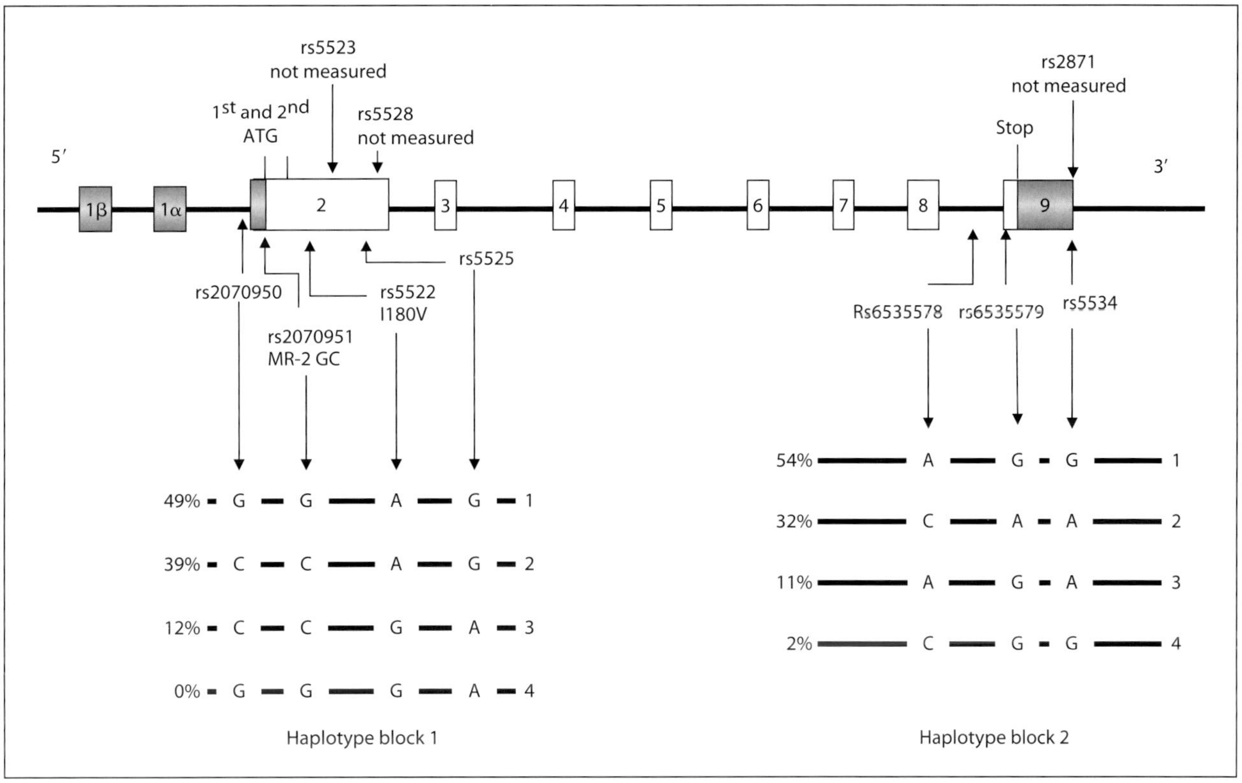

Fig. 2. The human MR gene. The human MR and GR genes have a similar organization, with exon 2 to exon 9 being translated. Corticosteroid receptor proteins can roughly be divided in three functional parts: a transcriptional active domain (exon 2), the DNA-binding domain (exons 3–4) and the ligand binding domain (exons 5–9), the latter also harboring transactivational capacity. The arrows indicate some important single nucleotide polymorphisms (SNPs). Haplotypes are depicted in two blocks, one in the 5′ promoter region and the other in the 3′ region including exon 9 and the 3′ untranslated region (UTR). Haplotype frequencies are indicated at the right. For the MR, the -2G/C and the I180V were found to affect transactivation in vitro. Haplotypes 1, 2a, 3 and 4 were also tested in vitro, showing profound differences. Moreover, the -2C SNP did not affect mRNA transcription, but induces increased MR expression in vitro, associating with increased transactivational capacity.

the genes being formed by exons 2–9 (see fig. 2) [3, 9, 10]. In order to disclose common genetic variability in the human MR gene, we screened the exonic regions of the human MR gene by direct sequencing. Fifty anonymous EDTA blood samples were obtained from the general physician laboratory in Leiden, the Netherlands. No new SNPs were identified in the exonic regions and we constructed haplotypes, resulting in two independent blocks, the first block 5′ in and around the first exonic regions and the second block around exon 9, located in the 3′ region of the human MR gene (fig. 2).

Table 1. MR gene variants associate with neuroticism in patients with depression

5′ promoter region MR gene (block 1): alleles 1–3

Block 1	0 copies	1 copy	2 copies	p
Allele 1	44.1 (±7.9)	43.6 (±8.4)	45.0 (±7.2)	n.s.
Allele 2	42.7 (±8.0)	44.7 (±7.8)	46.3 (±7.3)	n.s.
Allele 3	45.3 (±7.4)	41.1 (±8.5)	37.3 (±7.1)	0.02

Exon 9/3′ untranslated region MR gene (block 2): alleles 1–3

Block 2	0 copies	1 copy	2 copies	p
Allele 1	38.6 (±8.9)	45.5 (±7.6)	44.8 (±6.5)	n.s.
Allele 2	44.5 (±7.0)	45.4 (±7.3)	35.1 (±10.1)	0.02
Allele 3	44.1 (±8.1)	44.0 (±7.2)		n.s

Indicated are neuroticism scores (using the NEO-FFI) in patients carrying 0, 1 or 2 copies of the haplotypes from block 1 (59 promoter region; upper panel) or block 2 (exon 9/3′ untranslated region; lower panel). In block 1, allele 3 (or haplotype 3, harboring the MR-I1801V, allele frequency ~12%) associated with significantly decreased neuroticism scores, with a gene dose effect. In block 2, allele 2 (frequency ~32%) significantly associated with decreased neuroticism scores in homozygotes.

In vitro Testing

Initial computer screenings revealed that rs2070951 is located in the first Kozak sequence of the gene, the region which determines translation of the protein. In addition, rs2871 is located in a DNA sequence prone for loop formation, this might have consequences for the stability of the mRNA. Other SNPs were not located at predicted transcription binding sites, regulatory regions or micro-RNA target regions.

A high-frequency SNP designated MR-I180V (rs5522, allele frequency 12%) is located in exon 2, changing the primary structure of the MR protein. The SNP results in an isoleucine to valine (minor allele) amino acid change in the N-terminal domain of the protein and influences the transactivation capacity of the MR in vitro [11], as was shown before.

A second SNP tested in the MR is MR-2G/C (rs2070951, allele frequency 50%) located two nucleotides before the first translational start site of the MR. The C-allele of this SNP probably creates a stronger translation start site resulting in more MR protein expression compared to the G allele [12]. Furthermore, in vitro studies demonstrated a change in transactivational capacity, using cortisol as the ligand, by this

G/C variant [13]. We also tested the combinations to these 5′ SNPs, as they occur in the three common 5′ haplotypes. Testing of haplotypes is important since haplotypes constitute the naturally occurring stretches DNA organisms possess. Haplotype 2 (frequency 39%), containing the -2C and the isoleucine-180 (major allele) displayed highest activity, slightly more than haplotype 3 (frequency 12%, -2C and the valine-180 (minor allele). Lower activity was observed with haplotype 1 (frequency 49%, -2G and isoleucine-180). Remarkably, much lower activity was seen with haplotype 4 (frequency <1‰, -2G and valine-180), which again indicates the importance of 'normal' MR activity [14].

To summarize, we described MR SNPs, constructed haplotypes and tested in vitro not only SNPs but also their combinations (haplotypes). Results indicated profound effects on MR protein expression and transactivational activity. These findings indicate that these common MR gene SNPs and haplotypes are functional and suggest that they have in vivo consequences.

Neuroendocrine Regulation

Pharmacological studies in both humans and rodents have shown that the brain MR is involved in tonic inhibition of plasma corticosteroid concentrations [3]. In rats, stress-induced increases in plasma corticosterone were further enhanced when animals were pretreated with spironolactone, an MR antagonist. Previously we have shown that the MR-I180V is associated with enhanced cortisol and autonomic responses to an acute psychosocial stressor (the Trier Social Stress Test (TSST)) in young male twins [11]. The TSST involves a public-speaking and mental arithmetic task in front of an audience and camera, with components of unpredictability, social evaluation and lack of control. More recently, we tested neuroendocrine responses following the TSST in a group of middle-aged elementary school teachers [14]. Both males and females showed highest cortisol and heart beat response when being carrier of the MR gene haplotype 2. These data indicate that in humans the MR is indeed involved in both autonomic and HPA axis reactivity.

In addition, we tested the influence of MR genetic variants on 'basal' non-stress levels of cortisol in humans. Initially, we found an association with lower morning concentrations of plasma cortisol and the -2C variant [15]. However, this study involved only one sampling time at vaguely defined time points; the samples were obtained before 10:00 a.m. Furthermore, the cohort was composed of very old people, between 85 and 90 years of age, and 2/3 being women. In an additional study, we tested if the cortisol awaking rise (CAR) did associate with MR gene variants. The CAR is a well-defined response, consisting of an increase in cortisol levels starting just after awakening with highest levels after around 30 min returning to the starting levels at around 60 min [16]. However, neither the two SNPs nor the haplotype were significantly related to awakening cortisol responses. In contrast, the two in vitro functional

MR gene polymorphisms, the MR-I180V and MR-2G/C were found to be associated with the CAR after a low dose of dexamethasone (0.25 mg) administered the previous evening [13]. This effect is gender-specific and, importantly, was also observed with the haplotypes of both SNPs. These findings indicate that associations with MR variants seem to depend on changing conditions, as for instance is imposed by a challenge with dexamethasone.

Together, the data clearly indicate that the human MR is involved at multiple levels in the regulation of the stress response and HPA axis reactivity.

Behavioral Traits

Extensive animal studies have demonstrated MR-mediated effects on behavior. MR blockade clearly influenced coping with stressful challenges, observed as a change in the initial search strategy for an escape route in the Morris water maze [7]. Accordingly, transgenic mice with increased levels of MR in the forebrain showed decreased anxiety-like behavior and enhanced memory [17–19]. In humans, MR blockade impaired e.g. selective attention and mental flexibility, behavioral responses that represent the initial behavioral response to a stressor [20].

By using functional MR gene variants as 'genetic tools', we aim to test involvement of the MR in human behavioral traits and psychiatric disorders. As described above, these MR SNPs and concurrent haplotypes are functional and modify neuroendocrine responses, making it likely they will also influence behavioral traits. Initially, we found an association between feelings of depression and the MR-I180V in a cohort consisting of very old individuals [15]. Carriers of the valine-180 (minor allele) showed higher levels with the Geriatric Depression Scale compared to non-carriers. Furthermore, Ising et al. [21] described an association between the valine-180 (minor allele) and higher ACTH responses and anxiety during the second TSST performed in this group.

Next, we wanted to test for associations with personality, mood and anxiety disorders in a cohort containing both healthy controls (n = 50) and patients (n = 100). We have examined a cohort of 100 patients (mean age 33.1, SD 11.2 years, 65 females) with a depressive and/or anxiety disorder (DSM-IV) that were recruited from the outpatient department of the mental health center Rivierduinen in Leiden, the Netherlands. In addition, all subjects completed the Dutch translation of the Brief Symptom Inventory (BSI) and the Mood and Anxiety Symptom Questionnaire (MASQ) [22]. The total BSI score generates an overall measure of psychopathological symptom severity. The MASQ measures three scales of general distress: depressive symptoms (12 items), anxious symptoms (11 items) and mixed symptoms (15 items), it has an anxiety-specific scale (anxious arousal, 17 items) and depression-specific scale (anhedonic depression, 22 items). For both BSI and MASQ (sub)scales, higher scores reflect more symptoms. Neuroticism was assessed with the subscale of the NEO Five Factor Inventory (NEO-FFI).

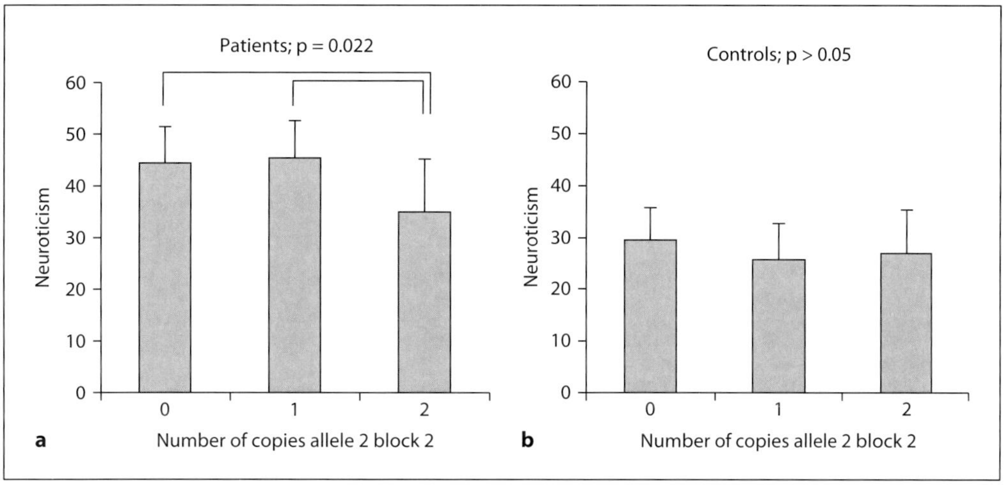

Fig. 3. MR gene variants in exon 9/3′ UTR associate with neuroticism. Association of MR allele 2 haplotype block 2 and neuroticism scores in depressed patients (**a**) and in control individuals (**b**). No associations are observed in controls who show much lower levels of neuroticism. In patients however a significant association was detected, with lowest levels in homozygotes.

Concerning psychopathology, the genotype distribution of MR SNPs or haplotype frequencies were not significantly different between the patient and control groups ($\chi^2 = 5.0$, $p = 0.167$). The total score on the BSI was significantly higher in the patients compared to the controls (patients 1.31 ± 0.69 vs. controls 0.18 ± 0.18) while also on the eight different subscales the patients scored significantly higher. There were no significant associations between the haplotypes and BSI total scores or subscale scores. General distress, anhedonic depression, and anxious arousal measured with the MASQ were also not associated with MR haplotypes or single SNPs.

In the patients there was a significant association between both MR haplotype blocks and neuroticism; patients carrying one copy of allele 3 of haplotype block 1 (MR valine-180 (minor allele) carriers) had lower scores than non-carriers and the scores were even lower in the patients carrying two copies of allele 3 ($p = 0.04$). In block 2, homozygote carriers of allele 2 had lower scores of neuroticism compared to patients without allele 2 or only one copy of allele 2 ($p = 0.02$; fig. 3). There were no significant effects of the MR haplotypes on neuroticism scores in healthy individuals ($p > 0.05$; fig. 3).

Neuroticism was associated with both haplotype blocks 1 and 2 in the patient group; this is the first study showing that genetic variation in the MR is associated with neuroticism. The effect of the MR was only observed in the patient group, which as a total had significantly higher neuroticism scores as compared to the control subjects. A depressive state amplifies the personality profile and the scores can be interpreted as an accurate reflection of the current condition of the person. Therefore it is likely that the observed effect is only visible in this group with amplified scores. Previous studies demonstrated heritability of neuroticism, associations between cortisol and neuroticism, involvement of the MR in behavior and a role for genetic variation in stress responsiveness [16, 23].

The association between haplotype 3 in block 1 carrying the MR valine-180 minor allele and less neuroticism was unexpected, since this SNP was also associated with feelings of depression and anxiety following the TSST. This finding could be explained from differences in cohorts, such as age, gender, specific testing, cultural differences or genetic background. Moreover, testing has been performed in a relative small cohort, so findings have to be taken with caution.

In block 2, allele 2 is the allele associated with neuroticism. The SNPs in this block are not yet tested for functionality but computer screenings revealed that one of the SNPs, rs2781, promotes loop formation of the mRNA and this will probably alter the mRNA stability, leading to differences in MR protein regulation. The profound effects of the -2G/C, changing MR protein expression, on neuroendocrine regulation is supporting this hypothesis.

These data show that common variants of the human MR gene are associated with changes in neuroticism scores. This finding is of interest since the neuroticism as a personality trait associates with anxious and fearful feelings with enhanced sensitivity to stress and hyperarousal, which are all risk factors for stress-related mental disorders. Hence, high neuroticism correlates with elevated baseline cortisol levels and a blunted cortisol response to stress. However, the exact conditions that contribute to this finding require further study on the role of the MR variants in the regulation of emotional and stress reactivity under these conditions.

Conclusions

We have described common haplotypes of the human MR, which showed different transactivational capacities in vitro. This indicates that in the human population, common MR gene variants occur with potential effects on neuroendocrine regulation and behavioral functioning. In line with this prediction, we found associations with different levels of HPA axis control and with autonomic output. Moreover, we describe significant associations with feelings of depression and the behavioral trait neuroticism, although the behavioral associations show inconsistencies and are preliminary, thus requiring further study. The inconsistencies might result from differences between the cohorts. Alternatively, MR expression is highly dynamic and specific MR gene variant-environmental interactions are very well possible. This is of interest in view of the recent evidence for rapid membrane actions in addition to the classical gene-mediated actions. Hence, the role of the MR in these environmental interactions has in common that it is important for the control of the initial reaction to the stressor involving mechanisms underlying appraisal processes as well as emotional and violent responses. In conclusion, we have reported here new findings on common functional human MR gene variants which reveal these variants as previously unknown factors associated with neuroticism conferring vulnerability to stress-related mental disorders, such as depression and the posttraumatic stress syndrome.

Acknowledgements

This work was supported by a grant from the Royal Netherlands Academy of Arts and Sciences, the Nederlandse Hersenstichting and psychiatric hospital Rivierduinen. We would like to thank Ellen Klaassens and Gerthe Veen for providing the cohort consisting of depressed patients and healthy controls.

References

1 De Kloet ER, Joëls M, Holsboer F: Stress and the brain: from adaptation to disease. Nat Rev Neurosci 2005;6:463–475.
2 Funder JW: Mineralocorticoid receptors: distribution and activation. Heart Fail Rev 2005;10:15–22.
3 DeRijk RH: Single nucleotide polymorphisms related to HPA axis reactivity Neuroimmunomodulation 2009;16:340–352.
4 Pascual-Le Tallec L, Lombes M: The mineralocorticoid receptor: a journey exploring its diversity and specificity of action. Mol Endocrinol 2005;19:2211–2221.
5 Tasker JG, Di S, Malcher-Lopes R: Rapid glucocorticoid signaling via membrane-associated receptors. Endocrinology 2006;147:5549–5556.
6 Joëls M, Karst H, Derijk R, de Kloet ER: The coming out of the brain mineralocorticoid receptor. Trends Neurosci 2008;31:1–7.
7 Oitzl MS, Champagne DL, van der Veen R, de Kloet ER: Brain development under stress: hypotheses of glucocorticoid actions revisited. Neurosci Biobehav Rev 2010;34:853–866.
8 Caspi A, Hariri AR, Holmes A, Uher R, Moffitt TE: Genetic sensitivity to the environment: the case of the serotonin transporter gene and its implications for studying complex diseases and traits. Am J Psychiatry 2010;167:509–527.
9 DeRijk RH, de Kloet ER: Corticosteroid receptor genetic polymorphism and stress responsivity. Endocrine 2005;28:263–270.
10 De Kloet ER, DeRijk RH, Meijer OC: Therapy Insight: is there an imbalanced response of mineralocorticoid and glucocorticoid receptors in depression? Nat Clin Pract Endocrinol Metab 2007;3:168–179.
11 DeRijk RH, Wüst S, Meijer OC, Zennaro M-C, Federenko IS, Hellhammer DH, Giacchetti G, Vreugdenhil E, Zitman FG, de Kloet ER: A common polymorphism in the mineralocorticoid receptor modulates stress responsiveness. J Clin Endocrinol Metab 2006;91:5083–5089.
12 van Leeuwen N, Caprio M, Blaya C, Fumeron F, Sartorato P, Ronconi V, Giacchetti G, Mantero F, Fernandes-Rosa FL, Simian C, Peyrard S, Zitman FG, Penninx BW, de Kloet ER, Azizi M, Jeunemaitre X, DeRijk RH, Zennaro MC: Hypertension 2010; 56:995–1002. Epub 2010.
13 Van Leeuwen N, Kumsta R, Entringer S, de Kloet ER, Zitman FG, DeRijk RH, Wüst S: Functional MR gene variation influences the cortisol awakening response after dexamethasone. Psychoneuroendocrinology 2009;35:339–349.
14 Van Leeuwen N, Bellingrath S, de Kloet ER, Zitman FG, Kudielka BM, DeRijk RH, Wust S: Human mineralocorticoid receptor (MR) gene haplotypes modulate MR expression and transactivation: implication for the stress response. Psychoneuroendocrinology 2010. In press.
15 Kuningas M, DeRijk RH, Westendorp RG, Jolles J, Slagboom EP, van Heemst D: Mental performance in old age is dependent on cortisol and genetic variance in the mineralocorticoid and glucocorticoid receptors. Neuropsychopharmacology 2007;32:1295–1301.
16 Wilhelm I, Born J, Kudielka BM, Schlotz W, Wust S: Is the cortisol awakening rise a response to awakening? Psychoneuroendocrinology 2007;32:358–366.

17 Lai M, Horsburgh K, Bae SE, Carter RN, Stenvers DJ, Fowler JH, Yau JL, Gomez-Sanchez CE, Holmes MC, Kenyon CJ, Seckl JR, MacLeod MR: Forebrain mineralocorticoid receptor overexpression enhances memory, reduces anxiety and attenuates neuronal loss in cerebral ischaemia. Eur J Neurosci 2007;25: 1832–1842.
18 Rozeboom AM, Akil H, Seasholtz AF: Mineralocorticoid receptor overexpression in forebrain decreases anxiety-like behavior and alters the stress response in mice. Proc Natl Acad Sci USA 2007;104:4688–4693.
19 Berger S, Wolfer DP, Selbach O, Alter O, Erdmann G, Reichard HM, Chepkova AN, Welzl H, Haas HL, Lipp H-P, Schütz G: Loss of the limbic mineralocorticoid receptor impairs behavioral plasticity. Proc Natl Acad Sci USA 2006;103:195–200.
20 Otte C, Moritz S, Yassouridis A, Koop M, Madrischewski AM, Wiedemann K, Kellner M: Blockade of the mineralocorticoid receptor in healthy men: effects on experimentally induced panic symptoms, stress hormones, and cognition. Neuropsychopharmacology 2007;32:232–328.
21 Ising M, Depping AM, Siebertz A, Lucae S, Unschuld PG, Kloiber S, Horstmann S, Uhr M, Muller-Myhsok B, Holsboer F: Polymorphisms in the FKBP5 gene region modulate recovery from psychosocial stress in healthy controls. Eur J Neurosci 2008;28:389–398.
22 De Beurs E, Hollander-Gijsman ME, Helmich S, Zitman FG: The tripartite model for assessing symptoms of anxiety and depression: psychometrics of the Dutch version of the mood and anxiety symptoms questionnaire. Behav Res Ther 2006;45:1609–1617.
23 Portella MJ, Harmer CJ, Flint J, Cowen P, Goodwin GM: Enhanced early morning salivary cortisol in neuroticism. Am J Pschiatry 2005;162:807–809.

Roel H. DeRijk
Department of Medical Pharmacology
Leiden Amsterdam Center for Drug Research (LACDR), Leiden University
Einsteinweg 55, NL–2333 CC Leiden (The Netherlands)
Tel. +31 71 5276229, Fax +31 71 5274715, E-Mail r.h.de.rijk@lacdr.leidenuniv.nl

Ghizzoni L, Cappa M, Chrousos G, Loche S, Maghnie M (eds): Pediatric Adrenal Diseases.
Endocr Dev. Basel, Karger, 2011, vol 20, pp 149–160

Adrenoleukodystrophy

Marco Cappa[a] · Carla Bizzarri[a] · Catello Vollono[b] ·
Anna Petroni[c] · Sebastiano Banni[d]

[a]Dipartimento Pediatrico Universitario Ospedaliero, Unità Operativa Complessa di Endocrinologia e Diabetologia, [b]Dipartimento di Neuroscienze, Unità Operativa Complessa di Neurologia, Ospedale Bambino Gesù, Roma; [c]Dipartimento di Scienze Farmacologiche, Università di Milano, Milano, e [d]Dipartimento di Biologia Sperimentale, Università di Cagliari, Cagliari, Italy

Abstract

X-linked adrenoleukodystrophy (ALD) is caused by mutations in the *ABCD1* gene that encodes a protein of the peroxisomal membrane named ALDP. Mutations in ALDP result in elevated levels of very long chain fatty acids (VLCFA) and reduced VLCFA oxidation in peroxisomes. Three main phenotypes are seen in affected males. The childhood cerebral form manifests usually between ages 4 and 8 years. It initially resembles attention deficit disorder or hyperactivity. Progressive central demyelination with impairment of cognition, behavior, vision, hearing, and motor function follow the initial symptoms and often lead to total disability within 2 years. The second phenotype, adrenomyeloneuropathy, manifests most commonly in the late twenties as progressive paraparesis, sphincter disturbances, sexual dysfunction, and often, impaired adrenocortical function; all symptoms are progressive over decades. The third phenotype, 'Addison disease only', presents with primary adrenocortical insufficiency between age 2 years and adulthood and most commonly by age 7.5 years, without evidence of neurologic abnormality. Approximately 50% of females who are carriers develop neurologic manifestations that resemble adrenomyeloneuropathy but have a later onset (age ≥35 years) and a milder disease. In this review, we will give an overview of the present understanding of ALD, and the implications of new diagnostics and treatment.

X-linked adrenoleukodystrophy (ALD) is the most frequent peroxisomal disorder with a minimum incidence ranging from 1:42,000 for hemizygous males only, to 1:17,000 for hemizygous males plus heterozygous females combined [1]. The corresponding gene defect maps to Xq28, the gene product is a peroxisomal membrane protein (ALDP) that is a member of the large family of ATP-binding cassette transporters: ABCD1. The physiologic role of ALDP is not yet understood. It is known that ALDP is localized in peroxisomal membranes in human fibroblasts, and that it has a role in very long chain fatty acid (VLCFA) metabolism [2]. As ALDP belongs to the ABC transporter superfamily, it seems plausible that ALDP acts as a transporter. This is indeed the case in yeast, but the role in mammalian cells is still uncertain [3]. The disorder primarily affects the nervous system and the steroid-producing tissues (adrenal cortex and gonads). The pathology in the nervous system is variable. It may be either an inflammatory demyelination (cerebral ALD) that usually begins in the parieto-occipital regions and may involve autoimmune mechanisms or a more slowly progressive axonopathy affecting the ascending and descending tracts of the spinal cord and to a lesser extent peripheral nerves adrenomyeloneuropathy (AMN) [4].

ALD has been diagnosed in all geographic regions and ethnic groups, and there is no evidence that the frequency varies with ethnic background. The underlying defect and the phenotypic variability present important questions in the understanding of the pathophysiology of this as other genetic diseases, and have implications for the assessment of treatment.

Diagnosis

Clinical Diagnosis

Initially, ALD was thought to have a single phenotype, consisting of the severe cerebral inflammatory demyelinating changes that lead to the progressive symptoms of cognitive impairment, gait disturbances, and visual loss followed by a vegetative state and ultimately death.

On the contrary, there are seven distinct clinical phenotypes of ALD described in males, ranging from the more severe cerebral forms, to AMN, and finally to asymptomatic persons or isolated adrenal insufficiency, without involvement of the central nervous system (CNS). Cerebral ALD is further divided into childhood, adolescent and adult form. Cerebral ALD is frequently associated with Addison's disease, but the primary adrenal insufficiency may precede, coexist or develop after neurological disturbances. Almost 70% of patients display Addison's disease and high ACTH levels usually represent the first sign of ongoing adrenal insufficiency. Childhood cerebral ALD patients usually develop symptoms from age 4 to 8 years, including visual and auditory disturbances, walking difficulties, which seem to be associated with a strong inflammatory reaction in the CNS, particularly in the white matter. The disease progresses rapidly and patients usually die approximately 2–5 years after onset of symptoms. AMN most commonly manifests in males during the late twenties as progressive paraparesis, sphincter disturbances, and sexual dysfunction; all symptoms worsen over decades, 30–50% of AMN patients manifest symptoms of inflammatory cerebral involvement. In male patients, AMN and cerebral ALD together account for more than 80% of clinical forms of ALD. Approximately 50% of women who are heterozygous for ALD develop an AMN-like syndrome, a higher proportion than what had been recognized in the past. The mean age at onset of symptoms in heterozygous females (37 ± 14.6 years, range 32–73 years) is later than in men with AMN (28 ± 9 years), clinical symptoms are milder and progression is slower. Only 3% of women heterozygous for ALD had inflammatory brain disease [5]. Adrenal insufficiency is extremely rare in female carriers. Moser et al. [4] proposed that the manifestations of ALD are subdivided into two main and pathologically distinct categories, inflammatory and non-inflammatory forms. AMN is the non-inflammatory form, characterized by distal axonopathy, but the affected subjects may develop a secondary and rapidly progressive inflammatory brain involvement. All the cerebral forms are inflammatory forms, associated with a rapidly progressive intensely inflammatory myelinopathy that usually begins in the parieto-occipital regions.

The diagnosis of ALD should be considered in four clinical settings: (1) Boys with symptoms of attention deficit disorder who also show signs of dementia, progressive behavioral disturbance, vision loss, difficulty in understanding spoken language, worsening handwriting, incoordination, or other neurologic disturbances. (2) Young or middle-aged men with progressive gait disorders, leg stiffness or weakness, abnormalities of sphincter control, sexual dysfunction, with or without adrenal insufficiency or cognitive or behavioral deficits. (3) All males with primary adrenocortical insufficiency, with or without evidence of other neurological abnormalities. (4) Middle-aged or older women with progressive paraparesis, abnormalities of sphincter control, and sensory disturbances mainly affecting the legs. It may be difficult to establish the correct diagnosis of ALD in a female with a negative family history.

Biochemical Diagnosis

The plasma concentration of VLCFA is abnormal in 99% of males with ALD, irrespective of age. An increased concentration of VLCFA in plasma and/or cultured skin fibroblasts is also evident in approximately 85% of affected females, but 15–20% of known carriers have a normal plasma concentration of VLCFA. Molecular genetic testing should be used to test females with a normal concentration of VLCFA if the disease-causing ABCD1 mutation has been identified in the family. Three parameters are analyzed for testing VLCFA: concentration of C26:0, ratio of C24:0 to C22:0, ratio of C26:0 to C22:0. All the three parameters are elevated in the majority of hemizygous males, though some variation is observed. Table 1 summarizes mean results for normal controls, affected males, and female carriers [6]. The VLCFA assay is performed in a limited number of laboratories worldwide.

Molecular Diagnosis

ALD is inherited in an X-linked manner. About 93% of index cases have inherited the ABCD1 mutation from one parent, 7% of individuals with ALD have a de novo mutation. Affected males transmit the ABCD1 mutation to all of their daughters and none of their sons. Female carriers have a 50% chance of transmitting the ABCD1 mutation in each pregnancy. Males who inherit the mutation will be affected; females who inherit the mutation will be carriers. It is appropriate to measure plasma VLCFA

Table 1. Plasma VLCFA values (μg/ml) in ALD

VLCFA	Normal	Males with ALD	Obligate female carriers
C26:0	0.23+0.09	1.30+0.45	0.68+0.29
C24:0/C22:0	0.84+0.10	1.71+0.23	1.30+0.19
C26:0/C22:0	0.01+0.004	0.07+0.03	0.04+0.02

concentration in the mothers of both affected males and carrier females and in the fathers of carrier females. When the disease-causing mutation has been identified in an affected family member, molecular genetic testing of the *ABCD1* gene can be used in the evaluation of the parents. PCR and sequence analysis of the coding region and flanking intronic regions identify mutations of the *ABCD1* gene in about 93% of hemizygous males or obligate heterozygote females. Deletion/duplication analysis identifies deletions/duplications not detectable by sequence analysis of genomic DNA; a variety of methods including Southern blot, quantitative PCR, real-time PCR, multiplex ligation-dependent probe amplification, and array CGH may be used. Most of the individuals without an ABCD1 mutation identified by sequence analysis have a gene deletion detected by Southern blot analysis [7, 8].

Prenatal Testing

Prenatal testing is possible for pregnancies of women who are carriers in whom the risk of having an affected male is 25% (or 50% if the fetus is known to be male). The usual procedure is to determine fetal sex by performing chromosome analysis on fetal cells obtained by chorionic villous sampling at approximately 10–12 weeks' gestation or by amniocentesis usually performed at approximately 15–18 weeks' gestation. If the karyotype is 46,XY and if the disease-causing mutation has been identified in a family member, DNA from fetal cells can be analyzed for the known disease-causing mutation. If the molecular genetic testing on fetal cells is not possible, VLCFA can be measured in cultured amniocytes or cultured chorionic villous cells. False negative test results have been reported, probably related to technical factors [9].

Considering that the phenotypic expression and prognosis of an affected male is an unpredictable variable, the justification of prenatal testing could be questionable and it is now a matter of discussion.

Neuroimaging

Brain MRI is always abnormal in neurologically symptomatic males with cerebral disease and often provides the first diagnostic lead. In approximately 85% of affected individuals, MRI shows a characteristic pattern of symmetric confluent hyperintensive lesions in T_2 and FLAIR images, with contrast enhancement at the advancing

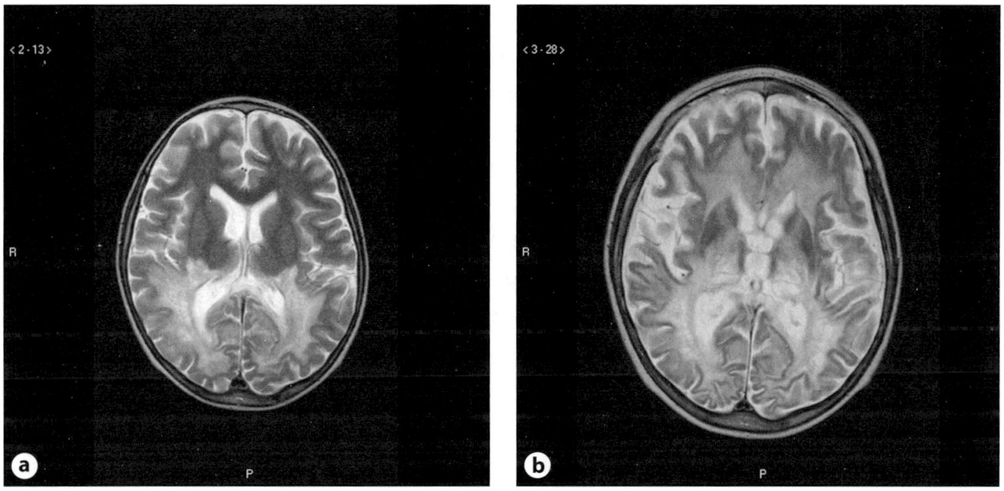

Fig. 1. **a** MRI scan of brain (T_2 image) at diagnosis in a child with childhood-onset ALD, showing diffuse high signal abnormalities in the periventricular white matter of both parietal and occipital lobes. **b** MRI scan performed 10 months later, showing significant advancing of the high signal abnormalities.

margin. Enhanced gadolinium contrast indicates that the demyelinating lesions are inflammatory and disrupt the blood-brain barrier (fig. 1).

The lesions usually first appear in the parieto-occipital region of both hemispheres, the progression of the lesions is unpredictable, it can be severe and rapid, mostly in the childhood cerebral form; milder and slower in adult forms or in AMN with secondary cerebral involvement.

Pathogenesis

ALDP contains 745 amino acids and is located in the peroxisomal membrane. It is a member of the ATP-binding cassette (ABC) protein transporter family. ABC transporters function as dimers of two related halves. ALDP represents only one of these halves and is referred to as a half-transporter. Binding of two half-transporters creates a functional transporter, whereby the two membrane domains form a channel through which the substrate is transported. Each of the ABC half-transporters contains a hydrophobic membrane domain with six membrane-spanning segments. The combination of the individual components may determine the specificity of the transporter. ALDP has been shown to be able to form a homodimer. However, the peroxisomal membrane contains three additional ABC transporters: PMP70, PMP69 (P70R), and the ALD-related protein (ALDR); it can also form heterodimers with some of these. ALDR, which has been mapped to 12q11, is of particular interest because it has 66%

identity to ALDP and can substitute for the function of ALDP in restoring the capacity of ALD fibroblasts to metabolize VLCFA. Variations in the interactions between ALDP and its homologues may influence the phenotypic expression of the disease [6, 8].

The physiologic function of ALDP is still unclear. It has been supposed that ALDP regulates the transport of very long chain acyl-CoA synthase (VLCS) through the membrane of peroxisome to the peroxisomal matrix [10]. When the ALDP is defective, the metabolism of VLCFA is impaired, particularly hexacosanoate (C26:0) which accumulates in brain lipids, mainly in cholesterol esters and gangliosides. Lipids containing an abnormally high proportion of VLCFA can act as triggers that initiate the cascade of inflammatory demyelination [11].

There is a phenotypic variation among family members with the same mutations [12], including a report of different clinical phenotypes in monozygotic twins [13]. Additionally, there is no correlation between the levels of VLCFA in plasma and fibroblasts and the severity of neurological disease [14]. The knockout mouse model does not develop neurological symptoms or demyelination at 1 year of age, despite the presence of VLCFA in the brain [15]. All these findings suggest that VLCFA accumulation is necessary but not sufficient for the pathogenesis of ALD.

First described by Schaumburg et al. [16], the affected white matter lesions in ALD are characterized by three distinct zones. These include an outermost zone that has active destruction of myelin with lack of inflammatory cells, a middle zone characterized by demyelination with axonal sparing and perivascular inflammatory cells, and an innermost zone demonstrating gliosis and scattered astrocytes with little to no inflammatory cells.

There is evidence for the involvement of all the different components of the immune system (cellular, cytokines, humoral, and complement) in the pathogenesis of the inflammatory cerebral lesions, but the results are not conclusive. One of the broad hypothesis for the pathogenesis of CNS lesions in ALD is that the VLCFA leads to myelin instability, followed by an inflammatory or immune-mediated process that destroys myelin [17].

Management and Treatment of Clinical Manifestations

Adrenal Insufficiency

The proportion of cases in which Addison disease is attributable to ALD is age-dependent. ALD is the most common cause of adrenal insufficiency in males before 15 years of age [18]. Adrenal function is abnormal in 90% of neurologically symptomatic ALD boys and in 70% of men with AMN. It is usually normal in female carriers. The most sensitive indicators of adrenal dysfunction are elevated plasma ACTH concentration and impaired rise of plasma cortisol concentration in response to administered ACTH [19]. Adrenal antibodies are absent [20]. Renin and aldosterone levels are usually normal, and patients rarely need mineralcorticoid therapy. In an early phase of the disease, basal ACTH levels may be elevated, concomitant basal cortisol concentrations are normal or slightly elevated, but not responsive to administered ACTH. In this condition, basal ACTH and cortisol levels tend to decrease and cortisol response to ACTH improves after starting a low VLCFA diet and Lorenzo's oil therapy. These findings suggest that adrenal insufficiency in ALD is due to ACTH resistance, probably related to VLCFA accumulation in the cellular membranes that causes a progressive impairment of the membrane functions and blocks ACTH receptors. In an early phase this damage could be reversible [M. Cappa, J Endocrinol Invest, in press].

Glucocorticoid replacement therapy is essential for ALD patients with adrenal insufficiency to prevent morbidity and mortality associated with adrenal crisis, but there is no evidence that this affects the neurologic problems associated with the disease.

For ALD male patients without overt Addison disease, a periodic evaluation of adrenocortical function is mandatory. Early identification of asymptomatic or minimally symptomatic at-risk males permits timely treatment of adrenal insufficiency.

Intravenous Immunoglobulin

Intravenous immunoglobulin (IVIG) has been tried as a possible therapeutic approach. A 15-year-old Japanese boy with cerebral ALD was treated with IVIG. His regimen was 100 mg/kg/day for 5 consecutive days, followed by 100 mg/kg/day every 2 weeks. Within 2 weeks of starting therapy, he began having improvement in his symptoms and no neurological deterioration was seen for 18 months after beginning treatment. Interestingly, VLCFA levels and brain CT scans showed no change [21]. In contrast, our group reported the results of a 1-year-long trial of IVIG treatment in 6 patients who were also on a restricted diet and Lorenzo's oil therapy and compared to a control group of 6 ALD patients who only received diet and Lorenzo's oil therapy. After 2 months, all patients had normal VLCFA levels but both groups continued to have similar progression of neurologic symptoms and MRI findings [22]. Of particular interest is the significantly different disease progression at long-term follow-up in the two groups. All patients in the IVIG group died within 4 years from the start of treatment, whereas all subjects in the Lorenzo's oil-only group are still alive 16 years after that [M. Cappa, unpubl. data]. The issue is intriguing and the reasons of the significant difference in clinical progression need to be clarified.

Lorenzo's Oil

Lorenzo's oil, a mixture of monounsaturated oleic acid ($C_{18:1}$) in triglyceride form (glyceryl-trioleate) and monounsaturated erucic acid ($C_{22:1}$) in triglyceride form (glyceryl-trierucate) at the 1:4 proportion, combined with a VLCFA-poor diet, can normalize VLCFA levels in patients after 4 weeks of therapy [23]. Lorenzo's oil barely passes the blood-brain barrier, leaving open the question whether it is really effective in the CNS. It has been demonstrated that Lorenzo's oil does not change the progressive nature of the disease in patients who have developed cerebral disease, but a recent study suggested that it may significantly reduce the chance of developing cerebral disease in asymptomatic boys with normal brain MRI results [24].

Hematopoietic Stem Cell Transplantation

Hematopoietic stem cell transplantation (HSCT) can halt the progression of the disease in patients with inflammatory cerebral involvement. Outcome after HSCT is directly related to neurological status at the time of transplantation. Patients with mild neurologic impairment and MRI involvement seem to have the best outcomes [25]. The international experience of HSCT in cerebral ALD from 1982 to 1999 has been recently reported. 126 patients underwent HSCT at 43 different transplant centers, and complete data were available for 94 patients. At a median follow-up of 11 months, full or partial donor engraftment was noted in 86% of patients. The incidence of severe acute graft-versus-host-disease was 12%, and a significant difference was not seen between donor type. The 5-year survival for related donor HSCT was 64% and for unrelated donor HSCT it was 53%. Pre-HSCT disease status was linked to overall outcome, as 5-year survival significantly worsened with increasing number of neurologic deficits and MRI severity scores. Performance IQ appears to be an excellent predictor of both neurocognitive function and survival after HSCT, with superior outcomes seen in patients with performance IQ >80. Due to the progressive nature of the disease, patients can often be expected to have further demyelination until stabilization occurs after HSCT. Because up to 50% of patients with normal MRI may not develop cerebral disease, HSCT is not recommended as a first-line treatment for asymptomatic patients [26]. Additionally, HSCT is clearly not even a therapeutic option for those patients who do not have a suitable HLA-matched related or unrelated donor. Once symptoms begin, the combination of the progressive nature of the disease with the typical search time of several months needed to try to identify an unrelated bone marrow donor, if no suitable HLA-matched family member is available, creates an exceedingly narrow window of opportunity for intervention. Advantages of using cord blood units for transplantation include the ability to tolerate more mismatches as well as quick availability of the units. These advantages are often offset by the longer time to engraftment, subsequent increase in infectious risks, and the decreased chance of having an adequate cell dose for patients weighing >20 kg [27].

Long-term follow-up data 5–10 years after HSCT show that cognitive status stabilizes or improves despite the continued presence of elevated VLCFA [28]. These data, along with the fact that a particular enzyme deficiency is not corrected in HSCT for ALD, suggest that the major mechanism of action of HSCT is the attainment of a new immune system and not reversal of a metabolic condition. The ALD protein participates in the peroxisomal degradation of VLCFA in oligodendrocytes and microglia, and deficiency of this protein disrupts myelin maintenance by these cells. The long-

Table 2. X-linked ALD-DRS level description

0	No difficulties
1	Mild learning or coordination difficulties from ALD; patient does not require support or intervention
2	Moderate learning, sensory, and/or neurologic abnormality; patient requires support or intervention in a few areas
3	Severe learning, sensory, and/or neurologic abnormality; patient requires support or intervention in many areas
4	Loss of cognitive ability and disorientation; patient requires constant supervision

term benefits of HSCT in ALD are probably due to the replacement of brain microglial cells by myelomonocytic cells derived from donor bone marrow.

Given the variability of the clinical progression and the wide phenotypic variation in ALD, HSCT should be reserved for ALD patients with early but definite evidence of cerebral disease at brain MRI. Various multidisciplinary scoring systems have been proposed to evaluate the extent of cerebral demyelination at MRI [29] and the degree of disability associated with the impairment of the different functional areas. In particular, the ALD disability rating scale (ALD-DRS) was developed at the University of Minnesota to provide insight into the quality of life of boys with cerebral ALD, and to describe different levels of disability and the assistance required for the disability (table 2). More than 50% of symptomatic patients do not meet the criteria for HSCT at the time of diagnosis [25].

Gene Therapy

The results of the first gene therapy trial in 2 patients with childhood cerebral ALD for whom there were no matched donors were reported in 2009 [30]. Autologous CD34+ cells were removed from the patients, genetically corrected ex vivo with a lentiviral vector encoding wild-type ABCD1, and then reinfused into the patients after they had received myeloablative treatment. Over a span of 24–30 months of follow-up, a significant polyclonal reconstitution was detected, with 9–14% of granulocytes, monocytes, and T and B lymphocytes expressing ALDP. These results strongly suggest that hematopoietic stem cells were transduced in the patients. Beginning 14–16 months after infusion of the genetically corrected cells, progressive cerebral demyelination stopped in both patients. Contrast enhancement of demyelinating lesions (indicating the inflammatory disruption of the blood-brain barrier) disappeared completely 12 months after transplantation in 1 patient.

If these results will be confirmed by larger studies, lentiviral-mediated gene therapy of hematopoietic stem cells in ALD could have less risks of mortality and provide more clinical benefits than HSCT. The fact that the neurologic benefits of gene therapy in ALD were comparable to those seen with HSCT supports further testing of this treatment in an extended series of ALD patients with cerebral demyelination who do not have an available HLA-matched donor or cord blood. Gene therapy might also be considered as a therapeutic option for adult ALD patients who develop cerebral demyelination, for whom the mortality risk of HSCT is about 40%.

A Novel Nutritional Approach

Lorenzo's oil barely passes the blood-brain barrier, leaving open the question whether it is really effective in the CNS. In an ongoing trial we have hypothesized that the addition to Lorenzo's oil of another fatty acid, that may enhance peroxisomal β-oxidation and passes the blood-brain barrier, could work synergistically with Lorenzo's oil. The possible candidate was conjugated linoleic acid (CLA). It is a high-affinity ligand of the peroxisome proliferator-activated receptor, a transcription factor for the genes involved in peroxisomal β-oxidation and thereby it is able to induce the acyl-CoA oxidase, a key enzyme for peroxisomal β-oxidation. Considering that the prompt incorporation of CLA into brain has been recently demonstrated in rats, its activity in the CNS should be ensured [31].

Preliminary analysis of the somatosensory-evoked potentials performed in 5 symptomatic female carriers before and after oral supplementation of Lorenzo's oil + CLA showed a slight reduction of absolute cortical latencies of the lower limbs. Similarly, the time intervals between peripheral stimulus and cortical response (interpeak latencies), appeared reduced after treatment [M. Cappa, submitted].

In conclusion, ALD is a disorder with a huge phenotypic spectrum. Up to now no clinical or biochemical markers can predict the evolution of the disease. No definite cure exists for this disease, but gene therapy and novel nutritional approaches could be promising for affected patients.

References

1 Bezman L, Moser AB, Raymond GV, Rinaldo P, Watkins PA, Smith KD, Kass NE, Moser HW: Adrenoleukodystrophy: incidence, new mutation rate, and results of extended family screening. Ann Neurol 2001;49:512.

2 Mosser J, Lutz Y, Stoeckel ME, Sarde CO, Kretz C, Douar AM, Lopez J, Aubourg P, Mandel JL: The gene responsible for adrenoleukodystrophy encodes a peroxisomal membrane protein. Hum Mol Genet 1994;3:265.

3 Kemp S, Wanders RJ: X-linked adrenoleukodystrophy: very long-chain fatty acid metabolism, ABC half-transporters and the complicated route to treatment. Mol Genet Metab 2007;90:268.

4 Moser HW, Mahmood A, Raymond GV: X-linked adrenoleukodystrophy. Nat Clin Pract Neurol 2007; 3:140.

5 Restuccia D, Di Lazzaro, Valeriani M, Oliviero A, La Pera D, Colosimo C, Burdi N, Cappa M, Bertini E, Di Biase A, Tonali P: Neurophysiological abnormalities in adrenoleukodystrophy carriers. Evidence of different degree of central nervous system involvement. Brain 1997;120:1139.

6 Steinberg S, Jones R, Tiffany C, Moser A: Investigational methods for peroxisomal disorders. Curr Protoc Hum Genet 2008; chapt 17: unit 17.6.

7 Montagna G, Di Biase A, Cappa M, Melone MA, Piantadosi C, Colabianchi D, Patrono C, Attori L, Cannelli N, Cotrufo R, Salvati S, Santorelli FM: Identification of seven novel mutations in ABCD1 by a DHPLC-based assay in Italian patients with X-linked adrenoleukodystrophy. Hum Mutat 2005; 25:222.
8 X-linked Adrenoleukodystrophy Database [www.x--ald.nl].
9 Wanders RJ, Mooyer PW, Dekker C, Vreken P: X-linked adrenoleukodystrophy: improved prenatal diagnosis using both biochemical and immunological methods. J Inherit Metab Dis 1998;21:285.
10 Mosser J, Douar AM, Sarde CO, Kioschis P, Feil R, Moser H, Poustka AM, Mandel JL, Aubourg P: Putative X-linked adrenoleukodystrophy gene shares unexpected homology with ABC transporters. Nature 1993;361:726.
11 Beutler B, Cerami A: The history, properties, and biological effects of cachectin. Biochemistry 1988; 27:7575.
12 Moser HW: Adrenoleukodystrophy: from bedside to molecular biology. J Child Neurol 1987;2:140.
13 Wilichowski E, Ohlenbusch A, Korenke GC, Hunneman DH, Hanefeld F: Identical mitochondrial DNA in monozygotic twins with discordant adrenoleukodystrophy phenotype. Ann Neurol 1998;43:835.
14 Moser H: Adrenoleukodystrophy: phenotype, genetics, pathogenesis and therapy. Brain 1997;120: 1485.
15 Kobayashi T, Shinnoh N, Kondo A, Yamada T: Adrenoleukodystrophy protein-deficient mice represent abnormality of very long chain fatty acid metabolism. Biochem Biophys Res Commun 1997; 232:631.
16 Schaumburg HH, Powers JM, Raine, CS, Suzuki K, Richardson EP Jr: Adrenoleukodystrophy. A clinical and pathological study of 17 cases. Arch Neurol 1975;32:577.
17 Ito M, Blumberg BM, Mock DJ, Goodman AD, Moser AB, Moser HW, Smith KD, Powers JM: Potential environmental and host participants in the early white matter lesion of adrenoleukodystrophy: morphologic evidence for CD8 cytotoxic T cells, cytolysis of oligodendrocytes, and CD1-mediated lipid antigen presentation. J Neuropathol Exp Neurol 2001;60:1004.
18 Aubourg P, Chaussain JL: Adrenoleukodystrophy: the most frequent genetic cause of Addison's disease. Horm Res 2003;59:104.
19 Cappa M, Cambiaso P, Del Balzo P, Benedetti S, Borrelli P: Hormonal evaluation in adrenoleukodystrophy; in Uziel G, Wandres RJA, Cappa M (eds): Adrenoleukodystrophy and Other Peroxisomal Disorders. Clinical, Biochemical, Genetic and Therapeutic Aspects. Amsterdam, Excerpta Medica, 1990, pp 81–92.
20 Laureti S, Falorni A, Volpato M, Casucci G, Picchio ML, Angeletti G, Luthman H, Brunetti P, Betterle C, Santeusanio F: Absence of circulating adrenal autoantibodies in adult-onset X-linked adrenoleukodystrophy. Horm Metab Res 1996;28:319.
21 Miike T, Taku K, Tamura T, Ohta J, Ozaki M, Yamamoto C, Sakai T, Antoku Y, Yadomi C: Clinical improvement of adrenoleukodystrophy following intravenous gammaglobulin therapy. Brain Dev 1989;11:134.
22 Cappa M, Bertini E, del Balzo P, Cambiaso P, Di Biase A, Salvati S: High dose immunoglobulin IV treatment in adrenoleukodystrophy. J Neurol Neurosurg Psychiatry 1994;57:69.
23 Rizzo WB, Leshner RT, Odone A, Dammann AL, Craft DA, Jensen ME, Jennings SS, Davis S, Jaitly R, Sgro JA: Dietary erucic acid therapy for X-linked adrenoleukodystrophy. Neurology 1989;39:1415.
24 Moser HW, Raymond GV, Lu SE, Muenz LR, Moser AB, Xu J, Jones RO, Loes DJ, Melhem ER, Dubey P, Bezman L, Brereton NH, Odone A: Follow-up of 89 asymptomatic patients with adrenoleukodystrophy treated with Lorenzo's oil. Arch Neurol 2005:62: 1073.
25 Peters C, Charnas LR, Tan Y, Ziegler RS, Shapiro EG, DeFor T, Grewal SS, Orchard PJ, Abel SL, Goldman AI, Ramsay NKC, Dusenbery KE, Loes DJ, Lockman LA, Kato S, Aubourg PR, Moser HW, Krivit W: Cerebral X-linked adrenoleukodystrophy: the international hematopoietic cell transplantation experience from 1982 to 1999. Blood 2004;104:881.
26 Moser HW, Raymond GV, Dubey P: Adrenoleukodystrophy: new approaches to a neurodegenerative disease. JAMA 2005;294:3131.
27 Copelan EA: Hematopoietic stem-cell transplantation. N Engl J Med 2006;354:1813.
28 Krivit W, Aubourg P, Shapiro E, Peters C: Bone marrow transplantation for globoid cell leukodystrophy, adrenoleukodystrophy, metachromatic leukodystrophy, and Hurler syndrome. Curr Opin Hematol 1999;6:377.
29 Loes DJ, Hite S, Moser H, Stillman AE, Shapiro E, Lockman L, Latchaw RE, Krivit W: Adrenoleukodystrophy: a scoring method for brain MR observations. AJNR Am J Neuroradiol 1994;15: 1761.

30 Cartier N, Hacein-Bey-Abina S, Bartholomae CC, Veres G, Schmidt M, Kutschera I, Vidaud M, Abel U, Dal-Cortivo L, Caccavelli L, Mahlaoui N, Kiermer V, Mittelstaedt D, Bellesme C, Lahlou N, Lefrère F, Blanche S, Audit M, Payen E, Leboulch P, l'Homme B, Bougnères P, Von Kalle C, Fischer A, Cavazzana-Calvo M, Aubourg P: Hematopoietic stem cell gene therapy with a lentiviral vector in X-linked adrenoleukodystrophy. Science 2009;326: 818.

31 Fa M, Diana A, Carta G, Cordeddu L, Melis MP, Murru E, Sogos V, Banni S: Incorporation and metabolism of c9,t11 and t10,c12 conjugated linoleic acid isomers in rat brain. Biochim Biophys Acta 2005;1736:61.

Marco Cappa
Unità Operativa Complessa di Endocrinologia e Diabetologia
Dipartimento Pediatrico Universitario Ospedaliero, Ospedale Bambino Gesù, IRCCS
Piazza S. Onofrio, 4, IT–00165 Roma (Italy)
Tel. +39 06 685 92508, Fax +39 06 695 93374, E-Mail marco.cappa@opbg.net

Ghizzoni L, Cappa M, Chrousos G, Loche S, Maghnie M (eds): Pediatric Adrenal Diseases.
Endocr Dev. Basel, Karger, 2011, vol 20, pp 161–172

Autoimmune Addison's Disease

Corrado Betterle · Luca Morlin

Clinical Immunology, Endocrine Unit, Department of Medical and Surgical Sciences, University of Padova, Padova, Italy

Abstract

Primary adrenocortical insufficiency, or Addison's disease (AD), results from an adrenal cortex hypofunction/dysfunction with a deficient production of glucocorticoids, mineralocorticoids and androgens, and with high levels of both ACTH and plasma renin activity. The prevalence of AD is 110–144 cases per million population in the developed countries. Autoimmune AD is the most frequent etiological form in adult patients, accounting for about 80% of cases, followed by post-tuberculosis AD in 10–15%, the remaining 5% being cases are due to vascular, neoplastic or rare genetic forms. Congenital adrenal hyperplasia is the most frequent form of AD in children and accounts for 72% of cases, whereas autoimmune AD is seen in around 10–15% of cases. The markers of autoimmune AD are adrenal cortex (ACA) or 21-hydroxylase autoantibodies (21-OHAbs) and they are present at diagnosis in more than 90% of cases. In autoimmune AD, the adrenal cortex is infiltrated by lymphocytes and plasma cells and the glands are sclerotic and reduced in volume. Autoimmune AD occurs mainly in middle-aged females, alone or associated with other (clinical, subclinical or potential) autoimmune diseases, giving rise to various forms of autoimmune polyglandular syndrome (type 1, 2 or 4). Replacement therapy with gluco- and mineralocorticoids is life-saving for patients with chronic adrenal insufficiency.

Primary Adrenocortical Insufficiency

The bilateral destruction or dysfunction of the adrenal cortex gives rise to primary adrenocortical insufficiency, or Addison's disease (AD), and the biochemical pattern of this condition is a deficient production of glucocorticoids, mineralocorticoids and androgens, associated with high levels of both ACTH and plasma renin activity [1].

Prevalence of AD

The prevalence of AD in Europe has increased from 40–70 cases per million population in the 1960s [2, 3] to 93–144 cases per million population by the end of the

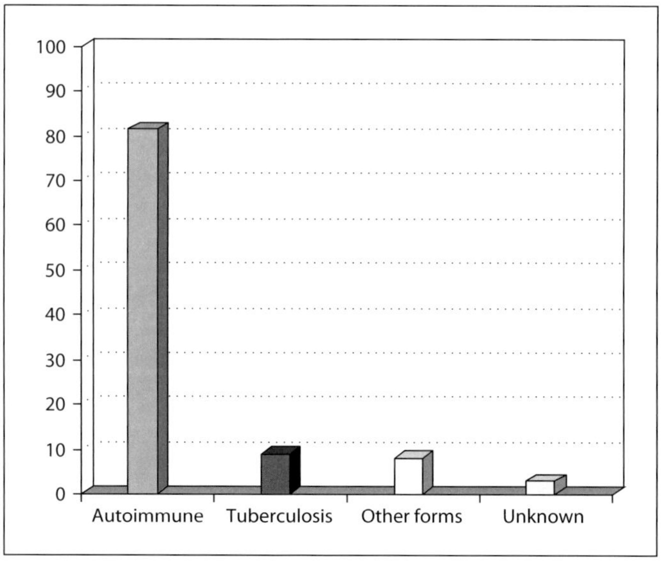

Fig. 1. Causes of AD in our series of 615 cases.

last century and in recent years [4–8], with a currently estimated incidence of 4.4–6 new cases per million population a year [7]. In Europe, tuberculosis was the most common cause of AD during the first half of the last century [9], but in recent years autoimmune AD has become the most frequent form of AD [8]. The increase in the frequency of AD over the last few decades, associated with a decline in the cases of tuberculosis, is indicative of the rising proportion of cases of autoimmune AD [10].

In our series of 615 patients with AD, collected from 1969 to 2009, the autoimmune form was diagnosed in 501 cases (81.5%), the tuberculosis-related form in 9%, and other causes (including infections, genetics, infiltrative diseases, primary and secondary cancers, drugs or surgery-induced hypoadrenalism) account for about 8% of cases (fig. 1).

In the pediatric population with AD, the frequency of the various forms of the disease differs considerably from the situation in the adult population, with a prevalence of the genetic forms. In a paper assessing 103 cases of AD seen at the Pediatric Endocrinology Service in Quebec (Canada) over 20 years (1981–2001), the most frequent etiology was congenital adrenal hyperplasia, accounting for 71.8% of cases, while other genetic forms accounted for another 5.9%, the autoimmune forms for only 12.7%, the remainder being of unknown cause [11].

Histopathology

In patients with autoimmune AD, the adrenal glands are small and weigh <1–2 g, and it is difficult to identify them on imaging or locate them during on autoptic examination.

The cortex is usually atrophic and infiltrated with mononuclear cells characterized by lymphocytes, sometimes with germinal centers, plasma cells and macrophages. The remaining adrenocortical cells are hyperplastic and surrounded by lymphocytes, while the cells of the medulla are normal and are not infiltrated. There are varying degrees of fibrosis. The majority of the infiltrating cells are T cells, with a CD4/CD8 ratio of 5–6/1, with only 5% of B cells. Almost half of them are positive for class II HLA [12].

Cellular Immunity

Increased levels of circulating Ia-positive T cells have been reported in patients with AD of recent onset [13], and a proliferative T-cell response to an 18- to 24-kDa molecular weight adrenal-specific protein fraction has been demonstrated [14]. A defective suppressor function of the human CD4+/CD25+ regulatory T cells has also been found in patients with autoimmune AD in the context of an autoimmune polyglandular syndrome (APS) type 2, but not in patients with autoimmune AD alone [15]. It was recently demonstrated that patients with autoimmune AD have circulating peripheral blood mononuclear cells that can be stimulated by 21-OH enzyme and specifically by amino acids 342–361 of the enzymatic molecule, possibly constituting a particular disease epitope represented by cells bearing HLA-DRB1*0404. These cells stimulated by 21-OH peptides are capable of producing IFN-γ and IL-2 in vitro [16].

Animal Models

An experimental adrenalitis was induced in rats by injecting adrenal extracts characterized by a mononuclear cell infiltration, with adrenal cortex insufficiency and circulating antibodies against adrenal tissue [17, 18], and the passive transfer of this disease was induced by lymphoid cells in these animals [19]. A spontaneous type of autoimmune AD was also reported in dogs [20–25] and in cats [21, 26–28], but the presence of adrenal cortex autoantibodies (ACA) was never described in these animals. NOD (non-obese diabetic) mice, which spontaneously develop type 1 diabetes mellitus, may also reveal a mononuclear cell infiltrate in the adrenal cortex, in the absence of overt signs of hypoadrenalism [29].

Humoral Immunity

Adrenal Cortex Autoantibodies (ACA)
Autoantibodies to adrenal cortex extracts were first demonstrated in 1957 in 25% of patients with 'idiopathic' AD [30]. Subsequently, using the classical indirect immunofluorescence technique on thin slides of adrenal cortical tissue, ACA were

demonstrated in 51% of patients with autoimmune AD [31]. These antibodies are organ- but not species-specific and are directed against a microsomal autoantigen located in the cytoplasm of all three cell layers of the adrenal cortex [31–33]. Using normal animal or human adrenal cortex tissues, the immunofluorescent method was employed for 30 years thereafter and was able to identify ACA in 61% of patients with autoimmune AD of varying duration and in about 90% of those with newly-diagnosed autoimmune AD, but in less than 2% of those with tuberculosis-related AD, as reviewed by Betterle [34].

In 1992, it was discovered that 21-hydroxylase (21-OH), an enzyme of the steroidogenic pathway [35–37], is the autoantigen recognized by ACA, and an immunoprecipitation assay using recombinant 21-OH labeled with ^{35}S-methionine was produced in 1995 to detect autoantibodies to 21-hydroxylase (21-OHAbs) [38–40], followed by an immunoprecipitation assay using recombinant human 21-OH labeled with ^{125}I [41]. These methods were able to identify 21-OHAbs in 78% of patients with unselected autoimmune AD, and in more than 90% of cases of recent onset, as well as in 1.9% of the normal population (reviewed by Betterle [34]).

Other Autoantibodies

Autoimmune AD may be preceded, accompanied, or followed by other clinical autoimmune diseases and, in this case, they combine to produce the clinical picture of APS type 1, 2, or 4. In cases of apparently isolated autoimmune AD, autoantibody screening (i.e., for thyroid peroxidase, thyroglobulin, gastric parietal cells, intrinsic factor, steroid-producing cells, islet cells, glutamic acid decarboxylase, second islet antigen, insulin, transglutaminase, etc.) can reveal the presence of one or more subclinical or potential autoimmune diseases [42].

A recent nationwide Norwegian study demonstrated that autoimmune AD can be associated with other clinical autoimmune diseases in 61% of cases [8]. Specifically, thyroid autoimmune diseases are found in 46.7%, type 1 diabetes mellitus in 12%, vitiligo in 11%, pernicious anemia in 10%, alopecia in 3.8%. In addition, if autoantibody screening was used in patients with apparently isolated AD, a further proportion of patients revealed subclinical or potential autoimmune associations. Taking this approach, 88% of the Norwegian population of autoimmune AD patients had one or more associated clinical, subclinical or latent autoimmune diseases, while only 12% of this population had AD alone.

In our population with autoimmune AD, the proportion of patients with one or more clinical, subclinical or latent autoimmune diseases was 82%, while only 18% had AD alone, which confirms the data of the Norwegian population. Specifically, among our patients, 61.9% had thyroid autoimmune diseases, 39.5% had gastric autoimmunity, 39% had premature ovarian failure, 23.8% had type 1 diabetes mellitus, 11% had chronic hypoparathyroidism, 9.5% had chronic candidiasis, 8.2% had vitiligo, 5.3% had celiac disease, 5.3% had alopecia, 4.2% had autoimmune hepatitis, 3.1% had cancer, 1.3% had malabsorption, and 1% had Sjögren's syndrome.

Table 1. Classification of APS

APS type 1	Chronic candidiasis + chronic hypoparathyroidism + AD
APS type 2	AD + thyroid autoimmune diseases and/or type 1 diabetes mellitus
APS type 3	Thyroid autoimmune diseases + other autoimmune diseases (excluding AD)
APS type 4	Combinations of two or more autoimmune diseases not falling into the above categories (e.g. AD and autoimmune gastritis, or vitiligo, or alopecia or celiac disease)

Another important point to remember is that, after autoimmune AD has been diagnosed in a given individual, it is important to investigate their first-degree relatives for any signs of clinical, subclinical or potential autoimmune diseases.

Different Clinical Forms of Autoimmune AD

APS were first classified by Neufeld and Blizzard [43, 44]. This classification includes four types of APS and AD may be one of the fundamental diseases in three of these (table 1). Based on the associated autoimmune diseases, patients with autoimmune AD may manifest APS type 1, 2, or 4, or they may have autoimmune AD alone [42], so autoimmune AD can manifest in four main clinical forms. In our personal experience of 501 patients with autoimmune AD, the frequency of the four different forms and their main characteristics are summarized in figure 2.

Genetic and Environmental Factors

The genetic profile seen in autoimmune AD varies greatly according to the clinical forms. APS type 1 is associated with mutations in the AIRE (acronym of AutoImmune REgulator) gene, located on the chromosome 21q22 [45, 46]. So far, more than 60 separate mutations associated with APS type 1 have been identified in the AIRE gene [47].

In Italian patients with AD and APS type 1, we identified specific, different AIRE gene mutations in the different regions [48–52]. In the other forms of autoimmune AD, a link has been demonstrated with the class II HLA genes on the chromosome 6. In particular, in Italian patients with APS type 2 and in patients with isolated autoimmune AD, there is evidence of an increased frequency of HLA-DR3 and/or DR4 and a reduction of DR1 and DR13 [42, 53]. As for patients revealing autoimmune AD with APS type 4, only a few cases have been identified so the data are not evaluable.

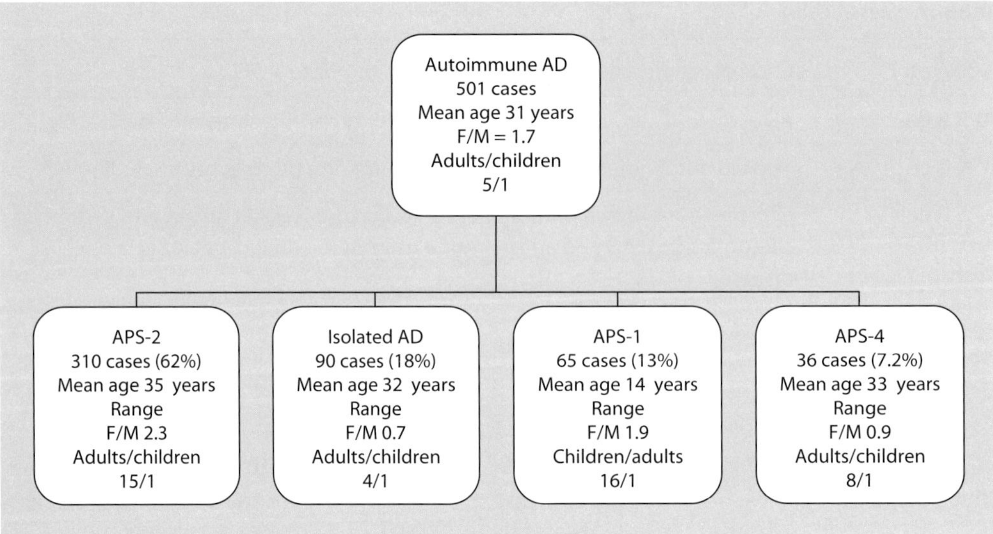

Fig. 2. Main features of autoimmune AD and their subtypes.

Norwegian patients with autoimmune AD revealed a link with HLA-DR3-DQ2/DR4-DQ8 [8]. An association with a microsatellite polymorphism of the CTLA-4 gene was demonstrated in English and Norwegian patients with autoimmune AD unassociated with APS type 1 [54, 55].

There are currently no data available on the pathogenetic environmental factors inducing AD, but we assume that any stressful event may be important and crucial in highlighting a latent AD. The involvement of environmental factors and their potential role in the natural history of autoimmune AD remains to be elucidated, however.

The Natural History of Autoimmune AD

Autoimmune AD is a chronic disorder with a long asymptomatic preclinical phase marked by the presence of ACA/21-OHAbs. In 1983, the predictive role of these autoantibodies was demonstrated when 4/9 ACA-positive patients with organ-specific autoimmunity developed overt AD within 1–31 months of follow-up [56]. The ACTH test made it possible to recognize five different stages (stages 0–4) of adrenal impairment in the natural history of autoimmune AD.

Specifically, stage 0 (a potential AD status) is characterized only by the presence of ACA and/or 21-OHAbs, with no impairment of ACTH, cortisol, plasma renin activity or aldosterone during the ACTH test. Stage 1 (the first level of subclinical AD) is marked by an increase in plasma renin activity: this impairment indicates that the zona glomerulosa is the first to be damaged, or is more vulnerable to the autoimmune assault. Stage 2 (the second level of subclinical AD) is denoted by a low cortisol response to ACTH. Stage 3 (the third level of subclinical AD) is characterized by an increase in basal ACTH with low cortisol levels. Stages 2 and 3 indicate that the zona fasciculata is probably the last to be attacked, or is more resistant to the autoimmune attack because it can protect itself by producing cortisone locally. Finally, stage 4 (clinical AD) features very high ACTH levels with very low cortisol levels and the appearance of the classical clinical signs of the disease [57–61]. Skin hyperpigmentation may occur only many months after the rise in ACTH levels.

Generally speaking, only a proportion of patients with ACA/21 OHAbs develop clinical AD, which is why a risk stratification model of AD has been developed and we have demonstrated that the risk of acquiring AD can be established from five simple variables: patient's age, autoantibody titers, adrenal functional status on the discovery of ACA, gender, and types of autoimmune disease involved [62]. Based on these five parameters, a mathematical model can predict the (low, medium or high) risk of AD developing in the future. Adrenal function should be evaluated every 6 months in patients judged to be at high risk and, in the event of any increase in plasma renin activity, replacement therapy can be started early to prevent a life-threatening adrenal crisis.

Pathogenesis of Autoimmune AD

The mechanisms by which the adrenal cortex is destroyed remain unclear. The disease does not seem to be antibody-mediated. In fact, although it has been demonstrated that ACA/21-OHAbs can block the enzymatic activity in vitro [63], they are not pathogenetic in vivo [64] and the passive transfer of ACAs/21-OHAbs from an addisonian mother to her fetus during pregnancy does not necessarily lead to a form of adrenal insufficiency in the newborn [65]. Moreover, the existence of ACTH receptor-blocking antibodies [66] was not confirmed [67]. The disease thus seems to be due to cell-mediated immunity. Further studies will provide more evidence of how autoantigen expression and immune dysregulation contribute to the onset of autoimmune AD.

Diagnosis of Autoimmune AD

Clinical Signs of AD

Autoimmune AD generally goes through a long, silent preclinical period marked by the presence of ACA/21-OHAbs. The clinical signs and symptoms do not appear until 90%, or more, of the adrenal cortex has been destroyed, though they obviously depend on the rate of progression of the adrenal cortex's destruction and also on superimposed stressful events [42]. The main clinical symptoms are: fatigue, lack of energy, anorexia, general malaise, weakness, gastric pain, nausea, vomiting, weight loss, myalgia, joint pain, dizziness, salt craving, and loss of libido (in women) [1].

The main clinical signs are: cutaneous and mucosal hyperpigmentation, postural hypotension, and loss of axillary and pubic hair [1]. In patients with autoimmune AD, moreover, the clinical evidence of other autoimmune diseases (such as vitiligo, alopecia, type 1 diabetes mellitus, hypogonadism, thyroid dysfunction, malabsorption, etc.) may precede, accompany, or follow the adrenal insufficiency.

Biochemical Tests

At diagnosis, patients have reduced serum levels of sodium, chloride and bicarbonate, while their potassium levels are high. Hyponatremia is found in all patients and is due to a loss of sodium in the urine and to an increase in plasma vasopressin, which impairs free water clearance. The aldosterone deficiency causes hyperkalemia in 50–70% of patients. For unknown reasons, mild or moderate hypercalcemia is recorded in 10–20% of patients [1]. Anemia is identified in 40–50%, with 10–15% of patients showing eosinophilia and lymphocytosis. At diagnosis the majority of patients with autoimmune AD are positive for ACA and/or 21-OHAbs.

Hormone Tests

AD is characterized by an increase in ACTH (normal values: 4–22 pmol/l) and a decrease in basal cortisol levels (normal values: 140–550 nmol/l). In the initial stages of autoimmune AD, plasma renin activity increases above the normal range and plasma aldosterone levels are generally within the normal range. Dehydroepiandrosterone sulfate is involved in adrenal failure, causing a pronounced androgen deficiency in women, resulting in the loss of axillary and pubic hair, dry skin, a reduced libido and an impaired wellbeing [1]. It may be necessary to perform an ACTH test (see above) in cases with no clear clinical signs of AD, but with ACAs/21-OHAbs. TSH levels increase in 30% of patients due to the loss of the inhibitory effect of cortisol on TSH secretion, so it is important to start adrenocortical therapy and only initiate thyroid hormone replacement therapy if the patient's TSH remains high [68].

Imaging

Adrenal imaging is unnecessary in patients who reveal ACA/21-OHAbs, but compulsory in those who do not [69]. In fact, bilateral enlargements occur with or without calcifications in cases of tuberculosis, cancer, infections and infiltrative or vascular disorders [1]. In the event of an anti-phospholipid syndrome being identified, imaging usually shows a larger volume of both adrenal glands due to bilateral hemorrhagic infarction [70].

Therapy

For patients with chronic adrenal insufficiency, therapy is life-saving [1]. Glucocorticoids (20–30 mg of hydrocortisone or 25–50 mg of cortisone acetate) are needed in two or three daily doses. The first dose is administered in the morning, the second in the early afternoon (about 6–8 h after the first), and the third in the early evening. A 24-hour urinary free-cortisol test is considered a good way to evaluate the adequacy of the cortisone intake [71, 72]. A single dose of 0.05–0.1 mg of fludrocortisone usually suffices for mineralocorticoid replacement. Patients should also be given a diet rich in sodium. Measuring blood pressure, serum electrolytes and plasma renin activity levels is useful to optimize the therapy [73]. Patients with AD should carry a medical identification card, outlining their current therapy and containing advice for dealing with emergency situations such as febrile illnesses, injury, vomiting, surgical procedures, dental extractions or pregnancies. In such situations, the glucocorticoid intake needs to be doubled or tripled, possibly via intramuscular or intravenous injection [74].

Acknowledgement

This study was supported in part by a grant from European Union Seventh Framework Program, the Euradrenal Project: Pathophysiology and Natural Course of Autoimmune Adrenal Failure in Europe. Grant Agreement No. 2008-201167.

References

1 Arlt W, Allolio B: Adrenal Insufficiency. Lancet 2003;361:1881.
2 Stuart-Mason A, Meade TW, Lee JAH, Morris JN: Epidemiological and clinical picture of Addison's disease. Lancet 1968;2:19.
3 Nerup J: Addison's disease – clinical studies. A report of 108 cases. Acta Endocrinol 1974;76:127.
4 Kong MF, Jeffocoate W: Eighty-six cases of Addison's disease. Clin Endocrinol (Oxf) 1994;41:757–761.
5 Willis AC, Vince FP: The prevalence of Addison's disease in Coventry, UK. Postgrad Med J 1997;73:286.
6 Laureti S, Vecchi L, Santeusanio F, Falorni A: Is the prevalence of Addison's disease underestimated? J Clin Endocrinol Metab 1999;84:1762.
7 Løvås K, Husebye ES: High prevalence and increasing incidence of Addison's disease in western Norway. Clin Endocrinol (Oxf) 2002;56:787.
8 Erichsen MM, Lovas K, Skinningsrud B, Wolff AB, Undlien DE, Svartberg J, Fougner KJ, Berg TJ, Bollerslev J, Mella B, Carlson JA, Erlich H, Husebye ES: Clinical, immunological, and genetic features of autoimmune primary adrenal insufficiency: observation from a Norwegian registry. J Clin Endocrinol Metab 2009;94:4882.
9 Dunlop D: Eighty-six cases of Addison's disease. Br Med J 1963;3:887.
10 Cooper GS, Stroehla BC: The epidemiology of autoimmune diseases. Autoimmun Rev 2002;2:119.
11 Perry R, Kecha O, Paquette J, Vliet GV, Deal C: Primary adrenal insufficiency in children: twenty years' experience at the Sainte-Justine Hospital, Montreal. J Clin Endocrinol Metab 2005;90:3243.
12 McNicol AM, Laidler P: The adrenal glands and extra-adrenal paraganglia; in: Lewis PD, (ed): Endocrine System. N.Y, Churchill Livingstone, 1996 pp 59–129.
13 Rabinowe SL, Jackson RA, Dluhy RG, Williams GH: Ia-positive T lymphocytes in recently diagnosed idiopathic Addison's disease. Am J Med 1984;77:597.
14 Freeman M, Weetman AP: T- and B-cell reactivity to adrenal antigens in autoimmune Addison's disease. Clin Exp Immunol 1992;88:275.
15 Krieghel MA, Lohmann T, Gabler C, Blank N, Kalden JR, Lorenz HM: Defective suppressor function of human CD4+ CD25+ regulatory T cells in autoimmune polyglandular syndrome type II. J Exp Med 2004;199:1285.
16 Bratland E, Skinningsrud B, Undlien DE, Mozes E, Husebye ES: T-cell responses to steroid cytochrome P450 21-hydroxylase in patients with autoimmune primary adrenal insufficiency. J Clin Endocrinol Metab 2009;94:5117.

17 Andrada JA, Skelton FR, Andrada EC, Milgrom F, Witebsky E: Experimental autoimmune adrenalitis in rats. Lab Invest 1968;19:460.
18 Irino T, Grollman A: Induction of adrenal insufficiency in the rat by sensitization with homologous tissue. Metabolism 1968;17:717.
19 Levine S, Wenk EJ: The production and passive transfer of allergic adrenalitis. Am J Pathol 1968;52:4120.
20 Harlton BW: Addison's disease in a dog. Vet Med Small Anim Clin 1976;71:285.
21 Kaufman J: Diseases of the adrenal cortex of dogs and cats. Mod Vet Pract 1984;65:513.
22 Little C, Marshall C, Downs J: Addison's disease in the dog. Vet Rec 1989;124:469.
23 Kintzer PP, Peterson ME: Diagnosis and management of primary spontaneous hypoadrenocorticism (Addison's disease) in dogs. Semin Vet Med Surg 1994;9:148.
24 Sadek D, Schaer M: Atypical Addison's disease in the dog: a retrospective survey 14 cases. J Am Anim Hosp Assoc 1996;32:159.
25 Dunn KJ, Herrtage ME: Hypocortisolaemia in a Labrador retriever. J Small Anim Pract 1998;39:90.
26 Peterson ME, Greco DS, Orth DN: Primary hypoadrenocorticism in ten cats. J Vet Intern Med 1989;3:55.
27 Tasker S, MacKay AD, Sparkes AH: Case report. A case of feline primary hypoadrenocorticism. J Felin Med Surg 1999;1:257.
28 Stonehewer J, Tasker S: Hypoadrenocroticism in a cat. J Small Anim Pract 2001;42:186.
29 Beales PE, Castri F, Valiani A, Rosignoli G, Buckley L, Pozzilli P: Adrenalitis in the non-obese diabetic mouse. Autoimmunity 2002;35:329.
30 Anderson JR, Goudie RB, Gray KG, Timbury GC: Autoantibodies in Addison's disease. Lancet 1957;1:1123.
31 Blizzard RM, Kyle M: Studies on adrenal antigens and autoantibodies in Addison's disease. J Clin Invest 1963;42:1653.
32 Irvine WJ, Barnes EW: Addison's disease, ovarian failure and hypoparathyroidism. Clin Endocrinol Metab 1975;4:379.
33 Sotsiu F, Bottazzo GF, Doniach D: Immunofluorescence studies on autoantibodies to steroid-producing cells, and to germline cells in endocrine disease and infertility. Clin Exp Immunol 1980;39:97.
34 Betterle C: Addison' disease and autoimmune polyglandular syndromes: in Geenen V, Chrosus G (eds): Immunoendocrinology in Health and Disease. New York, Dekker, 2004, chapt 24, pp 491–536.
35 Baumann-Antczak A, Wedlock N, Bednarek J, Kiso Y, Krishnan H, Fowler S, Rees Smith B, Furmaniak J: Autoimmune Addison's disease and 21- hydroxylase. Lancet 1992;340:429.
36 Bednarek J, Furmaniak J, Wedlock N, Kiso Y, Baumann-Antczak A, Fowler S, Krishnan H, Craft JA, Rees Smith B: Steroid 21-hydroxylase is a major autoantigen involved in adult onset autoimmune Addison's disease. FEBS Lett 1992;309:51.
37 Winqvist O, Karlsson FA, Kampe O: 21-Hydroxylase, a major autoantigen in idiopathic Addison's disease. Lancet 1992;339:1559.
38 Colls J, Betterle C, Volpato M, Rees Smith B, Furmaniak J: A new immunoprecipitation assay for autoantibodies to steroid 21-hydroxylase in Addison's disease. Clin Chem 1995;41:375.
39 Falorni A, Nikoshokov A, Laureti S, Grenbäck E, Hulting AL, Casucci G: High diagnostic accuracy for idiopathic Addison's disease with a sensitive radiobinding assay for autoantibodies against recombinant human 21-hydroxylase. J Clin Endocrinol Metab 1995;80:2752.
40 Chen S, Sawicka J, Betterle C, Powell M, Prentice L, Volpato M, Rees Smith B, Furmaniak J: Autoantibodies to steroidogenic enzymes in autoimmune polyglandular syndrome, Addison's disease, and premature ovarian failure. J Clin Endocrinol Metab 1996;83:2977.
41 Tanaka H, Perez MS, Powell M, Sandres JF, Sawicka J, Chen S, Prentice L, Asawa T, Betterle C, Volpato M, Rees Smith B, Furmaniak J: Steroid 21-hydroxylase autoantibodies: measurements with a new immunoprecipations assay. J Clin Endocrinol Metab 1997;82:1440.
42 Betterle C, Dal Pra C, Mantero F, Zanchetta R: Autoimmune adrenal insufficiency and autoimmune polyendocrine syndromes: autoantibodies, autoantigens, and their applicability in diagnosis and disease prediction. Endocr Rev 2002;23:327.
43 Neufeld M, Blizzard RM: Polyglandular autoimmune diseases; in Pinchera A, Doniach D, Fenzi GF, Baschieri L (eds): Symposium on Autoimmune Aspects of Endocrine Disorders. New York, Academic Press, 1980, pp 357–365.
44 Neufeld M, MacLaren NK, Blizzard RM: Two types of autoimmune Addison's disease associated with different polyglandular autoimmune syndromes. Medicine 1981;60:355.
45 Nagamine K, Peterson P, Scott HS, Kudoh J, Minoshima S, Heino M, Krohn KJE, Lalioti MD, Mullis PE, Antonarakis SE, Kawasaki K, Asakawa S, Ito F, Shimizu N: Positional cloning of the APECED gene. Nat Genet 1997;17:393.

46 The Finnish-German APECED Consortium: An autoimmune disease, APECED, caused by mutations in a novel gene featuring two PHD-type zinc-finger domains. Nat Genet 1997;17:399.
47 Mathis D, Benoist C: Aire. Annu Rev Immunol 2009;27:287–312.
48 Rosatelli MC, Meloni A, Devoto M, Cao A, Scott HS, Peterson P, Heino M, Krohn KJ, Nagamine K, Kudoh J, Shimizu N, Antonarakis SE: A common mutation in Sardinian autoimmune polyendocrinopathy-candidiasis-ectodermal dystrophy patients. Hum Genet 1998;103:428.
49 Cetani F, Barbesino G, Borsari S, Pardi E, Cianferotti L, Pinchera A, Marcocci C: A novel mutation of the autoimmune regulator gene in an Italian kindred with autoimmune polyendocrinopathy-candidiasis-ectodermal dystrophy, acting in a dominant fashion and strongly co-segregating with hypothyroid autoimmune thyroiditis. J Clin Endocrinol Metab 2001;86:4747.
50 Meloni, A, Perniola R, Faa V, Corvaglia E, Cao A, Rosatelli MC: Delineation of the molecular defects in the AIRE gene in autoimmune polyendocrinopathy-candidiasis-ectodermal dystrophy patients from Southern Italy. J Clin Endocrinol Metab 2002;87:841.
51 De Luca F, Valenzise M, Alaggio R, Arrigo T, Crisafulli G, Salzano G, Cervato S, Mariniello B, Lazzarotto F, Betterle C: Sicilian family with autoimmune polyendocrinopathy-candidiasis-ectodermal dystrophy (APECED) and lethal lung disease in one of the affected brothers. Eur J Pediatr 2008;167:1283.
52 Cervato S, Mariniello B, Lazzarotto F, Morlin L, Zanchetta R, Radetti G, De Luca F, Valenzise M, Giordano R, Rizzo D, Giordano C, Betterle C: Evaluation of the autoimmune regulator (AIRE) gene mutations in a cohort of Italian patients with autoimmune-polyendocrinopathy-candidiasis-ectodermal-dystrophy (APECED) and in their relatives. Clin Endocrinol (Oxf) 2009;70:421.
53 Betterle C, Lazzarotto F, Presotto F: Autoimmune polyglandular syndrome type 2: the tip of an iceberg? Clin Exp Immunol 2004;137:225.
54 Kemp EH, Ajjan RA, Husebye ES, Peterson P, Uibo R, Imrie H, Pearce SH, Watson PF, Weetman AP: A cytotoxic T-lymphocytic antigen-4 (CTLA-4) gene polymorphism is associated with autoimmune Addison's disease in English patients. Clin Endocrinol (Oxf) 1998;49:609.
55 Blomhoff A, Lie BA, Myhre AG, Kemp EH, Weetman AP, Akselsen HE, Huseby ES, Undlien DE: Polymorphisms in the cytotoxic T-lymphocyte antigen-4 gene region confer susceptibility to Addison's disease. J Clin Endocrinol Metab 2004;89:3474.
56 Betterle C, Zanchetta R, Trevisan A, Zanette F, Pedini B, Mantero F, Rigon F: Complement-fixing adrenal autoantibodies as a marker for predicting onset of idiopathic Addison's disease. Lancet 1983;1:1238.
57 Betterle C, Scalici C, Presotto F, Pedini B, Moro L, Rigon F, Mantero F: The natural history of adrenal function in autoimmune patients with adrenal autoantibodies. J Endocrinol 1988;117:467.
58 Betterle C, Volpato M, Rees Smith B, Furmaniak J, Chen S, Greggio NA, Sanzari M, Tedesco F, Pedini B, Boscaro M, Presotto F: I. Adrenal cortex and steroid 21-hydroxylase autoantibodies in adult patients with organ-specific autoimmune diseases: markers of low progression to clinical Addison's disease. J Clin Endocrinol Metab 1997;82:932.
59 Betterle C, Volpato M, Rees Smith B, Furmaniak J, Chen S, Zanchetta R, Greggio NA, Pedini B, Boscaro M, Presotto F: II. Adrenal cortex and steroid 21-hydroxylase autoantibodies in children with organ-specific autoimmune diseases: markers of high progression to clinical Addison's disease. J Clin Endocrinol Metab 1997;82:939.
60 Laureti S, Arvat E, Candeloro P, Di Vito L, Ghigo E, Santeusanio F, Falorni A: Low dose (1 g) ACTH test in the evaluation of adrenal dysfunction in preclinical Addison's disease. Clin Endocrinol (Oxf) 2000;53:107.
61 Betterle C, Coco G, Zanchetta R: Adrenal cortex autoantibodies in subjects with normal adrenal function. Best Pract Res Clin Endocrinol Metab 2005;19:85.
62 Coco G, Dal Pra C, Presotto F, Albergoni MP, Canova C, Pedini B, Zanchetta R, Chen S, Furmaniak J, Rees Smith B, Mantero F, Betterle C: Estimated risk for developing autoimmune Addison's disease in patients with adrenal cortex autoantibodies. J Clin Endocrinol Metab 2006;91:1637.
63 Furmaniak J, Kominami S, Asawa T, Wedlock N, Colls J, Rees Smith B: Autoimmune Addison's disease. Evidence for a role of steroid 21-hydroxylase autoantibodies in adrenal insufficiency. J Clin Endocrinol Metab 1994;79:1517.
64 Boscaro M, Betterle C, Volpato M, Fallo F, Furmaniak J, Rees Smith B, Sonino N: Hormonal responses during various phases of autoimmune adrenal failure: no evidence for 21-hydroxylase enzyme activity inhibition in vivo. J Clin Endocrinol Metab 1996;81:2801.
65 Betterle C, Dal Pra C, Pedini B, Zanchetta R, Chen S, Furmaniak J, Rees Smith B: Assessment of adrenal function and adrenal autoantibodies in a baby of a mother with autoimmune polyglandular syndrome type 2. J Endocrinol Invest 2004;27:618.

66 Wulfraat NM, Drexhage HA, Bottazzo GF, Wiersinga WM, Jeucken P, Van der Gaag R: Immunoglobulins of patients with idiopathic Addison's disease block the in vitro action of adrenocotropin. J Clin Endocrinol Metab 1989;69:231.
67 Wardle CA, Weetman AP, Mitchell R, Peers N, Robertson WR: Adrenocorticotropic hormone receptor-blocking immunoglobulins in serum from patients with Addison's disease: a re-examination. J Clin Endocrinol Metab 1993;77:750.
68 Orth DN, Kavacs WJ: The adrenal cortex; in Wilson JD, Foster DW, Kronenberg HM, Larse PR (eds): Williams Textbook of Endocrinology, ed 9. Philadelphia, Saunders, 1998, pp 517–664.
69 Doppman JL: Adrenal imaging; in De Groot LJ, Jameson JL (eds): Endocrinology, ed 4. Philadelphia, Saunders, 2001, vol 125, pp 1747–1766.
70 Presotto F, Fornasini F, Betterle C, Federspil G, Rossato M: Acute adrenal failure as the heralding symptom of primary antiphospholipid syndrome: report of a case and review of the literature. Eur J Endocrinol. 2005;153:507.
71 Burch WM: Urine free-cortisol determination: a useful tool in the management of chronic hypoadrenal states. JAMA 1982;247:2002.
72 Howlett TA: An assessment of optimal hydrocortisone replacement therapy. Clin Endocrinol (Oxf) 1997;46:263.
73 Williams GH, Dluhy RG: Disease of the adrenal cortex; in Fauci AS, Braunwald E, Isselbacher KJ, et al (eds): Harrison's Principles of Internal Medicine, ed 14. New York, McGraw-Hill, 1998, pp 2040–2042.
74 Oelkers W: Adrenal insufficiency. N Engl J Med 1996;335:1206.

Prof. Corrado Betterle
Endocrine Unit, Department of Medical and Surgical Sciences
Via Ospedale Civile, 105, IT–35100 Padova (Italy)
Tel. +39 049 821 4273, Fax +39 049 657 391, E-Mail corrado.betterle@unipd.it

Ghizzoni L, Cappa M, Chrousos G, Loche S, Maghnie M (eds): Pediatric Adrenal Diseases.
Endocr Dev. Basel, Karger, 2011, vol 20, pp 173–180

Optimal Glucocorticoid Therapy

Miguel Debono · Richard J. Ross

Academic Unit of Diabetes, Endocrinology and Metabolism, Department of Human Metabolism, University of Sheffield, Sheffield, UK

Abstract

The rhythmic regulation of human physiology and behaviour is controlled by a central endogenous clock located in the suprachiasmatic nucleus. Most tissues have peripheral clocks that oscillate in time with this central clock. How the central time keeper controls peripheral clocks is not established, however there is evidence to suggest that the cortisol rhythm is one important secondary messenger. Loss of the endogenous cortisol rhythm is associated with sleep disturbance, depression, and metabolic abnormalities. In adrenal insufficiency, current glucocorticoid replacement regimens cannot replace the normal circadian rhythm of cortisol, and patients have an increased mortality and impaired quality of life. We propose that reproducing circadian cortisol levels may improve quality of life in patients with adrenal insufficiency and we have been investigating the impact of circadian hydrocortisone replacement. Using Chronocort, a modified release preparation of hydrocortisone, we have demonstrated that it is possible to simulate the overnight rise in cortisol release and, in preliminary studies in patients with congenital adrenal hyperplasia, control morning androgen levels. Future studies are now required to determine whether Chronocort can improve quality of life in patients with adrenal insufficiency.

Cortisol has a distinct circadian rhythm; levels are low on going to sleep and rise between 02:00 and 04:00 h (fig. 1) [1, 2]. There is an increase in adrenocorticotropic hormone (ACTH) stimulating cortisol secretion just prior to waking, which may provide a signal for waking [3]. Levels then decline through the day. The regulation of cortisol release is through the hypothalamic-pituitary-adrenal axis. The central endogenous clock (pacemaker), located in the hypothalamic suprachiasmatic nuclei (SCN), directly drives the circadian rhythm of corticotrophin-releasing hormone in the paraventricular nucleus which in turn generates circadian rhythms in ACTH and cortisol. The circadian rhythm of cortisol is disrupted in many conditions associated with insomnia including major depression, sleep apnoea, chronic fatigue and in patients taking steroids [4].

Circadian clocks, driven by clock genes, oscillate with a period length (τ) of about 24 h. They require daily adjustment by the photoperiod in order to be synchronised to the 24-hour day. In addition to the master clock in the SCN, molecular oscillators exist in most peripheral cells and tissues of mammals. The phase of these peripheral clocks is reset by signals from the SCN pacemaker and serum can induce circadian gene expression suggesting there is some circulating secondary messenger [5]. Glucocorticoid hormones are attractive candidates for being a secondary messenger from central to peripheral clocks. Using dexamethasone it has been shown that peripheral oscillators can be phase delayed or phase advanced [6]. Appropriately timed, light and melatonin are capable of phase shifting the central SCN clock. The pineal gland hormone, melatonin, also has a circadian rhythm generated by the SCN, with high levels at night which decline to baseline during the day. The melatonin rhythm is considered a reliable marker of the central circadian clock [7]. Thus, current theory suggests that the central clock drives the rhythms of melatonin and cortisol and that these two hormones may regulate peripheral clocks and tissues including regulating sleep, physiology and metabolism.

Current Hydrocortisone Therapy in Adrenal Insufficiency

Since the first published report of the efficacy of cortisone in the treatment of rheumatoid arthritis in 1949, patients with adrenal insufficiency have been treated with glucocorticoid replacement. Hydrocortisone is now used in most centres around the world although only cortisone acetate or synthetic glucocorticoids such as prednisolone are available in some countries. Traditionally the dose of hydrocortisone was two thirds in the morning and one third in the evening to try and imitate the normal cortisol circadian rhythm; 20 mg on waking and 10 mg in the evening. Hydrocortisone has a short plasma half-life of 1.7 h and patients on this dosing regimen wake with undetectable cortisol levels, only achieving peak cortisol levels an hour after taking their morning dose of hydrocortisone, followed by very low levels of cortisol by mid-afternoon, making it impossible to imitate the normal cortisol circadian rhythm [8, 9]. Initial hydrocortisone doses (30 mg) were based on an overestimated value of the cortisol production rate of 12–15 mg/m^2/day resulting in over-replacement. The current recommended total daily hydrocortisone dose (15–20 mg) for a 70-kg patient is approximately 8.1 mg/m^2/day, which is similar to the estimated cortisol production rate measured using stable isotopic methodology [10, 11]. Based on these data, treatment for adults is generally given as a fixed total dose of between 15 and 25 mg hydrocortisone, 2 or 3 times daily [12].

Weight-Related Dosing

Body weight and body surface area are important variables determining hydrocortisone clearance. We examined the variables influencing cortisol levels in 20 cortisol-

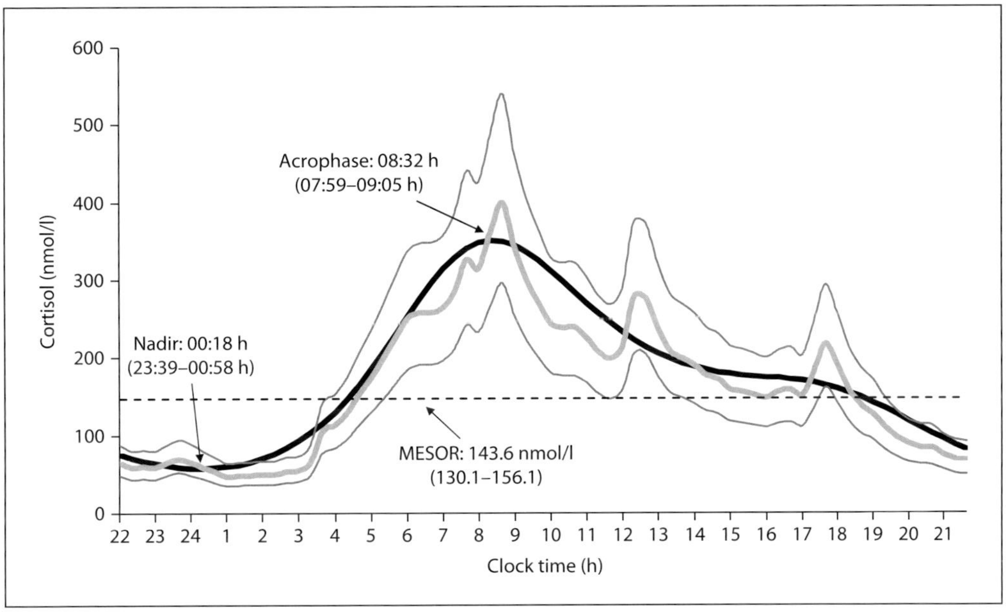

Fig. 1. Physiological cortisol circadian rhythm. Circadian rhythm of cortisol in 33 individuals with 20 min cortisol profiling. Peak cortisol levels are reached at around 08:30 h and nadir cortisol levels at around midnight. The peaks of cortisol at noon and around 18:00 h represent meal-induced cortisol stimulation [taken from Debono et al., JCEM 2009;94:1548–1554].

insufficient patients given oral immediate-release hydrocortisone either as a fixed or as a body surface area-adjusted dose in the fasted or fed state [9]. The results showed that weight-adjusted dosing decreased inter-patient variability in maximum cortisol concentration from 31 to 7%, and reduced overexposure to <5% (fig. 2). A weight-adjusted hydrocortisone dosage for the morning dose produces serum cortisol profiles associated with AUC (area under the curve) values similar to endogenous cortisol. Food intake prior to hydrocortisone ingestion prolonged its absorption half-life (mean ± SEM fed vs. fast 43 ± 12 vs. 15 ± 4 min). Consequently, 3 times daily, weight-related dosing, before food, was recommended as our preferred hydrocortisone treatment regimen [9].

Efficacy of Current Hydrocortisone Replacement Therapy

Patients with adrenal insufficiency have an excess mortality mainly from cardiovascular events and infections [13, 14]. Moreover, with current replacement regimens the majority of patients report impaired health-related quality of life [15, 16], and early morning fatigue [17, 18] with major socioeconomic health implications including

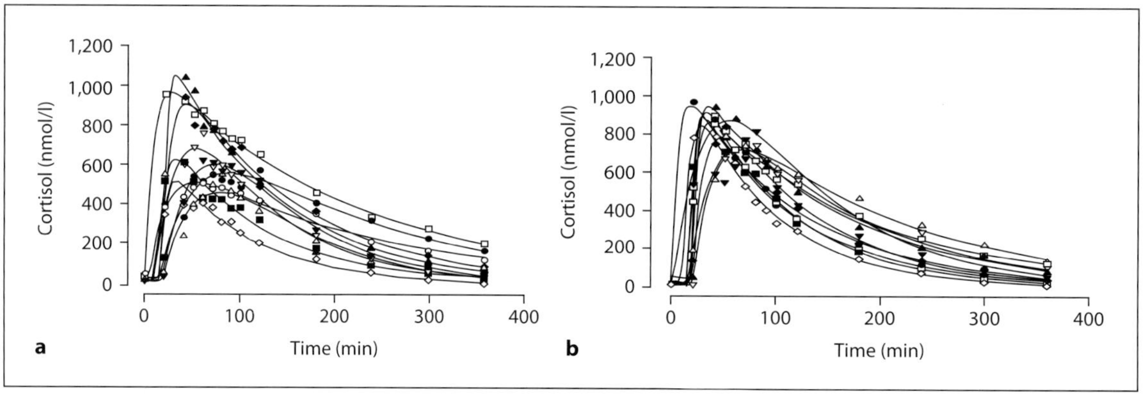

Fig. 2. **a** Serum cortisol concentrations after taking a fixed dose of 10 mg hydrocortisone. **b** Serum cortisol concentration in 10 fasted patients after taking a weight-adjusted (0.12 mg/kg) dose of hydrocortisone [taken from Mah et al., Clin Endocrinol 2004;61:367–375].

an inability to undertake employment. In a study on 79 patients, general health and vitality were both lower, whilst total fatigue, physical fatigue and mental fatigue were all higher than normal, and one quarter of patients in the 18–67 age range were out of work due to disablement compared to 10% in normal controls [18]. A poor quality of life has also been revealed in patients with secondary adrenal insufficiency [19]. Sleep disturbances (difficulties falling asleep, repeated awakenings, early morning awakenings) associated with increased daytime fatigue have been reported for patients with adrenal insufficiency [17], early morning awakening being the most common complaint.

Once-Daily Formulations of Hydrocortisone

One approach to improve hydrocortisone therapy has been to try and generate a once-daily medication. Duocort AB has generated a tablet which combines immediate- and extended-release characteristics with the intention to rapidly achieve physiological cortisol levels in the morning and sustain them over the day [20]. In a study to determine single-dose pharmacokinetics it was shown that the time to reach a clinically significant serum concentration of cortisol (>200 nmol/l) was within 25 min and a peak of 400–450 nmol/l was obtained within 50 min after the 20-mg tablet. Serum cortisol levels remained >200 nmol/l for around 6 h thereafter whereas all serum concentrations 18–24 h after intake were <50 nmol/l. However, this formulation does show variability in cortisol levels and although it may improve compliance it does not address the circadian early morning release of cortisol.

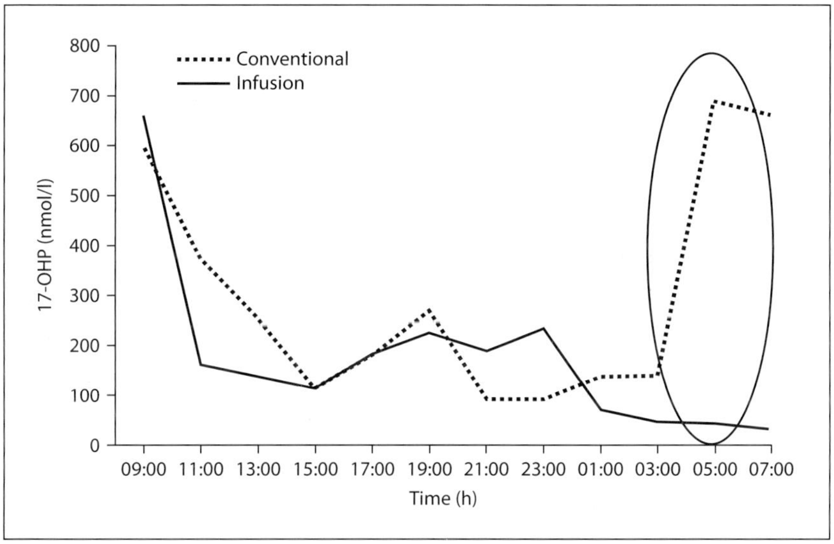

Fig. 3. Comparison of serum 17-OHP levels in 2 patients with CAH during conventional oral replacement therapy and during circadian intravenous infusion of hydrocortisone. The circled area highlights the clinically important difference in 17-OHP levels between the two treatments in the early morning hours of a day [taken from Merza et al., Clin Endocrinol 2006;65:45–50].

Circadian Cortisol Therapy

We and other groups have been investigating circadian hydrocortisone replacement using infusions to imitate the physiological cortisol rhythm. In a proof-of-concept study, intravenous infusions of hydrocortisone, delivered by a programmable pump, were used to assess whether circadian therapy could improve ACTH and androgen levels in patients with Addison's disease and congenital adrenal hyperplasia (CAH). The patients had very high plasma ACTH levels before their morning dose of conventional hydrocortisone treatment (68.5 ± 18.8 pmol/l). In contrast, ACTH levels were much lower after the circadian infusion (15.5 ± 9.9 pmol/l) ($p < 0.05$). Similarly, morning 17-hydroxyprogesterone (17-OHP) levels were 550 and 777 nmol/l after conventional treatment and 3 and 64 nmol/l after the circadian infusion [21] (fig. 3). These data support the notion that delivering physiological hydrocortisone replacement is likely to improve control in these patients. In another study, 7 patients with adrenal insufficiency received continuous subcutaneous hydrocortisone infusions. Morning ACTH levels were brought from high (>39 pmol/l) to near normal levels (<18 pmol/l). The glucocorticoid dose could be reduced in most patients and all were satisfied with the treatment. Two patients felt so much better that they wished to continue continuous subcutaneous hydrocortisone infusions permanently with positive evaluation of subjective health and no adverse events after 24 and 12 months respectively [22].

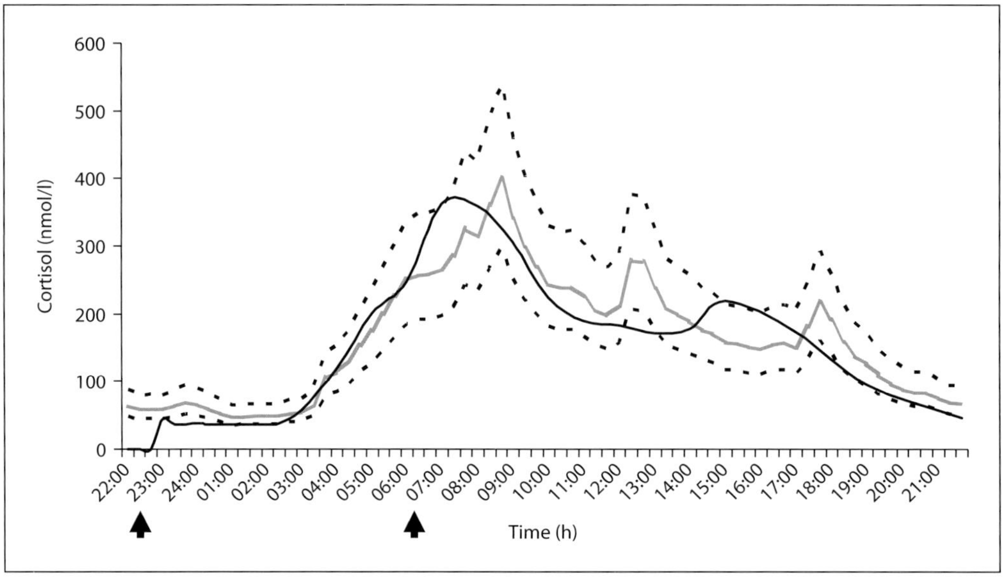

Fig. 4. Simulation of physiological cortisol rhythm using MR-HC, Chronocort (–––). The graph shows modelled concentration time profile obtained when giving 20 mg (15 + 5 mg) MR-HC at 23:00 h and 10 mg MR-HC at 07:00 h (- - -) superimposed on the physiological cortisol rhythm [taken from Debono et al., JCEM 2009;94:1548–1554].

Based on these observations it is clear that the circadian infusion of hydrocortisone can decrease morning ACTH and 17-OHP levels to near normal even in patients with poor biochemical control of their disease. The problem with hydrocortisone infusions are their lack of practicality. An alternative regimen is waking to take immediate-release oral hydrocortisone dose at 03:00 h, and such an approach resulted in a significant improvement in 17-OHP, testosterone and individual urinary 17-ketosteroids in 5 patients with CAH [23]. Although effective, this strategy is not practical for most patients and may result in sleep fragmentation.

A more practical solution is the generation of oral modified-release hydrocortisone formulations. We have developed Chronocort, a modified-release hydrocortisone tablet, that can be taken late at night and allows a delayed and then sustained hydrocortisone release that can simulate the overnight cortisol circadian rhythm, by allowing a rise in circulating cortisol starting in the early hours of the morning and peaking at the time of waking [24]. In a pilot study on 6 dexamethasone-suppressed healthy males (to suppress endogenous cortisol production), two formulations of the drug were tested. Visual inspection of the profiles associated with both formulations indicated a good match to the normal circadian overnight profile but lower cortisol levels than normal after 12:00 h on day 2. Subsequent detailed phase 1 studies in dexamethasone-suppressed normal individuals showed that Chronocort 15 mg demonstrated delayed and sustained release with a mean (SEM) C_{max} of 457 (38.4) nmol/l at 7.41 (0.57) h after drug. Bioavailability of Chronocort at 5, 10 and 15 mg was 100, 79, and 86% that of hydrocortisone. Modelling suggested that Chronocort 15–20 mg at 23:00 h and 10 mg at 07:00 h could reproduce physiological cortisol levels (fig. 4). In a pilot study, Chronocort 30 mg was given as a single night-time dose to 14 patients with CAH and compared to 3 times daily hydrocortisone given as 30 mg in a reverse circadian fashion [25]. Hydrocortisone therapy resulted in three cortisol peaks whereas Chronocort had a single cortisol peak at approximately 06:00 h. Chronocort resulted in significantly ($p < 0.001$) lower 24-hour, afternoon (12:00–20:00 h), and night time (20:00–04:00 h) cortisol compared to hydrocortisone. From 04:00 to 12:00 h, when physiological cortisol is highest, cortisol was higher on Chronocort than hydrocortisone ($p < 0.001$). Patients on Chronocort had significantly ($p < 0.05$) higher afternoon (12:00–20:00 h) 17-OHP, androstenedione and ACTH, but significantly ($p = 0.025$) lower 08:00 h 17-OHP. In conclusion, Chronocort represents a foundation for future drug development in the pursuit of physiological cortisol replacement. Overnight adrenal androgens were well controlled, but rose in the afternoon with once-daily dosing. Further studies are needed to determine the optimal dosing regimen and long-term clinical outcome.

Disclosure Statement

R.J.R. is a founding director and equity holder in Diurnal Ltd that is developing new hydrocortisone preparations for patients with CAH.

References

1 Krieger DT, Allen W, Rizzo F, Krieger HP: Characterization of the normal temporal pattern of plasma corticosteroid levels. J Clin Endocrinol Metab 1971;32:266–284.

2 Debono M, Ghobadi C, Rostami-Hodjegan A, Huatan H, Campbell MJ, Newell-Price J, Darzy K, Merke DP, Arlt W, Ross RJ: Modified-release hydrocortisone to provide circadian cortisol profiles. J Clin Endocrinol Metab 2009;94:1548–1554.

3 Born J, Hansen K, Marshall L, Molle M, Fehm HL: Timing the end of nocturnal sleep. Nature 1999;397:29–30.

4 Buckley TM, Schatzberg AF: On the interactions of the hypothalamic-pituitary-adrenal (HPA) axis and sleep: normal HPA axis activity and circadian rhythm, exemplary sleep disorders. J Clin Endocrinol Metab 2005;90:3106–3114.

5 Balsalobre A, Damiola F, Schibler U: A serum shock induces circadian gene expression in mammalian tissue culture cells. Cell 1998;93:929–937.

6 Balsalobre A, Brown SA, Marcacci L, Tronche F, Kellendonk C, Reichardt HM, Schutz G, Schibler U: Resetting of circadian time in peripheral tissues by glucocorticoid signaling. Science 2000;289:2344–2347.

7 Arendt J, Skene DJ: Melatonin as a chronobiotic. Sleep Med Rev 2000;9:25–39.

8 Derendorf H, Mollmann H, Barth J, Mollmann C, Tunn S, Krieg M: Pharmacokinetics and oral bioavailability of hydrocortisone. J Clin Pharmacol 1991;31:473–476.

9 Mah PM, Jenkins RC, Rostami-Hodjegan A, Newell-Price J, Doane A, Ibbotson V, Tucker GT, Ross RJ: Weight-related dosing, timing and monitoring hydrocortisone replacement therapy in patients with adrenal insufficiency. Clin Endocrinol (Oxf) 2004;61:367–375.

10 Linder BL, Esteban NV, Yergey AL, Winterer JC, Loriaux DL, Cassorla F: Cortisol production rate in childhood and adolescence. J Pediatr 1990;117:892–896.

11 Kerrigan JR, Veldhuis JD, Leyo SA, Iranmanesh A, Rogol AD: Estimation of daily cortisol production and clearance rates in normal pubertal males by deconvolution analysis. J Clin Endocrinol Metab 1993;76:1505–1510.
12 Ten S, New M, Maclaren N: Addison's disease 2001. J Clin Endocrinol Metab 2001;86:2909–2922.
13 Bergthorsdottir R, Leonsson-Zachrisson M, Oden A, Johannsson G: Premature mortality in patients with Addison's disease: a population-based study. J Clin Endocrinol Metab 2006;91:4849–4853.
14 Tomlinson JW, Holden N, Hills RK, Wheatley K, Clayton RN, Bates AS, Sheppard MC, Stewart PM: Association between premature mortality and hypopituitarism. West Midlands Prospective Hypopituitary Study Group. Lancet 2001;357:425–431.
15 Arlt W, Allolio B: Adrenal insufficiency. Lancet 2003;361:1881–1893.
16 Hahner S, Loeffler M, Fassnacht M, Weismann D, Koschker AC, Quinkler M, Decker O, Arlt W, Allolio B: Impaired subjective health status in 256 patients with adrenal insufficiency on standard therapy based on cross-sectional analysis. J Clin Endocrinol Metab 2007;92:3912–3922.
17 Lovas K, Husebye ES, Holsten F, Bjorvatn B: Sleep disturbances in patients with Addison's disease. Eur J Endocrinol 2003;148:449–456.
18 Lovas K, Loge JH, Husebye ES: Subjective health status in Norwegian patients with Addison's disease. Clin Endocrinol (Oxf) 2002;56:581–588.
19 Hahner S, Loeffler M, Arlt W, Koschker A, Weismann M, Fassnacht M, Allolio B: Quality of life in chronic adrenal insufficiency results of a cross-sectional survey in 256 patients and a population-based reference cohort. Endocr Soc 2006;P3–730.
20 Johannsson G, Bergthorsdottir R, Nilsson AG, Lennernas H, Hedner T, Skrtic S: Improving glucocorticoid replacement therapy using a novel modified-release hydrocortisone tablet: a pharmacokinetic study. Eur J Endocrinol 2009;161:119–130.
21 Merza Z, Rostami-Hodjegan A, Memmott A, Doane A, Ibbotson V, Newell-Price J, Tucker GT, Ross RJ: Circadian hydrocortisone infusions in patients with adrenal insufficiency and congenital adrenal hyperplasia. Clin Endocrinol (Oxf) 2006;65:45–50.
22 Lovas K, Husebye ES: Continuous subcutaneous hydrocortisone infusion in Addison's disease. Eur J Endocrinol 2007;157:109–112.
23 Moeller H: Chronopharmacology of hydrocortisone and 9α-fluorhydrocortisone in the treatment for congenital adrenal hyperplasia. Eur J Pediatr 1985; 144:370–373.
24 Newell-Price J, Whiteman M, Rostami-Hodjegan A, Darzy K, Shalet S, Tucker GT, Ross RJ: Modified-release hydrocortisone for circadian therapy: a proof-of-principle study in dexamethasone-suppressed normal volunteers. Clin Endocrinol (Oxf) 2008;68:130–135.
25 Verma S, Vanryzin C, Sinaii N, Kim MS, Nieman LK, Ravindran S, Calis KA, Arlt W, Ross RJ, Merke DP: A pharmacokinetic and pharmacodynamic study of delayed- and extended-release hydrocortisone (Chronocort) vs. conventional hydrocortisone (Cortef) in the treatment of congenital adrenal hyperplasia. Clin Endocrinol (Oxf) 2010;72:441–447.

Prof. Richard J. Ross
University of Sheffield, Room 112, Floor M, Royal Hallamshire Hospital
Glossop Road, Sheffield S10 2JF (UK)
Tel. +44 114 2712052, Fax + 44 114 2711863, E-Mail r.j.ross@sheffield.ac.uk

Ghizzoni L, Cappa M, Chrousos G, Loche S, Maghnie M (eds): Pediatric Adrenal Diseases.
Endocr Dev. Basel, Karger, 2011, vol 20, pp 181–186

Modulation of Glucocorticoid Metabolism by the GH-IGF-I Axis

Sebastian J.C.M.M. Neggers · Aart J. van der Lely

Department of Internal Medicine, Erasmus University MC, Rotterdam, The Netherlands

Abstract

Growth hormone (GH) can generate insulin-like growth factor-I production, provided that the liver encounters portal insulin as a permissive factor that switches on the liver sensitivity for GH. This phenomenon is important for a proper insight into the pathophysiology of diseases as type 1 and 2 diabetes which differ in portal insulin levels. Also, acromegaly and obesity can be better understood when this effect of insulin on liver sensitivity for GH is taken into account. Moreover, as all of these factors seem to influence activity of the 11β-hydroxysteroid dehydrogenase (11β-HSD) type 1 (and 2), an extensive knowledge on the interplay between them is crucial as nowadays treatment options for obesity using the 11β-HSD1 are emerging.

Much is known about the effects of growth hormone (GH) on insulin-like growth factor-I (IGF-I) generation [1]. This also accounts for the interplay between insulin and glucose [2]. Another well-studied system is the 11β-hydroxysteroid dehydrogenase system (11-HSD), which is responsible for the activation and deactivation of glucocorticoids in a tissue-specific manner [3, 4]. The interplay between these systems has been less well studied. Knowledge about this interplay is essential however in order to understand the important aspects of several disease conditions as obesity, diabetes and others. This report addresses some of the interactions between the systems as mentioned above.

Interplay between GH, IGF-I and Insulin

Mammals, including humans, are capable of prolonged fasting because they can recruit fuels from liver, muscle and fat when they run out of readily available carbohydrates

[5, 6]. Because of the hypoglycemic properties of IGF-I, a drop in portal insulin levels should result in a drop of IGF-I action and levels. Therefore, it is essential that insulin has some control over IGF-I action and production. Leung et al. [7] reported that insulin regulates hepatic GH receptor (GHR) biosynthesis and surface translocation in a reciprocal manner, with surface receptor availability the net result of the divergent effects. The divergent actions of insulin appear to be mediated by the mitogen-activated protein kinase and phosphatidylinositol 3-kinase pathways, respectively. So when insulin levels in the portal circulation are high, the liver becomes GH hypersensitive, while reduced portal insulin levels make the liver GH resistant. In this way, insulin can recruit IGF-I and IGF-I action (by reducing IGFBP-1 levels) to help insulin in its hypoglycemic task [8].

In conclusion, high portal insulin levels make the liver GH sensitive and, therefore, increase IGF-I levels, while low portal insulin levels reduce the sensitivity for GH of the liver and, therefore, reduce serum IGF-I levels.

Effects of GH on 11β-HSD1

The interconversion of hormonally active cortisol (F) and inactive cortisone (E) is catalyzed by two isozymes of 11β-HSD, an oxo-reductase converting E to F (11β-HSD1) and a dehydrogenase (11β-HSD2) converting F to E [9].

11β-HSD1 is important in mediating glucocorticoid-regulated glucose homeostasis and regional adipocyte differentiation [9]. Neither IGF-I nor GH has any effect on 11β-HSD2 activity, however [9]. Interestingly, GH seems to inhibit 11β-HSD1 activity through an IGF-I-mediated effect [9, 10]. This is noteworthy in the light of the fact that the liver lacks IGF-I receptors [11].

11β-HSD1 is widely expressed and increases glucocorticoid action through its unique ability to convert inactive glucocorticoids. The enzyme has roles in the normal hypothalamus-pituitary-adrenal axis, has been implicated in metabolic syndrome, and may modulate various aspects of the immune response [3].

According to Stewart et al. [12], the effects of the 11β-HSD isozymes are clinically very important. Patients with acromegaly show a reduction in hepatic-derived metabolites of cortisol/cortisone, and these levels return to normal when GH concentrations are normalized [12]. Conversely, patients with GH deficiency in the setting of hypopituitarism demonstrate an increased cortisol/cortisone metabolite ratio and reduction in circulating cortisol concentrations in patients on hydrocortisone replacement. Treatment with low-dose GH replacement reverses these abnormalities [12].

To date, pharmacological intervention with 11β-HSD1 inhibition in vivo has shown the full range of protective metabolic, anti-obesity, anti-inflammatory and anti-atherogenic effects with a growing body of evidence for beneficial effects in man [13].

Effects of Insulin on 11β-HSD1

Valsamakis et al. [14] studied glucocorticoid metabolism in 33 male subjects with type 2 diabetes mellitus and 38 normal controls. Circulating F:E ratios were elevated in the diabetic group. Interestingly, they found no differences in 11β-HSD1 activity between diabetic subjects and controls [14]. In addition, 11β-HSD1 activity was unaffected by BMI in diabetic subjects only, while in control subjects, increasing BMI was associated with impaired 11β-HSD1 activity [14]. The degree of inhibition correlated tightly with visceral fat mass. One of the reasons for the impaired E to F metabolism in obesity is that it may help preserve insulin sensitivity and prevent diabetes mellitus. Failure to down-regulate 11β-HSD1 activity in patients with diabetes may potentiate dyslipidemia, insulin resistance, and obesity. Inhibition of 11β-HSD1 may therefore represent a therapeutic strategy in patients with type 2 diabetes mellitus and obesity [14].

Liu et al. [15] measured the mean systolic blood pressure and the 11β-HSD activity and mRNA levels for both 11β-HSD1 and 11β-HSD2 in the kidney of streptozotocin (STZ)-diabetic female rats. Three weeks after injection of STZ, the mean systolic blood pressure of diabetic rats was elevated above that of normal rats. They observed that the renal 11β-HSD2 activity and level of mRNA expression were significantly decreased in diabetic rats. However, the treatment of rats with STZ did not decrease the levels of renal 11β-HSD1 activity and mRNA expression in diabetic rats. Insulin administered subcutaneously to diabetic rats for 2 weeks completely reversed the decrease in renal 11β-HSD2 activity and gene expression and prevented the elevation in blood pressure in the diabetic rat. These results indicate that alteration of renal 11β-HSD2 activity and gene expression may be primarily responsible for the changes in blood pressure of STZ-diabetic rats after early treatment with insulin [15].

In conclusion, insulin and IGF-I seem to be able to reduce HSD1 activity, while insulin might be able to increase HSD2 activity of the kidney.

Interplay between the GH-IGF-I Axis and the HSD System in Disease Conditions

Insulin-Dependent and -Independent Diabetes

Type 1 diabetes mellitus is associated with decreased IGF-I, and increased GH and IGF-BP1 levels [16]. Since the liver is the major source of IGF and IGFBP production, Shishko et al. [16] studied whether levels of IGF-I and IGFBPs differ when insulin is infused into the portal or peripheral vascular system. Diabetic patients with conventional subcutaneous insulin therapy have low IGF-I compared with patients with continuous subcutaneous insulin infusion and especially intraportal insulin infusion. In their study, the mean elevated GH level before insulin administration was decreased by especially portal insulin therapy [16]. Their results show the role of glycemic control and the route of insulin administration in the normalization of IGF-1, IGFBP-1 and GH up to non-diabetic controls in patients with recent-onset type 1 diabetes mellitus [16].

Therefore, in most of the IDDM patients who administer their daily insulin subcutaneously, GH levels will remain high and IGF-I levels relatively low. These high GH levels will prevent an increased HSD1-dependent activation of cortisol production which is in line with reports in which type 2 DM and not type 1 DM is accompanied by increased F over E levels [14].

Because of the higher sensitivity of the liver in type 2 DM for GH [7], IGF-I levels are usually relatively high while GH levels are low [1, 17]. In obesity, global 11β-HSD1 activity as measured by urinary corticosteroid metabolite analysis is impaired. This predominantly reflects a decrease in hepatic activity and this is thought to be a compensatory mechanism to preserve insulin sensitivity and decrease hepatic glucose output. In contrast, this decrease in activity does not occur in type 2 diabetics, suggesting that failure to down-regulate 11β-HSD1 activity may contribute to the development of insulin resistance. In the subcutaneous adipose depot, studies have predominantly shown 11β-HSD1 expression and activity is positively correlated to obesity and insulin resistance [18].

Acromegaly

As described above, GH and insulin resistance may exert opposite effects on the conversion of E to F by 11β-HSD1. Mondok et al. [19] determined F and E concentrations in 562 plasma samples obtained from acromegalic patients during an active phase (76 patients) and after cure of the disease (68 patients). In addition, they studied whether type 2 diabetes mellitus or impaired glucose tolerance, frequently associated with active acromegaly, could influence plasma F and E levels in these patients. Interestingly, they found that plasma F concentrations were similar in patients with active acromegaly and in those who were cured, independent of the treatment modality (pituitary surgery, irradiation and/or medical therapy) [19]. However, plasma E levels were significantly higher in patients with active compared to those with cured acromegaly, resulting in a lower F/E ratio in patients with active disease. They also observed significantly lower plasma E levels and higher F/E ratios in active acromegalic patients with type 2 diabetes mellitus or impaired glucose tolerance compared to those with normal carbohydrate metabolism. Plasma F concentrations were similar in these two groups, however [19]. These findings indicate that disease activity exerts a significant impact on 11β-HSD in acromegalic patients, which is further modified with altered carbohydrate homeostasis.

As nowadays acromegalic patients are sometimes treated with the GHR antagonist pegvisomant, it is interesting to investigate the effects of this specific treatment on HSD activity, which is GH-dependent. Trainer et al. [20] studied this influence of pegvisomant on cortisol metabolism in patients with active acromegaly. Seven patients were studied at baseline and again after a mean of 46 weeks of treatment. The tetrahydrocortisol/tetrahydrocortisone ratio increased from a pretreatment mean value of 0.64 ± 0.21 to 0.98 ± 0.26 ($p < 0.02$) and in the 6 patients in whom serum IGF-I normalized, the ratio rose from 0.66 ± 0.23 to 1.01 ± 0.26 ($p < 0.04$).

Their data strongly suggest that blockade of GH action with pegvisomant in patients with acromegaly is associated with reversal of the inhibition of 11β-HSD1 and correction of cortisol metabolism. Moreover, they also observed that in active acromegaly, cortisol clearance is accelerated and that this is also reversed by pegvisomant. This is further evidence of the efficacy of pegvisomant in the management of acromegaly and has important implications for determining optimum pegvisomant dosages, as endogenous GH levels are difficult to measure because of cross-reactivity with pegvisomant in the available assays [21].

Conclusions

The many factors that influence 11β-HSD 1 and 2 activities are important as they enable us the make assumptions and predictions on in which direction cortisol activation and/or inactivation will change in many disease conditions in which we intervene with a multitude of (medical) interventions. This is especially true for diabetes, obesity and GH-related disorders as acromegaly and GH deficiency. The expanding knowledge on the tissue-specific regulation of HSD activity will surely lead to promising theories that might include future treatment options in these important diseases that affect many patients.

References

1 Moller N, Jorgensen JO: Effects of growth hormone on glucose, lipid, and protein metabolism in human subjects. Endocr Rev 2009;30:152–177.

2 Quinn SM, Baur LA, Garnett SP, Cowell CT: Treatment of clinical insulin resistance in children: a systematic review. Obes Rev 2010;11:722–730.

3 Cooper MS, Stewart PM: 11β-Hydroxysteroid dehydrogenase type 1 and its role in the hypothalamus-pituitary-adrenal axis, metabolic syndrome, and inflammation. J Clin Endocrinol Metab 2009;94: 4645–4654.

4 Ahmed A, Saksena S, Sherlock M, Olliff SP, Elias E, Stewart PM: Induction of hepatic 11β-hydroxysteroid dehydrogenase type 1 in patients with alcoholic liver disease. Clin Endocrinol (Oxf) 2008;68:898–903.

5 Moller N, Vendelbo MH, Kampmann U, Christensen B, Madsen M, Norrelund H, Jorgensen JO: Growth hormone and protein metabolism. Clin Nutr 2009; 28:597–603.

6 Salgin B, Marcovecchio ML, Humphreys SM, Hill N, Chassin LJ, Lunn DJ, Hovorka R, Dunger DB: Effects of prolonged fasting and sustained lipolysis on insulin secretion and insulin sensitivity in normal subjects. Am J Physiol 2009;296:E454–E461.

7 Leung KC, Doyle N, Ballesteros M, Waters MJ, Ho KK: Insulin regulation of human hepatic growth hormone receptors: divergent effects on biosynthesis and surface translocation. J Clin Endocrinol Metab 2000;85:4712–4720.

8 Frystyk J, Delhanty PJ, Skjaerbaek C, Baxter RC: Changes in the circulating IGF system during short-term fasting and refeeding in rats. Am J Physiol 1999;277:E245–E252.

9 Moore JS, Monson JP, Kaltsas G, Putignano P, Wood PJ, Sheppard MC, Besser GM, Taylor NF, Stewart PM: Modulation of 11β-hydroxysteroid dehydrogenase isozymes by growth hormone and insulin-like growth factor: in vivo and in vitro studies. J Clin Endocrinol Metab 1999;84:4172–4177.

10 Li RS, Nakagawa Y, Liu YJ, Fujisawa Y, Sai S, Nakanishi T, Chapman KE, Seckl JR, Ohzeki T: Growth hormone inhibits the 11β-hydroxysteroid dehydrogenase type 1 gene promoter activity via insulin-like growth factor I in HepG2 cells. Horm Metab Res 2000;40:286–288.
11 Nakae J, Kido Y, Accili D: Distinct and overlapping functions of insulin and IGF-I receptors. Endocr Rev 2001;22:818–835.
12 Stewart PM, Toogood AA, Tomlinson JW: Growth hormone, insulin-like growth factor-I and the cortisol-cortisone shuttle. Horm Res 2001;56(suppl 1):1–6.
13 Tomlinson JW, Durrani OM, Bujalska IJ, Gathercole LL, Tomlins PJ, Reuser TT, Rose GE, Curnow SJ, Stewart PM, Walker EA, Rauz S: The role of 11β-hydroxysteroid dehydrogenase 1 in adipogenesis in thyroid-associated ophthalmopathy. J Clin Endocrinol Metab 2010;95:398–406.
14 Valsamakis G, Anwar A, Tomlinson JW, Shackleton CH, McTernan PG, Chetty R, Wood PJ, Banerjee AK, Holder G, Barnett AH, Stewart PM, Kumar S: 11β-Hydroxysteroid dehydrogenase type 1 activity in lean and obese males with type 2 diabetes mellitus. J Clin Endocrinol Metab 2004;89:4755–4761.
15 Liu YJ, Nakagawa Y, Ohzeki T: Gene expression of 11β-hydroxysteroid dehydrogenase type 1 and type 2 in the kidneys of insulin-dependent diabetic rats. Hypertension 1998;31:885–889.
16 Shishko PI, Dreval AV, Abugova IA, Zajarny IU, Goncharov VC: Insulin-like growth factors and binding proteins in patients with recent-onset type 1 (insulin-dependent) diabetes mellitus: influence of diabetes control and intraportal insulin infusion. Diabetes Res Clin Pract 1994;25:1–12.
17 Yuen KC, Dunger DB: Therapeutic aspects of growth hormone and insulin-like growth factor-I treatment on visceral fat and insulin sensitivity in adults. Diabetes Obes Metab 2007;9:11–22.
18 Gathercole LL, Stewart PM: Targeting the prereceptor metabolism of cortisol as a novel therapy in obesity and diabetes. J Steroid Biochem Mol Biol 2010;122:21–27.
19 Mondok A, Varga I, Glaz E, Szucs N, Toth M, Patocs A, Beko G, Racz K: 11β-Hydroxysteroid dehydrogenase activity in acromegalic patients with normal or impaired carbohydrate metabolism. Steroids 2009; 74:725–729.
20 Trainer PJ, Drake WM, Perry LA, Taylor NF, Besser GM, Monson JP: Modulation of cortisol metabolism by the growth hormone receptor antagonist pegvisomant in patients with acromegaly. J Clin Endocrinol Metab 2001;86:2989–2992.
21 Bidlingmaier M, Freda PU: Measurement of human growth hormone by immunoassays: current status, unsolved problems and clinical consequences. Growth Horm IGF Res 2010;20:19–25.

A.J. van der Lely, MD
Department of Internal Medicine, Section of Endocrinology, Erasmus University MC
PO Box 2040, NL–3000 CA Rotterdam (The Netherlands)
Tel. +31 10 703 2862, Fax +31 10 703 3639, E-Mail a.vanderlelij@erasmusmc.nl

Ghizzoni L, Cappa M, Chrousos G, Loche S, Maghnie M (eds): Pediatric Adrenal Diseases.
Endocr Dev. Basel, Karger, 2011, vol 20, pp 187–193

Effects of Glucocorticoids on the Growth Plate

Julian C. Lui · Jeffrey Baron

Developmental Endocrinology Branch, Program in Developmental Endocrinology and Genetics, Eunice Kennedy Shriver National Institute of Child Health and Human Development, National Institutes of Health, Bethesda, Md., USA

Abstract

Glucocorticoids have a direct, inhibitory effect on the growth plate, as demonstrated by in vivo and organ culture studies. Glucocorticoids slow longitudinal bone growth by inhibiting chondrocyte proliferation, hypertrophy, and cartilage matrix synthesis. The molecular mediators of these effects are poorly understood. Glucocorticoids also delay growth plate senescence. The decreased rate of senescence appears to be a consequence of the growth inhibition and, in particular, may occur because glucocorticoids slow proliferation of stem-like cells in the resting zone and therefore conserve the limited proliferative capacity of these cells. This slowing of senescence appears to explain the phenomenon of catch-up growth following transient glucocorticoid exposure. After the exposure, the growth plate is less senescent, and therefore grows more rapidly than is normal for age. Glucocorticoids cause growth inhibition and subsequent catch-up growth not only in terms of longitudinal bone growth at the growth plate but also in terms of cross-sectional bone growth at the periosteum. Whether the underlying mechanisms are analogous to those at the growth plate is not known.

In mammals, longitudinal bone growth occurs at the growth plate, a cartilaginous structure present in tubular bones and vertebrae. The growth plate is composed of three principal layers: the resting zone, the proliferative zone, and the hypertrophic zone. The resting zone contains progenitor chondrocytes capable of forming new clones of proliferative chondrocytes [1]. In the proliferative zone, these clones of chondrocytes are arranged in columns parallel to the long axis of the bone. Within each column, the cells proliferate rapidly. The chondocytes that are located farthest from the end of the bone in these columns stop dividing and start enlarging to become hypertrophic chondrocytes. In addition, chondrocytes throughout the growth plate produce cartilage, the composition of which varies in the different zones. In isolation, this cell proliferation, cell hypertrophy, and cartilage matrix synthesis would lead to progressive widening of the cartilaginous growth plate. However, simultaneously, the hypertrophic zone is invaded by blood vessels and bone cells, which remodel the newly formed cartilage into bone tissue. The net result is that new bone is progressively created at the bottom of the growth plate, resulting in longitudinal bone growth.

The rate of longitudinal bone growth at the growth plate is governed by a system of endocrine signals including growth hormone, IGF-I, thyroid hormone, glucocorticoids, androgens and estrogens. A likely evolutionary purpose for this endocrine regulation is to conserve nutrients in times of adversity. If a child is undernourished, circulating IGF-I and thyroid hormone decline while glucocorticoid levels rise. In adolescents, undernutrition also causes decline in sex steroids. These endocrine changes suppress bone growth, presumably to conserve nutrients for vital functions. However, as discussed below, during these times of adversity, growth is not simply lost irreversibly. Instead the growth plate is able to conserve much of its growth capacity until conditions improve, thus preserving, at least in part, the individual's adult stature.

Glucocorticoids provide an important component of this endocrine system that governs growth plate function. When glucocorticoid levels rise above physiological concentrations, longitudinal bone growth is potently inhibited. Although some of this effect may be mediated by indirect effects involving other endocrine signals, there is also an important direct, local effect of glucocorticoids on the growth plate. Evidence for a direct effect comes from studies in vivo and in vitro. In vivo, dexmethasone, a potent synthetic glucocorticoid, infused directly into the proximal tibial growth plates of young rabbits caused a 77% inhibition in growth compared to the contralateral, vehicle-treated growth plates [2]. In vitro, addition of dexamethasone to the culture medium caused a similarly severe inhibition of growth in cultured mouse fetal metatarsal bones and in rat fetal and postnatal metatarsal bones [3]. This inhibition results from decreased proliferation and hypertrophy. In addition, dexamethasone may stimulate apoptosis of terminal hypertrophic chondrocytes [4]. Whether this last effect contributes to the growth inhibition is unclear.

Because longitudinal bone growth results from chondrocyte proliferation, hypertrophy, and cartilage matrix secretion, the growth-inhibitory effects of glucocorticoids might result from inhibition of any of these three processes. In fact, there is evidence for an effect on each. Studies using fetal rat metatarsal bones in organ culture indicate a suppressive effect on proliferation and hypertrophy [3]. Similarly, glucocorticoid hinders the proliferative capacity of chondrocytes in vitro [5, 6]. Furthermore, studies in primary cell culture systems suggest a stimulatory effect on proteoglycan synthesis at low concentration but a suppressive effect at high concentration [7].

The molecular mechanisms by which glucocorticoids suppress longitudinal bone growth are not well understood. Several studies have explored the possibility that glucocorticoids may act by modulating local components of the growth hormone-IGF-I system. In cultured rat growth plate chondrocytes, glucocorticoids can downregulate expression of growth hormone receptor, IGF-I, and IGF1R [8]. However, other studies suggest that glucocorticoids increase expression of IGF1R mRNA [9] in porcine-cultured chondrocytes and increase GHR mRNA in vivo in rabbit growth plate [10]. These disparate findings could reflect species differences but may be related to the experimental system chosen; when growth plate chondrocytes are removed from the context of the growth plate and placed in isolated cell culture, their behavior may change substantially [11] presumably because the chondrocytes are no longer subject to the complex array of paracrine signals from neighboring cells or to the normal cell-matrix interactions. Addition of IGF-I to the culture medium can partially compensate for the growth impairment of glucocorticoids, but the cellular mechanisms appear to be different with glucocorticoids inhibiting proliferation primarily and IGF-I stimulating hypertrophy primarily [12]. Thus, it is not clear to what extent glucocorticoids might inhibit growth plate function by modulating the local components of the GH-IGF-I system.

Other studies of the molecular mechanisms have taken an unbiased approach, using microarray to assess global changes in gene expression due to glucocorticoid exposure [13, 14]. In cultured growth plate chondrocytes, addition of dexamethasone to the culture medium significantly changes expression of hundreds of genes including many cytokines and growth factors [14]. However, the functional role of these changes remains to be established as does the applicability to the growth plate in vivo.

Glucocorticoids above physiological levels inhibit not only growth plate function but also growth plate senescence. Growth plate senescence is a developmental process that occurs during juvenile postnatal life [15]. With increasing age, the growth plate progressively involutes. The overall height of the growth plate decreases because of a decrease in the number of resting, proliferative, and hypertrophic chondrocytes [16]. Individual hypertrophic chondrocytes do not grow as large in an older animal and the chondrocyte columns become more widely spaced, with less intervening matrix [16]. In addition, growth plates undergo a progressive loss of function with age, primarily because of a decrease in the rate of chondrocyte proliferation [16]. Thus, growth plate senescence explains why linear growth progressively slows in postnatal life and eventually ceases.

Glucocorticoid excess slows the programmed senescence of the growth plate in young rabbits [17]. This slowing of senescence has also been observed with other growth-inhibiting conditions, including hypothyroidism [18] and tryptophan deficiency [19], suggesting that senescence is driven by the process of growth, rather than age itself. Each of these experiments involved a similar conceptual design in which growth was inhibited transiently in young animals and then the growth-inhibiting condition was withdrawn. After the animals had recovered, the growth plates were examined to determine whether the preceding period of growth inhibition had delayed the program of senescence. In these experiments the growth-inhibiting conditions did delay the progression of multiple structural markers of senescence, such as the decline in growth plate height, functional markers such as the decline in proliferation in the proliferative zone, and molecular markers, such as the decline in IGF-2 mRNA expression.

The mechanism by which growth drives growth plate senescence is not known. One possibility is that chondrocytes in the growth plate might contain a cell-cycle counter. This counter could be achieved if some characteristic of the cell were progressively increased or decreased with each successive cell division and this characteristic affected the subsequent behavior of the cell. While it is unclear whether a cell-cycle counter exists in the growth plate, there is evidence suggesting that a counter exists in some other mammalian tissues. For example, removal of pancreatic progenitor cells in mice during early development results in a smaller pancreas [20], implying that these progenitor cells undergo a defined number of cell divisions before proliferation ceases. If such a cell-division counter is responsible for growth plate senescence, it seems likely that it would reside in the resting zone chondrocytes. Cell-cycle counting would not occur in hypertrophic chondrocytes because these cells are terminally differentiated and do not proliferate. Cell-cycle counting in the proliferative zone would not readily provide an explanation for progression of senescence. Proliferative zone chondrocytes replicate transiently but then undergo hypertrophic differentiation, then apoptosis. Thus these cells remain in the growth plate only transiently and therefore are unlikely to store long-term information about the growth history of the growth plate. In contrast, the resting zone chondrocytes, which serve as progenitor cells, are thought to persist in the growth plate for the long term. This line of reasoning suggests that senescence might be a function of the number of cell divisions undergone by the resting zone chondrocytes [21].

Specifically, resting zone chondrocytes might have a finite proliferative capacity which is gradually exhausted, causing depletion of these cells. In vivo labeling experiments in rabbits support this concept. Resting zone chondrocytes do become depleted with age, both quantitatively depleted, that is, the number of these cells decreases with age, and also qualitatively depleted, that is, the replication rate of these cells declines with age [21]. Furthermore, administration of dexamethasone to juvenile rabbits slows the replication rate of the resting zone chondrocytes and slows their numerical depletion [21]. Thus, taken together, the findings support the concept that glucocorticoids slow senescence by inhibiting resting zone proliferation and consequently slowing the depletion of these stem-like cells. If this model is correct, then resting zone chondrocytes are not true stem cells, because one of the defining characteristics of a stem cell is the ability to proliferate indefinitely, or at least as long as the lifespan of the organism. Consequently, these resting zone chondrocytes have been labeled stem-like cells [15].

The effect of glucocorticoid excess on growth plate senescence may provide an explanation for the phenomenon of catch-up growth. Catch-up growth is defined as a growth rate greater than is normal for chronological age following a period of growth inhibition [22]. Catch-up growth has been observed after a variety of growth-inhibiting conditions, including glucocorticoid excess, in humans and other mammals [22]. It has been suggested that catch-up growth might be caused by a neuroendocrine mechanism within the central nervous system that senses the child's body size, compares that actual body size to an age-appropriate set point, and then adjusts the growth rate accordingly [23]. However, more recent data are not compatible with this neuroendocrine hypothesis. Dexamethasone was infused locally into the growth plates of juvenile rabbits for 4 weeks causing growth inhibition [24]. After the dexamethasone infusion ended, local catch-up growth was observed solely within the dexamethasone-treated growth plates. This anatomic specificity is not readily explained by any systemic mechanism but instead suggests that the mechanism responsible for catch-up growth resides within the growth plates themselves [24]. Furthermore, the growth pattern in the growth plates that had previously received dexamethasone suggested that the catch-up growth occurred because of a delay in the normal senescent decline in the growth rate. Similar patterns have been reported in catch-up growth in rabbits following systemic glucocorticoid treatment [17], in rats following hypothyroidism [18], and in children with celiac disease placed on a gluten-free diet [25]. In each case, once the growth-inhibiting condition resolved, the ensuing growth rate appeared to be a right shift of the normal senescent decline in growth rate. Furthermore, in the animal studies, the magnitude of the time shift was similar to the delay in the other markers of growth plate senescence. Thus, taken together, the findings support the hypothesis that catch-up growth occurs because growth-inhibiting conditions slow growth plate senescence. Consequently, after the growth-inhibiting condition resolves, the growth plates are less senescent and therefore behave like younger growth plates, growing more rapidly and for a longer period of time than is normal for age, causing catch-up growth.

Catch-up growth following glucocorticoid exposure occurs not only in longitudinal bone growth but also in cross-sectional growth [26]. Dexamethasone administration to juvenile rabbits slows periosteal bone formation, resulting in a thinner bone shaft. However, after the dexamethasone treatment was stopped the rate of periosteal bone formation in these rabbits exceeded that of controls, resulting in catch-up growth in bone width [26]. Whether this catch-up growth at the periosteal surface involves mechanisms analogous to those at the growth plate is unknown.

In summary, glucocorticoid in excess of physiological levels can inhibit longitudinal bone growth by a direct, local action at the growth plate. This slowing of growth appears to involve an inhibition of chondrocyte proliferation, hypertrophy, and cartilage matrix secretion. Glucocorticoid excess also slows growth plate senescence. Senescence is slowed by other growth-inhibiting conditions, suggesting that senescence is not a function solely of age but rather of growth itself. One attractive hypothesis is that stem-like cells in the resting zone have a finite proliferative capacity, and, as this capacity is exhausted, growth in the growth plate progressively slows. Glucocorticoid excess therefore may slow senescence by slowing proliferation in these cells, thereby conserving their proliferative capacity.

Acknowledgement

This work was supported by the Intramural Research Program of the Eunice Kennedy Shriver National Institute of Child Health and Human Development, NIH.

References

1 Abad V, Meyers JL, Weise M, Gafni RI, Barnes KM, Nilsson O, Bacher JD, Baron J: The role of the resting zone in growth plate chondrogenesis. Endocrinology 2002;143:1851.
2 Baron J, Huang Z, Oerter KE, Bacher JD, Cutler GB Jr: Dexamethasone acts locally to inhibit longitudinal bone growth in rabbits. Am J Physiol 1992;263: E489.
3 Chagin AS, Karimian E, Sundstrom K, Eriksson E, Savendahl L: Catch-up growth after dexamethasone withdrawal occurs in cultured postnatal rat metatarsal bones. J Endocrinol 2010;204:21.
4 Chrysis D, Ritzen EM, Savendahl L: Growth retardation induced by dexamethasone is associated with increased apoptosis of the growth plate chondrocytes. J Endocrinol 2003;176:331.
5 Guenther HL, Felix R, Fleisch H: Comparative study of deflazacort, a new synthetic corticosteroid, and dexamethasone on the synthesis of collagen in different rat bone cell populations and rabbit articular chondrocytes. Calcif Tissue Int 1984;36:145.
6 Robson H, Anderson E, Eden O, Isaksson O, Shalet S: Glucocorticoid pretreatment reduces the cytotoxic effects of a variety of DNA-damaging agents on rat tibial growth-plate chondrocytes in vitro. Cancer Chemother Pharmacol 1998;42:171.
7 Kato Y, Gospodarowicz D: Stimulation by glucocorticoid of the synthesis of cartilage-matrix proteoglycans produced by rabbit costal chondrocytes in vitro. J Biol Chem 1985;260:2364.
8 Jux C, Leiber K, Hugel U, Blum W, Ohlsson C, Klaus G, Mehls O: Dexamethasone impairs growth hormone (GH)-stimulated growth by suppression of local insulin-like growth factor (IGF)-I production and expression of GH- and IGF-I-receptor in cultured rat chondrocytes. Endocrinology 1998;139: 3296.
9 Smink JJ, Koedam JA, Koster JG, van Buul-Offers SC: Dexamethasone-induced growth inhibition of porcine growth plate chondrocytes is accompanied by changes in levels of IGF axis components. J Endocrinol 2002;174:343.
10 Heinrichs C, Yanovski JA, Roth AH, Yu YM, Domene HM, Yano K, Cutler GB Jr, Baron J: Dexamethasone increases growth hormone receptor messenger ribonucleic acid levels in liver and growth plate. Endocrinology 1994;135:1113.
11 Mancilla EE, De LF, Uyeda JA, Czerwiec FS, Baron J: Effects of fibroblast growth factor-2 on longitudinal bone growth. Endocrinology 1998;139:2900.
12 Mushtaq T, Bijman P, Ahmed SF, Farquharson C: Insulin-like growth factor-I augments chondrocyte hypertrophy and reverses glucocorticoid-mediated growth retardation in fetal mice metatarsal cultures. Endocrinology 2004;145:2478.
13 Owen HC, Roberts SJ, Ahmed SF, Farquharson C: Dexamethasone-induced expression of the glucocorticoid response gene lipocalin 2 in chondrocytes. Am J Physiol 2008;294:E1023.
14 James CG, Ulici V, Tuckermann J, Underhill TM, Beier F: Expression profiling of dexamethasone-treated primary chondrocytes identifies targets of glucocorticoid signalling in endochondral bone development. BMC Genomics 2007;8:205.
15 Nilsson O, Baron J: Fundamental limits on longitudinal bone growth: growth plate senescence and epiphyseal fusion. Trends Endocrinol Metab 2004; 15:370.
16 Weise M, De-Levi S, Barnes KM, Gafni RI, Abad V, Baron J: Effects of estrogen on growth plate senescence and epiphyseal fusion. Proc Natl Acad Sci USA 2001;98:6871.
17 Gafni RI, Weise M, Robrecht DT, Meyers JL, Barnes KM, De-Levi S, Baron J: Catch-up growth is associated with delayed senescence of the growth plate in rabbits. Pediatr Res 2001;50:618.
18 Marino R, Hegde A, Barnes KM, Schrier L, Emons JA, Nilsson O, Baron J: Catch-up growth after hypothyroidism is caused by delayed growth plate senescence. Endocrinology 2008;149:1820.
19 Forcinito P, Andrade AC, Finkielstain GP, Baron J, Nilsson O, Lui JC: Growth-Inhibiting Conditions Slow Growth Plate Senescence. J Endocrinol, in press.
20 Stanger BZ, Tanaka AJ, Melton DA: Organ size is limited by the number of embryonic progenitor cells in the pancreas but not the liver. Nature 2007; 445:886.

21 Schrier L, Ferns SP, Barnes KM, Emons JA, Newman EI, Nilsson O, Baron J: Depletion of resting zone chondrocytes during growth plate senescence. J Endocrinol 2006;189:27.
22 Boersma B, Wit JM: Catch-up growth. Endocr Rev 1997;18:646.
23 Tanner JM: Regulation of growth in size in mammals. Nature 1963;199:845.
24 Baron J, Klein KO, Colli MJ, Yanovski JA, Novosad JA, Bacher JD, Cutler GB Jr: Catch-up growth after glucocorticoid excess: a mechanism intrinsic to the growth plate. Endocrinology 1994;135:1367.
25 Emons JA, Boersma B, Baron J, Wit JM: Catch-up growth: testing the hypothesis of delayed growth plate senescence in humans. J Pediatr 2005;147:843.
26 Gafni RI, McCarthy EF, Hatcher T, Meyers JL, Inoue N, Reddy C, Weise M, Barnes KM, Abad V, Baron J: Recovery from osteoporosis through skeletal growth: early bone mass acquisition has little effect on adult bone density. FASEB J 2002;16:736.

Jeffrey Baron
Developmental Endocrinology Branch, Program in Developmental Endocrinology and Genetics
Eunice Kennedy Shriver National Institute of Child Health and Human Development
National Institutes of Health, CRC, Room 1-3330, 10 Center Drive, MSC-1103, Bethesda, MD 20892-1103 (USA)
Tel. +1 301 496 6312, Fax +1 301 402 0574, E-Mail jeffrey.baron@nih.gov

Ghizzoni L, Cappa M, Chrousos G, Loche S, Maghnie M (eds): Pediatric Adrenal Diseases.
Endocr Dev. Basel, Karger, 2011, vol 20, pp 194–201

Growth Hormone Treatment in Children on Chronic Glucorticoid Therapy

Martin O. Savage[a] · Dominique Simon[b] · Paul C. Czernichow[c]

[a]Department of Endocrinology, William Harvey Research Institute, Barts and the Royal London School of Medicine and Dentistry, London, UK; [b]Hôpital Robert Debré, and [c]Hôpital Necker, Paris, France

Abstract

Growth failure is common in paediatric disorders associated with inflammation and nutritional deficiency. Examples are Crohn's disease (CD), juvenile idiopathic arthritis (JIA) and cystic fibrosis. Several factors contribute to the abnormal growth, notably excess pro-inflammatory cytokine production and nutritional deficiency, both of which can disturb the GH-IGF-I axis. Typically, growth hormone (GH) secretion is normal, whereas IGF-I production is compromised. Against this background of a disturbed IGF system, glucorticoid therapy, particularly in JIA, and to some extent in CD is frequently initiated to suppress inflammation and provide symptomatic relief. The availability of GH therapy and its efficacy in non-GH-deficient disorders has led to a number of trials in children with glucocorticoid-dependent diseases. In JIA, GH in pharmacological dosages of 0.066 and 0.047 mg/kg/day was effective in preventing the rapid decrease in height SDS values in the acute phase of the disease and in increasing adult height after long-term therapy respectively. In CD, fewer children are steroid-dependent because of widespread use of enteral nutrition, non-steroidal immunosuppressive agents and anti-tumour necrosis factor therapy. However, two trials of GH, using similar doses to those above, have demonstrated anabolic effects with increase in height velocity, bone mineral density and fat free mass compared to untreated controls. Collaboration between paediatric endocrinologists and other subspecialists will improve the opportunity for successful GH therapy. Treatment should be initiated before or during early puberty.

Growth failure is seen in children with a number of chronic illnesses, particularly those associated with nutritional deficit and active inflammation. Examples are inflammatory bowel disease, notably Crohn's disease (CD) [1, 2], juvenile idiopathic arthritis (JIA) [3] and cystic fibrosis. When the inflammation is causing severe and constant symptoms and appears to be resistant to non-steroidal medications, treatment with glucocorticoids may be initiated and is often used in a long-term maintenance regimen. This is particularly true in JIA, where the symptoms are painful and debilitating [3]. Paediatric CD is now managed less with steroid therapy than previously, but

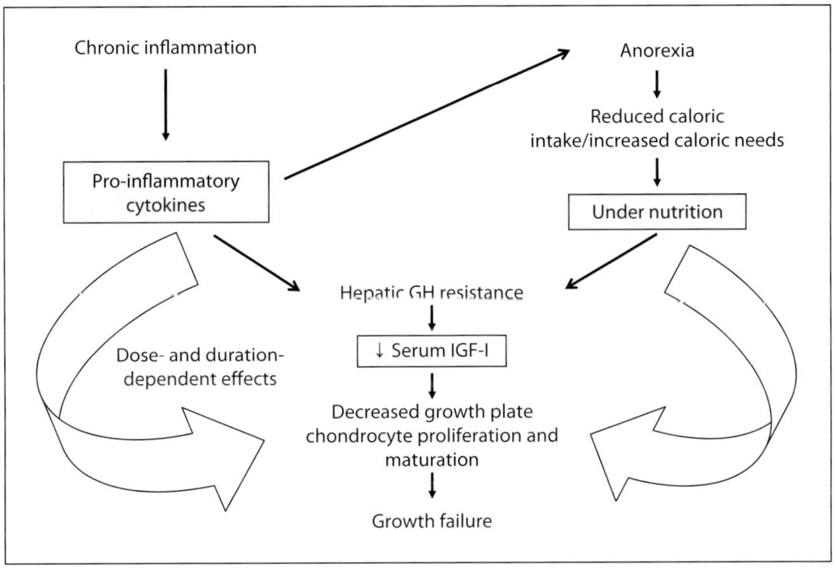

Fig. 1. Mechanisms contributing to growth failure in children with chronic inflammatory disorders.

resistance to non-steroidal immunosuppressive therapy may necessitate its use [1]. Other conditions such as nephrotic syndrome and rarely asthma may be treated with long-term steroids, however the growth disturbance prior to steroid initiation is less severe. Consequently, this chapter will concentrate on JIA and CD.

The availability of recombinant human growth hormone (GH) for treatment of non-GH-deficient growth disorders has led to its regulatory approval in conditions such as Turner syndrome and short stature related to birth weight and length small for gestational age. It was inevitable that studies of the efficacy of GH to promote growth in children with chronic diseases should be performed. Some such studies have been performed in children on chronic steroid therapy.

Disturbance of the GH-IGF-I Axis in CD

The function of the GH-IGF-I axis depends on finely tuned mechanisms which can be influenced by non-endocrine factors seen in states of chronic inflammation and nutritional deficiency. This axis is influenced by the presence of catabolism, increased energy expenditure, deficient nutritional support and inflammatory activity [4]. The two major influences are pro-inflammatory cytokines and nutritional deficiency (fig. 1). Inflammatory cytokines, notably tumour necrosis factor-α, interleukin-6 and interleukin-1β, have been shown to decrease liver GH receptor numbers and disturb STAT5b phosphorylation and IGF-I gene transcription [4]. Nutritional deficit can impair GH secretion and increase protease activity to influence the stability of the ternary complex and reduce the affinity of IGF-I for its carrier proteins, IGFBP-3 and acid-labile subunit. In vitro studies performed in the rat model with colitis induced by trinitrobenzenesulphonic acid have demonstrated an effect of inflammation on IGF-I generation and on linear growth [5]. Changes in the GH-IGF-I axis combine to create a state of relative GH resistance, which is manifested both in juvenile chronic arthritis and CD by decreased or subnormal serum IGF-I concentrations [6, 7] and by impaired linear growth.

This pathophysiologic state is further complicated by the introduction of glucocorticoid therapy, which may further impair GH secretion and also affect growth plate function. So it is in the context of significantly disturbed growth physiology that GH therapy attempts to induce its anabolic actions in a positive way to improve the growth of the child.

Growth Disturbance in JIA

The growth deficit in children with systemic JIA is considerable, due to a combination of the effects of the primary condition and the consequences of long-term glucocorticoid therapy, which remains the cornerstone of management [3, 6]. A significant loss of height (–2.7 ± –1.5 SD) was reported in the early active phase of the disease, which was also correlated with the duration of prednisone therapy [3]. Long-term growth is also severely compromised. Forty percent of children treated with steroids for more than 2 years had a final height below –2 SD and 80% had final heights below their mid-parental target heights [3]. Exposure to increased cytokine production and steroid therapy adversely affects growth plate function. Cytokines in particular restrict recovery of growth plate chondrogenesis [8] and steroids decrease chondrogenesis and chondrocyte differentiation [9].

GH Therapy in Children with JIA

In view of the severity of clinical growth failure in JIA, attempts to improve linear growth using GH therapy in the short term and also to increase adult height are certainly justified in the context of controlled clinical trials. This approach has been adopted in two centres, namely by Dr Dominique Simon in the group of Professor Paul Czernichow at Hôpital Robert Debré in Paris and by Dr Susanne Bechtold in the University Children's Hospital in Munich. Simon et al. [10] have studied the effect of GH therapy relatively early in the course of the disease in order to attempt to prevent the serious decrease in height SDS values, which no doubt contributes to the poor long-term height outcome. This GH 'prevention' study enrolled 30 patients, at mean age of 4.9 years, 12–15 months after initiation of steroid therapy and as soon

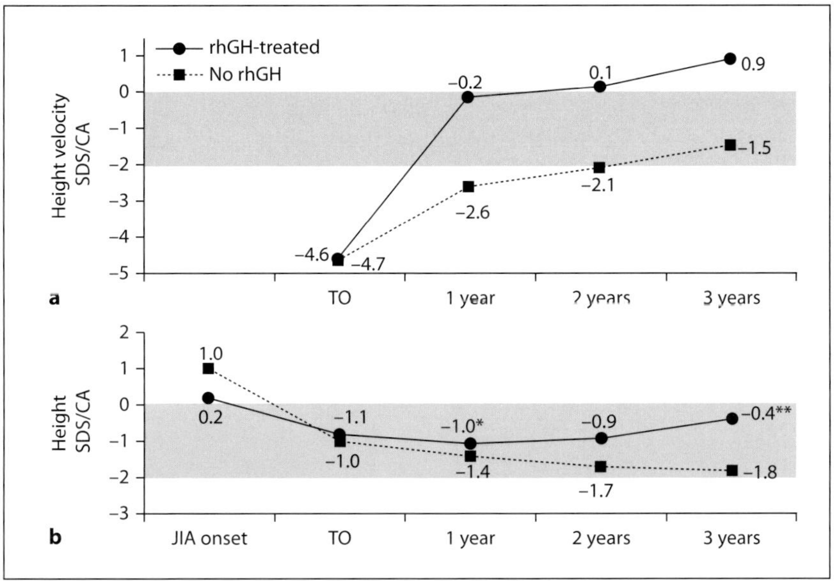

Fig. 2. Height SDS at onset of JIA. Changes in height velocity SDS/CA (**a**) and height SDS/CA (**b**) with and without rhGH group over 3 years. Changes were compared between the two groups over 3 years. Changes in height SDS versus baseline (**b**) were significantly different between the two groups in the first year of the study (* p = 0.01) and in the third year of the study (** p = 0.04). The height SDS gain over the 3-year study was significantly different between the rhGH group and the control group (mean height SDS change, +0.37 ± 1.5 and –0.96 ± 1.2, respectively) [10].

as they demonstrated steroid dependency. Patients were randomised to receive either no GH therapy or GH in a dose of 0.066 mg/kg/day (0.46 mg/kg/week) for a period of 3 years. The two groups were evenly matched and interestingly before GH therapy, serum IGF-I correlated negatively with serum C-reactive protein levels suggesting a suppressive effect of cytokines on IGF-I production.

The results showed that over 3 years the mean change in height SDS was +0.37 ± 1.5 in the GH-treated group compared with –0.96 ± 1.2 in the non-GH group, the difference being significant (p = 0.04). Height velocity also increased significantly in the GH-treated group, remaining low in the non-GH group (fig. 2). Lean mass also increased in the GH-treated subjects and its change was proportional to the change in height SDS, both in the GH-treated and in the non-GH-treated subjects. The choice of a supraphysiological dose of GH was justified by the ability of this regimen to normalise height velocity and thereby prevent the loss of height seen in the untreated group. All of the GH-treated subjects except two received glucocorticoid therapy throughout the 3 years of the study and following the therapeutic requirements in this disorder, other immunosuppressive agents were also prescribed.

The group of Simon [11] have recently reported their experience of analysing glucose metabolism in children treated on long-term glucorticoids in combination with

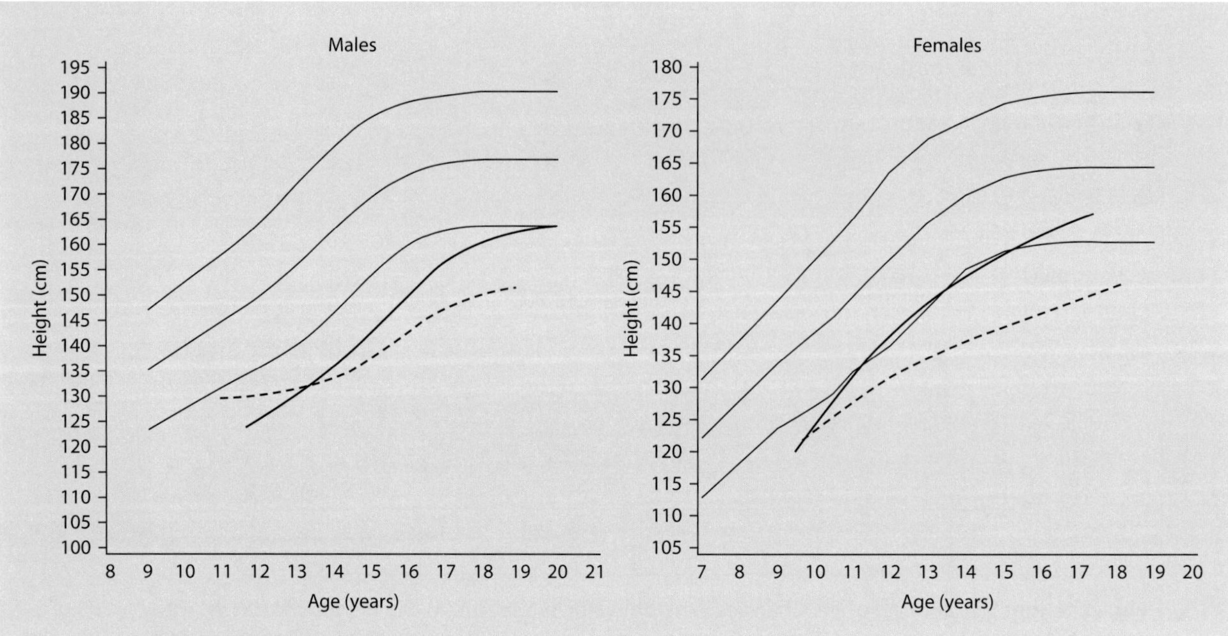

Fig. 3. Synchronized mean growth curves during GH treatment in 13 children and adolescents (solid line) with JIA in comparison with 18 control patients (dashed line) not treated with GH, according to sex; reference lines are for 97th, 50th, and 3rd percentiles, respectively [12].

GH. They report that GH treatment in JIA children decreased insulin sensitivity but had only modest effects on glucose tolerance. A significant number of patients (37%) had some degree of glucose intolerance and transient diabetes was present in some subjects (5%). Close monitoring by oral glucose tolerance testing is strongly recommended before and during GH treatment, particularly during puberty and relapses [11].

The studies performed by Dr Bechtold also demonstrated an impressive beneficial effect of pharmacological doses of GH prescribed in patients until final height. A randomised controlled study enrolled 31 growth-retarded children with chronic JIA, all of whom had received long-term glucocorticoid therapy at initiation of GH treatment [12]. The subjects were randomised to receive either GH 0.047 mg/kg/day (0.33 mg/kg/week), i.e., a slightly lower dose than in Simon's study, or no GH. After a mean treatment period of 6.7 years, the 13 GH-treated subjects reached a mean final height of –1.6 SD compared with the non-GH-treated subjects who had a final height of –3.4 SD ($p = 0.009$). Change in height SDS from baseline was also considerably greater ($p = 0.001$) in the GH-treated subjects (fig. 3). Dr Bechtold's group have also demonstrated increases and normalisation of total bone and muscle cross-sectional area at final height in the same population of GH-treated patients [13].

Growth Disturbance in CD

Growth is frequently disturbed in paediatric CD [2]. This contrasts with ulcerative colitis, where linear growth is usually normal. Height is often subnormal at the time of diagnosis and the degree of abnormality is related to the length of symptoms before diagnosis [14]. The presence of jejunal disease at diagnosis also predisposes to additional height loss. Despite expert treatment in a specialist paediatric gastroenterology department, growth velocity is frequently subnormal [2]. As indicated above, GH secretion is usually normal and serum IGF-I may be at or below the lower limit of normal indicating some degree of GH resistance [15]. Studies of growth until adult height of patients diagnosed with paediatric CD show a long-term height deficit with 25% failing to reach their parental target [16].

There have been changes in the management strategy of paediatric CD during the past 20 years. Glucocorticoids used to be the mainstay of therapy, however these have largely been replaced by the use of enteral nutrition and steroid-sparing immunosuppressive agents such as azathioprine [2]. Anti-tumour necrosis factor agents such as infliximab have also been used and have been shown to improve growth [17].

GH Therapy in Children with CD

There are relatively few publications on the effect of GH on children with CD, but the published series do include patients on glucocorticoid therapy. The group of Mauras [18] published a pilot study demonstrating anabolic effects of GH in 10 children. All had been on prednisone prior to and during treatment with GH. GH was prescribed in a pharmacological dose of 0.05 mg/kg/day. Results of GH therapy showed that body composition changed favourably with increased fat free mass ($p = 0.001$) and decreased percent fat mass ($p = 0.001$) after 4 months of treatment. Linear growth velocity increased from 3.5 ± 0.4 to 7.7 ± 0.9 cm/year after 6 months of combined prednisone/GH ($p = 0.001$). IGF-I and IGFBP-3 also increased. There was no change in overall disease activity. Consequently, it was concluded that GH therapy was able to induce anabolic effects in glucocorticoid-dependent children with CD.

The other definitive multi-centre study in the USA was performed by Heyman et al. [19]. In this study, 2 out of the 10 patients enrolled were on glucocorticoid therapy. The dose of GH was 0.043 mg/kg/day for 12 months. Results were compared with an untreated group. As in the previous study, growth velocity increased significantly, with a change of $+5.3 \pm 3.4$ cm/year. Height SDS also increased over 12 months by 0.76 ± 0.4 compared with 0.16 ± 0.4 in the control group as did bone mineral density ($p = 0.003$ vs. baseline) (fig. 4).

A key factor in all studies of growth promotion is the stage of puberty when the GH therapy was initiated. Treatment of subjects in late puberty does not induce significant acceleration in growth [2]. Consequently, children should be treated either

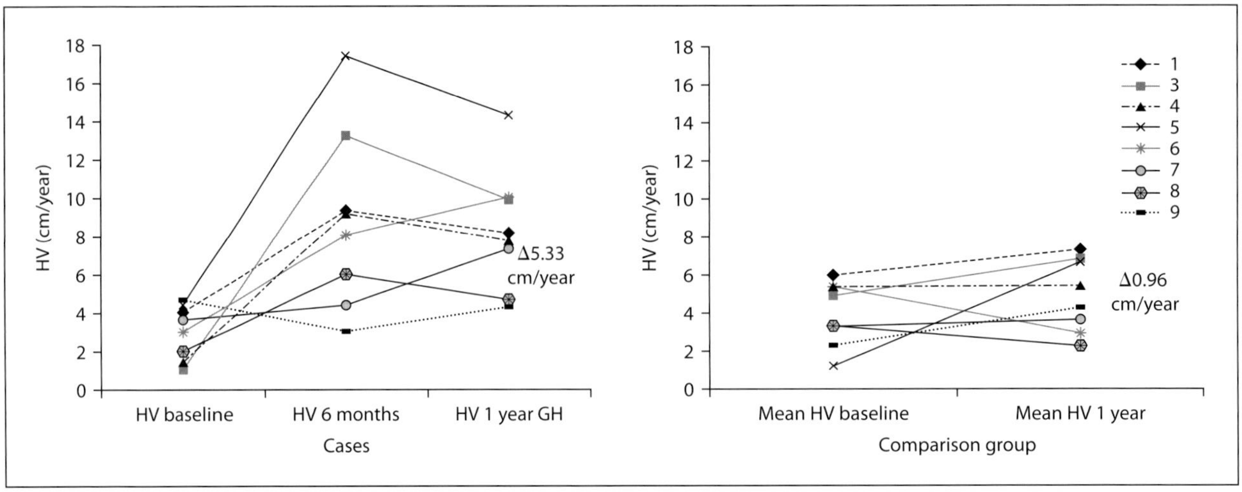

Fig. 4. Height velocity (HV) in 8 paediatric patients with CD at baseline and after 6 and 12 months of GH treatment and in 24 paediatric patients with CD not treated with GH (each line represents the mean of 3 matched control patients for each patient) [19].

before puberty or at the onset of puberty so they can benefit from catch-up growth and if remission of the primary disease is successful, the patient's own puberty growth spurt can provide additional height benefit [2].

Conclusions

The collaborations between departments of paediatric endocrinology and their counterparts in rheumatology or gastroenterology provide a challenge, which can enormously benefit the clinical outcome of the patients and the education of the specialists. Endocrine involvement can strengthen the auxological approach to growth assessment and also contribute to the understanding of potential hormonal abnormalities that might be influencing disturbances of growth and puberty. Successful collaboration has been demonstrated by the groups whose data have been included in this chapter and they are to be complimented for this.

It appears that GH therapy can improve growth in the presence of active inflammation and its management with glucocorticoids. This was particularly impressive in the prevention of the serious growth failure occurring early in JIA. The management of severe chronic inflammatory disorders is in itself a major challenge, particularly with fluctuations of disease activity. Nevertheless, GH therapy should be considered at adjunctive therapy in situations of severe growth failure and the temptation to wait until late puberty to initiate GH therapy should be resisted as the later treatment is started, the more mature the patient will be and the less advantageous will be the long-term benefit.

References

1 Heuschkel R, Salvestrini C, Beattie RM, Hildebrand H, Walters TD, Griffiths A: Guidelines for the management of growth failure in childhood inflammatory bowel disease. Inflamm Bowel Dis 2008;14:839–849.

2 Walters TD, Griffiths AM: mechanism of growth impairment in pediatric Crohn's disease. Nat Rev Gastroenterol Hepatol 2009;6:513–523.

3 Simon D, Fernando C, Czernichow PC, Prieur AM: Linear growth and final height in patients with systemic juvenile chronic arthritis treated with long-term glucocorticoids. J Rheumatol 2002;29:1296–1300.

4 Thissen J-P: How proinflammatory cytokines may impair growth and cause muscle wasting. Horm Res 2007;67(suppl 1):64–70.

5 Sawczenko A, Azooz O, Paraszczuk J, Idestrom M, Croft NM, Savage MO, Ballinger AB, Sanderson IR: Intestinal inflammation-induced growth retardation acts through IL-6 in rats and depends on the -174 IL-6 G/C polymorphism in children. Proc Natl Acad Sci USA 2005;102:13260–13265.

6 Davies UM, Jones J, Reeve J, Camacho-Hübner C, Charlett A, Ansell BM, Preece MA, Woo P: Juvenile rheumatoid arthritis: effects of disease activity and recombinant human growth hormone on insulin-like growth factor 1, insulin-like growth factor binding proteins 1 and 3 and osteocalcin. Arthritis Rheum 1997;40:332–340.

7 Street ME, De Angelis GL, Camacho-Hübner C, Giovanelli G, Ziveri MA, Bacchini PL, Bernasconi S, Sansebastiona G, Savage MO: Relationships between serum IGF-I, IGFBP-2, interleukin-1β and interleukin-6 in inflammatory bowel disease. Horm Res 2004;61:159–164.

8 Macrae VE, Wong SC, Farquarson C, Ahmed SF: Cytokine actions in growth disorders associated with pediatric chronic inflammatory diseases. Int J Mol Med 2006;18:1011–1018.

9 Mushtaq T, Ahmed SF: The impact of corticosteroids on growth and bone health. Arch Dis Child 2002;87:93–96.

10 Simon D, Prieur A-M, Quartier P, Ruiz JC, Czernichow P: Early recombinant human growth hormone treatment in glucocorticoid-treated children with juvenile idiopathic arthritis: a 3-year randomized study. J Clin Endocrinol Metab 2007;92:2567–2573.

11 Bismuth E, Chevenne D, Czernichow PC, Simon D: Moderate deterioration in glucose tolerance during high-dose growth hormone therapy in glucocorticoid-treated patients with juvenile idiopathic arthritis. Horm Res Paediatr 2010;73:465–472.

12 Bechtold S, Ripperger P, Dalla Pozza R, Bonfig W, Häfner R, Michels H, Schwarz HP: Growth hormone increases final height in patients with juvenile idiopathic arthritis: data from a randomized controlled study. J Clin Endocrinol Metab 2007;92:3013–3018.

13 Bechtold S, Ripperger P, Dalla Pozza R, Roth J, Häfner R, Michels H, Schwarz HP: Dynamics of body composition and bone in patients with juvenile idiopathic arthritis treated with growth hormone. J Clin Endocrinol Metab 2010;95:178–185.

14 Sawczenko A, Sandhu B: Presenting features of inflammatory bowel disease in Great Britain and Ireland. Arch Dis Child 2003;88:995–1000.

15 Wong SC, Smyth A, McNeill E, Galloway PJ, Hassan K, McGrogan P, Ahmed SF: The growth hormone-insulin-like growth factor-1 axis in children and adolescents with inflammatory bowel disease and growth retardation. Clin Endocrinol (Oxf) 2010;73:220–228.

16 Sawczenko A, Ballinger AB, Savage MO, Sanderson IR: Clinical features affecting final adult height in patients with pediatric-onset Crohn's disease. Pediatrics 2006;118:124–129.

17 Walters TD, Gilman AR, Griffiths AM: Linear growth improves during infliximab therapy in children with chronically active severe Crohn's disease. Inflamm Bowel Dis 2007;4:424–430.

18 Mauras N, George D, Evans J, Milov D, Abrams S, Rini A, Welch S, Haymond MW: Growth Hormone has anabolic effects in glucocorticoid-dependent children with inflammatory bowel disease: a pilot study. Metabolism 2002;51:127–135.

19 Heyman MB, Garnett EA, Wojcicki J, Gupta N, Davis C, Cohen SA, Gold B, Kirschner BS, Baldassano RN, Ferry GD, Winter HS, Kaplan S: Growth hormone treatment of growth failure in pediatric patients with Crohn's disease. J Pediatr 2008;153:651–658.

Prof. M.O. Savage
Department of Endocrinology, William Harvey Research Institute
Charterhouse Square, London EC1M 6BQ (UK)
Tel. +44 780 308 4491, Fax +44 20 7882 6234, E-Mail m.o.savage@qmul.ac.uk

Ghizzoni L, Cappa M, Chrousos G, Loche S, Maghnie M (eds): Pediatric Adrenal Diseases.
Endocr Dev. Basel, Karger, 2011, vol 20, pp 202–211

Ancient History of Congenital Adrenal Hyperplasia

Maria I. New

Mount Sinai School of Medicine, New York, N.Y., USA

Abstract

Although there are many erudite reports on the history of endocrinology and endocrine disorders, the history of congenital adrenal hyperplasia has not been published. I have tried to review ancient as well as modern history of CAH.

Most endocrinologists regard Giuseppe Marzo as the first patient described with congential adrenal hyperplasia (CAH) (fig. 1). The cadaver of Giuseppe Marzo in figure 1 was described by de Crecchio, who was conducting the autopsy. He published it in the journal *Morgagni* in 1865;7:154. The description indicated marked surprise at finding uterus and fallopian tubes in this patient, who had very large adrenals. The story of the patient, Giuseppe Marzo, was that he was originally assigned as a female called Giuseppina (Josephine), but then was found to have genital ambiguity and was reassigned as a male, called Giuseppe. He was reared as a boy, and as an adult joined the noble household in which his father worked. Giuseppe fell in love with a maid, proposed marriage, and produced his birth certificate, which said Giuseppina. It was reported that she ran away, and that Giuseppe died of a broken heart. Possibly he died of a salt-wasting crisis. Dr. de Crecchio wrote in very flowery language the remarkable size of the adrenals.

However, it is possible that a case of CAH was indirectly reported in 800 AD, when a female pope was said to be elected. This remarkable event is described in a book by D'Onofrio, entitled *La Papessa Giovanna (Pope Joan)* (fig. 2).

The legend of a female pope is first recorded in a manuscript by the friar Jean de Mailly from about 1250 AD. During the late Middle Ages and Reformation, dozens

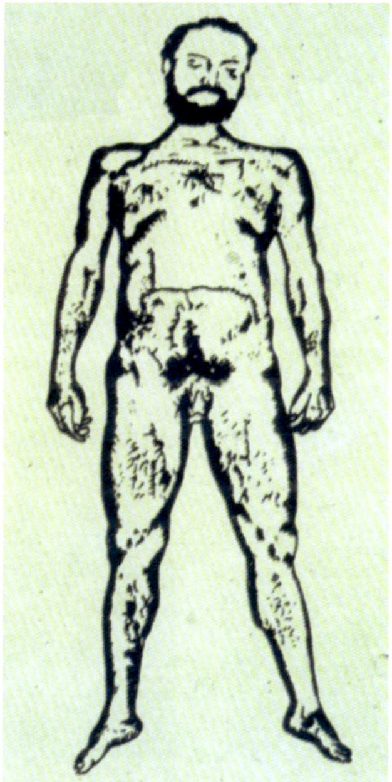

Fig. 1. Giuseppe Marzo from de Crecchio L. Sopra un caso di apparenze virili in una donna. Morgagni 1865;7:154.

of people wrote about this scandal, many of them Franciscan and Dominican friars or Protestants, and their stories were widely believed. The most popular version was a bestseller all over Europe for hundreds of years after its publication around 1265. This appeared in Martin Polanus' *Chronicon Pontificum et Imperatum*:

'After. . . Leo, John Anglicus, born at Mainz, was pope for 2 years, 7 months, and 4 days, and died in Rome, after which there was a vacancy in the papacy of 1 month. It is claimed that this John was a woman, who as a girl had been led to Athens dressed in the clothes of a man by a certain lover of hers. There she became proficient in a diversity of branches of knowledge, until she had no equal, and afterwards in Rome, she taught the liberal arts and had great masters among her students and audience. A high opinion of her life and learning arose in the city, and she was the choice of all for pope. While pope, however, she became pregnant by her companion. Through ignorance of the exact time when the birth was expected, she was delivered of a child while in procession from St. Peter's to the Lateran, in a narrow lane between the Coliseum and St. Clement's church. After her death, it is said that she was buried in that same place. The Lord Pope always turns aside from the street and it is believed by many that this is done because of abhorrence of the event. Nor is she placed on the list of the holy pontiffs, both because of her female sex and on account of the foulness of the matter.'

Boccaccio, the Italian humanist, relished the tale and included it in his book, *De Claris Mulieribus*, published circa 1350. This was the first recorded attribution of a woman's name, Gilberta, to John Anglicus (Pope John VIII). There are various accounts of how Pope Joan died; some authors have her torn limb from limb by an angry crowd; others say she was dragged by horses through the streets of Rome. Petrarch (1304–1376), evidently thinking of the plagues of Revelation, added that

Fig. 2. In 800 AD a female Pope was proposed.

after her death it rained blood for 3 days and nights and locusts with six wings and powerful teeth appeared in France.

Is There Any Evidence That Pope Joan Existed?

Is there any factual basis for the fantastic stories about Pope Joan? To answer this question, it helps to have some idea of the historical framework. In 800, on Christmas Day, Charlemagne was crowned Holy Roman Emperor by Pope Leo III in thanks for his help in staving off the Lombards who periodically attacked Rome and for defending him from enemies among the Roman nobility, who had physically attacked him. But after Charlemagne's death in 814, the papacy fell increasingly under the sway of the impious and powerful Roman nobility. The existence of this corruption is central to understanding the birth of the legend. Joan's papacy was placed by Martin Polanus and most of her

Fig. 3. Porphyry chair with perforated seat dating to classical times. Used in the installation of Popes 1099–1513. Discontinued by saintly Hadrian VI. Used to check the new pope's sex after Joan?

other chroniclers in the years 855–857, between Popes Leo IV (847–855?) and Benedict III (?855–858). Jean de Mailly and some others put her around 1100. But all of these accounts of Pope Joan were written long after she was said to have lived-in most cases 400 years after. The existence of a female pope is not mentioned in any private letters or official communications or histories at the time(s) of the alleged papacy.

Another avenue for this legend might have been the Christian literary tradition of the female saint in male disguise, which derives from the early legend of St. Pelagia the Penitent. Whatever the first source of the Pope Joan legend, turmoil within the Church continued for centuries, culminating in the Protestant Reformation, and so the story lived on, along with growing controversy as to its truth.

Is there any solid evidence that Pope Joan ever lived? There are three facts to weigh. The first is this: In the nave of the Cathedral of Siena is a series of busts representing each of the legitimate popes. The head of Johannes VIII, Femina de Anglia, was placed between those of Leo IX and Benedict III until 1600, when Pope Clement VIII ordered it removed, and it was recycled as the head of Pope Zacchary. Was this a cover-up?

Second, there is the famous 'Sedia Gestatoria,' a red marble chair, now in the Vatican museum, which has a perforated seat (fig. 3). It is known that this chair was used in the ceremonies of installation of the popes between 1099 (Paschall II) and 1513 (Leo X). It is rumored that it was used to ascertain the gender of popes-elect as a direct result of the scandal of a real Pope Joan. The 'junior cleric present' purportedly palpated the genitals of the papal candidate through the hole in the bottom of the chair, and, if he found what he expected, shouted in a loud voice, *'Testiculos habet et bene pendentes. (He has well-formed testicles).'* To which all the clerics replied, *'Habe ova noster papa (Our father is virile)'*. Finally, there is the fact that around the time that the legend of Joan was spread by Martin Polanus' book, formal papal processions began to turn off the most direct route between their home in The Lateran Palace and St. Peter's to avoid the place where Joan was said to have given birth. On route, there is a shrine presumably on the spot where Pope Joan gave birth. I walked that route, and at the shrine, there was a pregnant woman praying. I asked her to whom she was praying, and why. She answered, 'I am praying to Pope Joan for me to have a healthy baby'.

I think that the story of Pope Joan must be a legend because it is very hard for me to believe that this fantastic event could have occurred in the 9th century without a contemporary written record. Yet despite the unlikelihood of its truth, this tale has continued to be retold and even believed up to the present day. It has been the vehicle for both ribald mockery and serious accusation of the Church. It has been used for feminist propaganda, as a basis of romantic fiction and drama, and in card games such as the Tarot, which has a card, the High Priestess, first identified with Pope Joan in the famous deck made for the Visconti-Sforza family by Bonifacio Bembo in the 1400s. My brother-in-law is one of the chief librarians of the Library of Congress, and when I was preparing this lecture, I asked him to send me everything he had on Pope Joan. I was amazed to find that there are hundreds of publications about Pope Joan. Many are written by historians sorting through the evidence, but the story has a life of its own.

Real Patients That Could Have Been Pope Joan

If Pope Joan was a legend, why as a civilization have we not lost interest in her after all these centuries? Why do many people persist in believing that her story could have been true? The fact is, on a biological basis, it might have been. I have known a man who if he had lived in the 9th century could have been Pope Joan.

The very masculinized person shown in figure 4a is a patient with CAH whose karyotype is 46,XX and who was never treated. In figure 4b is the recent picture of a female subject who chose to take testosterone to become masculinized. She changed her birth certificate to that of a male, and married a female who had had 2 children by a former husband. The couple wanted a child and so the husband decided to stop taking testosterone and demasculinize. He began menstruating and became pregnant by artificial insemination.

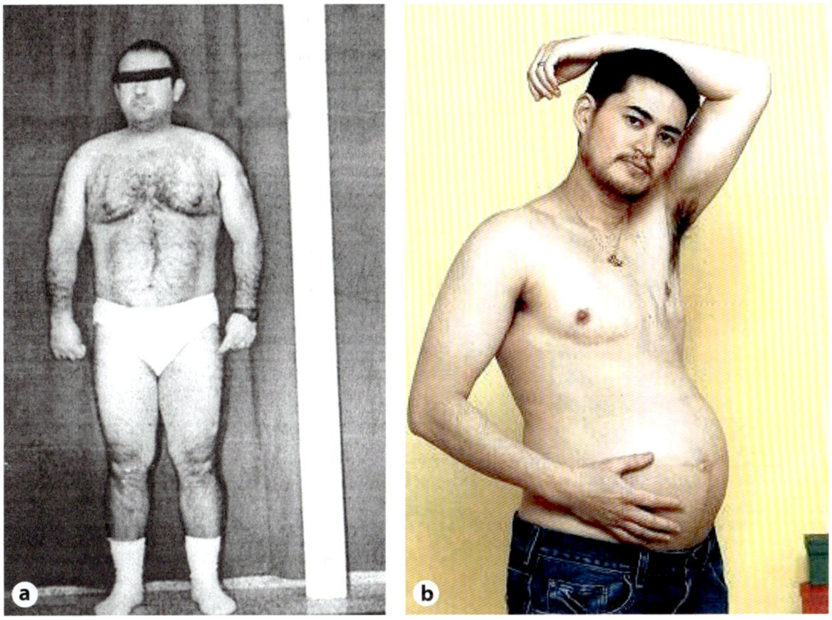

Fig. 4. **a** A very masculinized person with CAH whose karyotype is 46,XX and who was never treated. **b** Recent picture of a female subject who chose to take testosterone to become masculinized (see text for details).

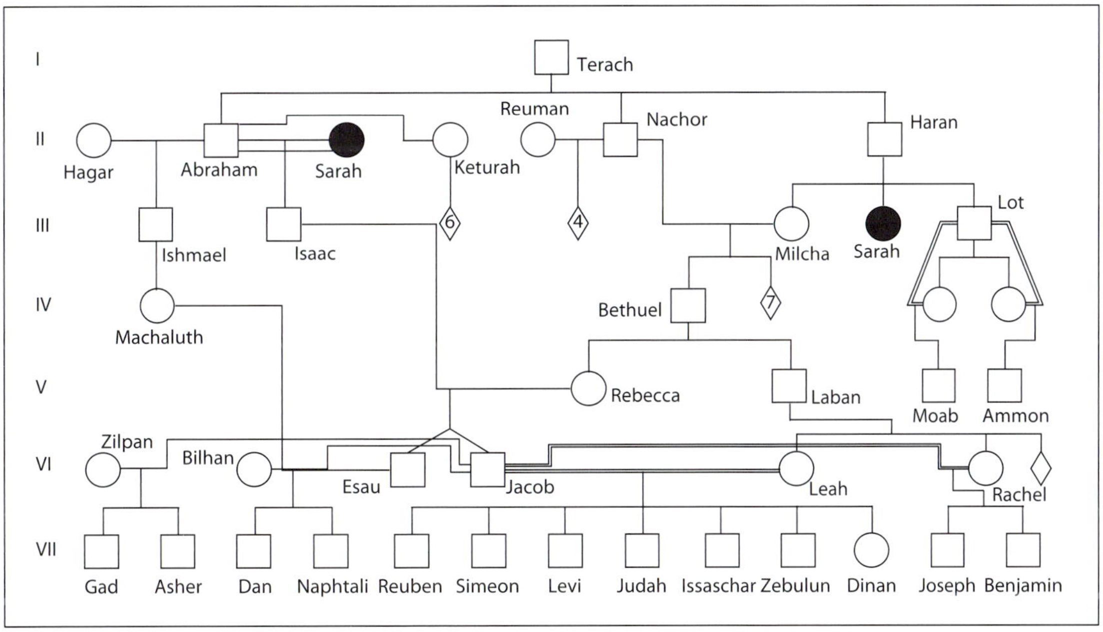

Fig. 5. The origin of the Twelve Tribes of Israel beginning with Terach, the father of Abraham, and proceeding to Jacob with his 12 sons. Note the high degree of consanguinity as indicated by the double lines. Tradition also states that Sarah and Abraham were related.

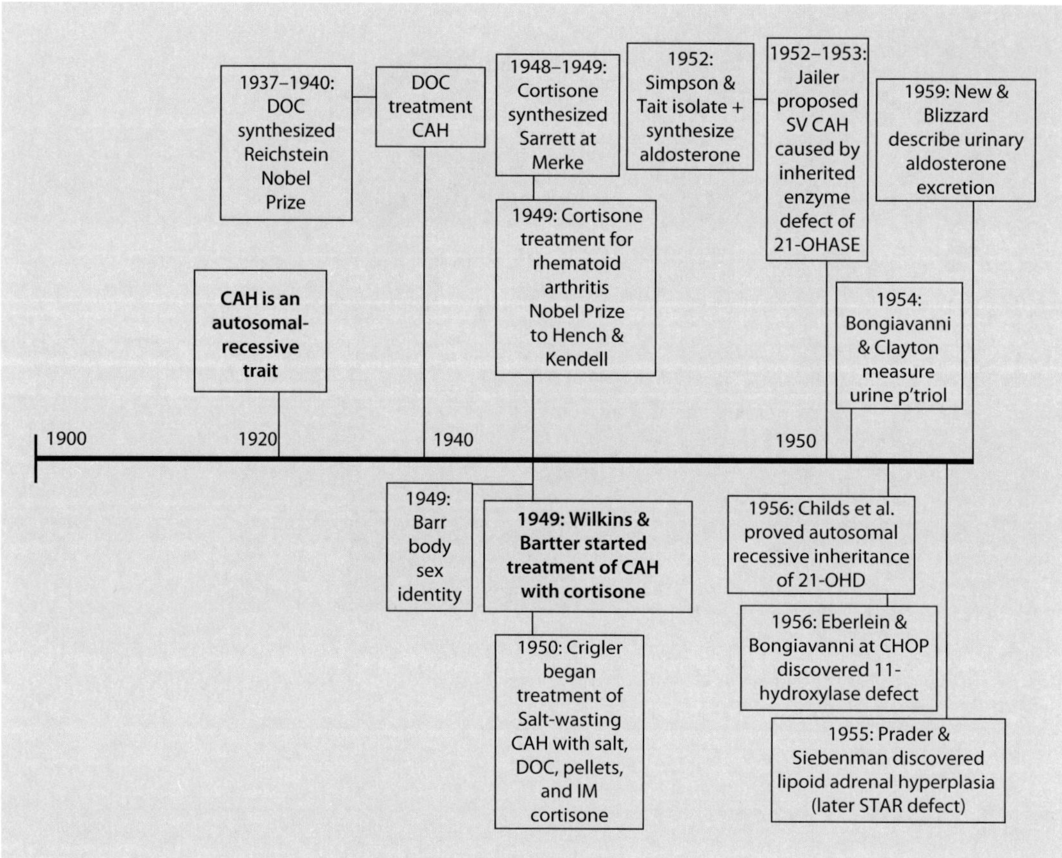

Fig. 6. Modern history of CAH: the timeline indicates that the modern history begins in the 20th century and continues into the 21st century.

In light of such cases, it is possible to imagine the story of Joan would be that Pope John VIII was a female pseudohermaphrodite who had a homosexual liaison, got pregnant, and delivered a child. The endocrinological version of the legend is that Pope Joan had a recognizable syndrome, CAH.

And Even Before 800 AD

The pedigree of the ancient biblical family shows consanguinity and infertility (fig. 5). Abraham married Sarah, who was the daughter of his brother, Haran. Haran died at an early age, and as was the custom, Abraham married the daughter was left orphaned. Further, it was said that Sarah was infertile and told Abraham to take the slave girl Hagar, by whom Abraham has a son called Ishmael. And it is said that at the age of 99, Sarah prayed to God, and gave Abraham a son called

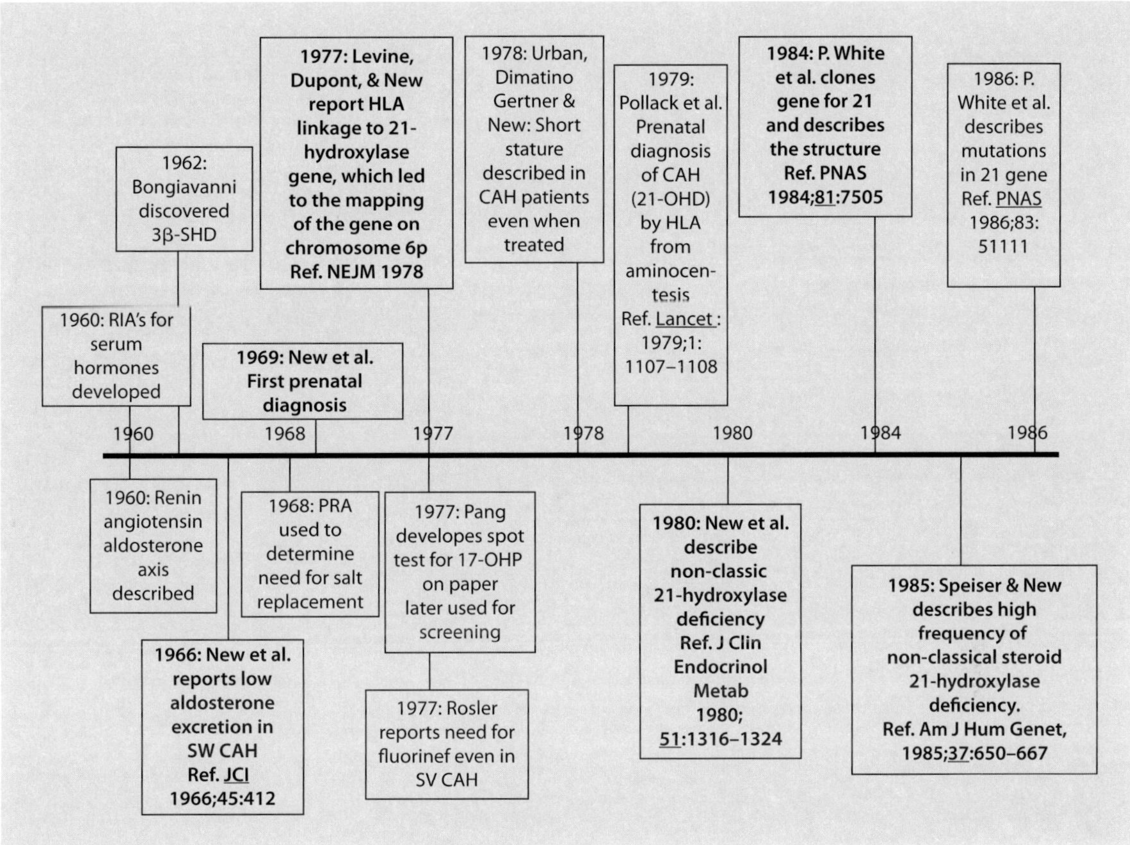

Fig. 6. Continued.

Isaac. Thus, in this ancient pedigree there was consanguinity and infertility, and I may speculate that this ancient family may have suffered from the non-classical form of CAH.

It is no wonder that the story of Pope Joan continues to resurface inspiring art and controversy. Whether or not she really lived and had a recognizable syndrome, Pope Joan is a symbol of the fascination and tragedy of gender ambiguity, and as such is likely to remain in the public eye.

Modern History of CAH

The modern history of CAH is depicted in figure. 6. The timeline indicates that the modern history begins in the 20th century, and continues into the 21st century.

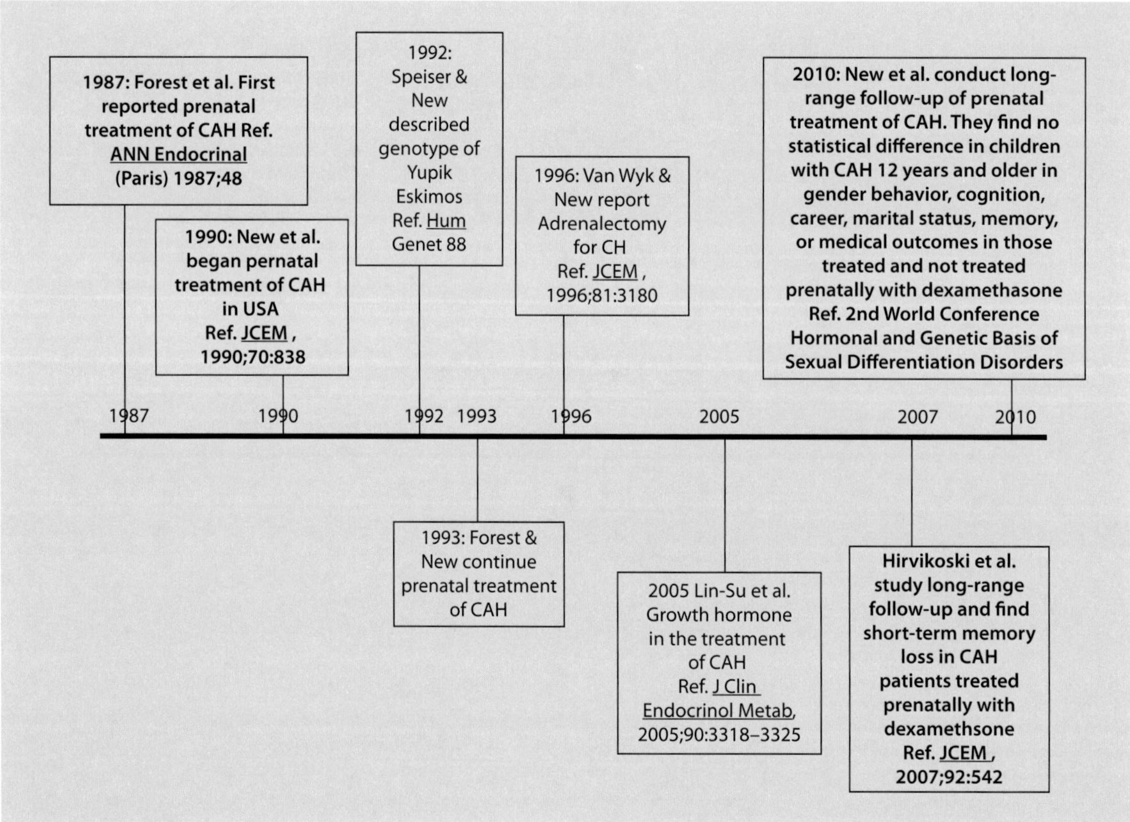

Fig. 6. Continued.

Future of CAH Investigation

The author proposes the following areas for future CAH investigation: (1) gene therapy, (2) non-invasive prenatal diagnosis and treatment, (3) advance understanding of psychoendocrinological factors, (4) improve surgical repair of genital ambiguity, and (5) discover the basis for the high-frequency of non-classical 21-hydroxylase deficiency.

Acknowledgement

The author wishes to acknowledge and warmly thank Mr. Ernest Schimizzi of Video Voice in New York City for his gift of the video of Michael Anderson's film, 'Pope Joan'. The author would also like to acknowledge the help of Elizabeth Kitzinger, Zachary Lerner, and Amina Ahmad in the preparation of this manuscript.

Suggested Reading

New MI: Pope Joan: a recognizable syndrome. Trans Am Clin Climatol Assoc 1993;104:104.

New MI, Kitzinger, ES: President's Address: Pope Joan: a recognizable syndrome. J Clin Endocrinol Metab 1993;76:3.

Pardoe R, Pardoe D: The Female Pope: The Mystery of Pope Joan. Wellingborough/UK, Crucible, 1988.

Strayer JR (ed): Dictionary of the Middle Ages. New York, Scribners, 1989, p 346.

Prof. Maria I. New
Pediatric Endocrinology, Genetics, and Genomics
1 Gustav L. Levy Place, Mount Sinai School of Medicine
Box 1200, New York, NY10029 (USA)
Tel. +1 212 241 7671, Fax +1 212 241 5405, E-Mail maria.new@mssm.edu

Author Index

Subject Index